LONDON MATHEMATICAL SOCIETY LECTURE NOTE SERIES

Managing Editor: Professor N.J. Hitchin, Mathematical Institute,
University of Oxford, 24–29 St Giles, Oxford OX1 3LB, United Kingdom

The titles below, and earlier volumes in the series, are available from booksellers
or from Cambridge University Press at www.cambridge.org

119 Triangulated categories in the representation theory of finite-dimensional algebras, D. HAPPEL
121 Proceedings of *Groups - St Andrews 1985*, E. ROBERTSON & C. CAMPBELL (eds)
128 Descriptive set theory and the structure of sets of uniqueness, A.S. KECHRIS & A. LOUVEAU
130 Model theory and modules, M. PREST
131 Algebraic, extremal & metric combinatorics, M.-M. DEZA, P. FRANKL & I.G. ROSENBERG (eds)
141 Surveys in combinatorics 1989, J. SIEMONS (ed)
144 Introduction to uniform spaces, I.M. JAMES
146 Cohen-Macaulay modules over Cohen-Macaulay rings, Y. YOSHINO
148 Helices and vector bundles, A.N. RUDAKOV *et al*
149 Solitons, nonlinear evolution equations and inverse scattering, M. ABLOWITZ & P. CLARKSON
150 Geometry of low-dimensional manifolds 1, S. DONALDSON & C.B. THOMAS (eds)
151 Geometry of low-dimensional manifolds 2, S. DONALDSON & C.B. THOMAS (eds)
152 Oligomorphic permutation groups, P. CAMERON
153 L-functions and arithmetic, J. COATES & M.J. TAYLOR (eds)
155 Classification theories of polarized varieties, TAKAO FUJITA
158 Geometry of Banach spaces, P.F.X. MÜLLER & W. SCHACHERMAYER (eds)
159 Groups St Andrews 1989 volume 1, C.M. CAMPBELL & E.F. ROBERTSON (eds)
160 Groups St Andrews 1989 volume 2, C.M. CAMPBELL & E.F. ROBERTSON (eds)
161 Lectures on block theory, BURKHARD KÜLSHAMMER
163 Topics in varieties of group representations, S.M. VOVSI
164 Quasi-symmetric designs, M.S. SHRIKANDE & S.S. SANE
166 Surveys in combinatorics, 1991, A.D. KEEDWELL (ed)
168 Representations of algebras, H. TACHIKAWA & S. BRENNER (eds)
169 Boolean function complexity, M.S. PATERSON (ed)
170 Manifolds with singularities and the Adams-Novikov spectral sequence, B. BOTVINNIK
171 Squares, A.R. RAJWADE
172 Algebraic varieties, GEORGE R. KEMPF
173 Discrete groups and geometry, W.J. HARVEY & C. MACLACHLAN (eds)
174 Lectures on mechanics, J.E. MARSDEN
175 Adams memorial symposium on algebraic topology 1, N. RAY & G. WALKER (eds)
176 Adams memorial symposium on algebraic topology 2, N. RAY & G. WALKER (eds)
177 Applications of categories in computer science, M. FOURMAN, P. JOHNSTONE & A. PITTS (eds)
178 Lower K- and L-theory, A. RANICKI
179 Complex projective geometry, G. ELLINGSRUD *et al*
180 Lectures on ergodic theory and Pesin theory on compact manifolds, M. POLLICOTT
181 Geometric group theory I, G.A. NIBLO & M.A. ROLLER (eds)
182 Geometric group theory II, G.A. NIBLO & M.A. ROLLER (eds)
183 Shintani zeta functions, A. YUKIE
184 Arithmetical functions, W. SCHWARZ & J. SPILKER
185 Representations of solvable groups, O. MANZ & T.R. WOLF
186 Complexity: knots, colourings and counting, D.J.A. WELSH
187 Surveys in combinatorics, 1993, K. WALKER (ed)
188 Local analysis for the odd order theorem, H. BENDER & G. GLAUBERMAN
189 Locally presentable and accessible categories, J. ADAMEK & J. ROSICKY
190 Polynomial invariants of finite groups, D.J. BENSON
191 Finite geometry and combinatorics, F. DE CLERCK *et al*
192 Symplectic geometry, D. SALAMON (ed)
194 Independent random variables and rearrangement invariant spaces, M. BRAVERMAN
195 Arithmetic of blowup algebras, WOLMER VASCONCELOS
196 Microlocal analysis for differential operators, A. GRIGIS & J. SJÖSTRAND
197 Two-dimensional homotopy and combinatorial group theory, C. HOG-ANGELONI *et al*
198 The algebraic characterization of geometric 4-manifolds, J.A. HILLMAN
199 Invariant potential theory in the unit ball of $\mathbf{C}^n$, MANFRED STOLL
200 The Grothendieck theory of dessins d'enfant, L. SCHNEPS (ed)
201 Singularities, JEAN-PAUL BRASSELET (ed)
202 The technique of pseudodifferential operators, H.O. CORDES
203 Hochschild cohomology of von Neumann algebras, A. SINCLAIR & R. SMITH
204 Combinatorial and geometric group theory, A.J. DUNCAN, N.D. GILBERT & J. HOWIE (eds)
205 Ergodic theory and its connections with harmonic analysis, K. PETERSEN & I. SALAMA (eds)
207 Groups of Lie type and their geometries, W.M. KANTOR & L. DI MARTINO (eds)
208 Vector bundles in algebraic geometry, N.J. HITCHIN, P. NEWSTEAD & W.M. OXBURY (eds)
209 Arithmetic of diagonal hypersurfaces over finite fields, F.Q. GOUVÉA & N. YUI
210 Hilbert $C^*$-modules, E.C. LANCE
211 Groups 93 Galway / St Andrews I, C.M. CAMPBELL *et al* (eds)
212 Groups 93 Galway / St Andrews II, C.M. CAMPBELL *et al* (eds)
214 Generalised Euler-Jacobi inversion formula and asymptotics beyond all orders, V. KOWALENKO *et al*
215 Number theory 1992–93, S. DAVID (ed)
216 Stochastic partial differential equations, A. ETHERIDGE (ed)
217 Quadratic forms with applications to algebraic geometry and topology, A. PFISTER

218 Surveys in combinatorics, 1995, PETER ROWLINSON (ed)
220 Algebraic set theory, A. JOYAL & I. MOERDIJK
221 Harmonic approximation, S.J. GARDINER
222 Advances in linear logic, J.-Y. GIRARD, Y. LAFONT & L. REGNIER (eds)
223 Analytic semigroups and semilinear initial boundary value problems, KAZUAKI TAIRA
224 Computability, enumerability, unsolvability, S.B. COOPER, T.A. SLAMAN & S.S. WAINER (eds)
225 A mathematical introduction to string theory, S. ALBEVERIO *et al*
226 Novikov conjectures, index theorems and rigidity I, S. FERRY, A. RANICKI & J. ROSENBERG (eds)
227 Novikov conjectures, index theorems and rigidity II, S. FERRY, A. RANICKI & J. ROSENBERG (eds)
228 Ergodic theory of $\mathbf{Z}^d$ actions, M. POLLICOTT & K. SCHMIDT (eds)
229 Ergodicity for infinite dimensional systems, G. DA PRATO & J. ZABCZYK
230 Prolegomena to a middlebrow arithmetic of curves of genus 2, J.W.S. CASSELS & E.V. FLYNN
231 Semigroup theory and its applications, K.H. HOFMANN & M.W. MISLOVE (eds)
232 The descriptive set theory of Polish group actions, H. BECKER & A.S. KECHRIS
233 Finite fields and applications, S. COHEN & H. NIEDERREITER (eds)
234 Introduction to subfactors, V. JONES & V.S. SUNDER
235 Number theory 1993–94, S. DAVID (ed)
236 The James forest, H. FETTER & B. GAMBOA DE BUEN
237 Sieve methods, exponential sums, and their applications in number theory, G.R.H. GREAVES *et al*
238 Representation theory and algebraic geometry, A. MARTSINKOVSKY & G. TODOROV (eds)
240 Stable groups, FRANK O. WAGNER
241 Surveys in combinatorics, 1997, R.A. BAILEY (ed)
242 Geometric Galois actions I, L. SCHNEPS & P. LOCHAK (eds)
243 Geometric Galois actions II, L. SCHNEPS & P. LOCHAK (eds)
244 Model theory of groups and automorphism groups, D. EVANS (ed)
245 Geometry, combinatorial designs and related structures, J.W.P. HIRSCHFELD *et al*
246 $p$-Automorphisms of finite $p$-groups, E.I. KHUKHRO
247 Analytic number theory, Y. MOTOHASHI (ed)
248 Tame topology and o-minimal structures, LOU VAN DEN DRIES
249 The atlas of finite groups: ten years on, ROBERT CURTIS & ROBERT WILSON (eds)
250 Characters and blocks of finite groups, G. NAVARRO
251 Gröbner bases and applications, B. BUCHBERGER & F. WINKLER (eds)
252 Geometry and cohomology in group theory, P. KROPHOLLER, G. NIBLO & R. STÖHR (eds)
253 The $q$-Schur algebra, S. DONKIN
254 Galois representations in arithmetic algebraic geometry, A.J. SCHOLL & R.L. TAYLOR (eds)
255 Symmetries and integrability of difference equations, P.A. CLARKSON & F.W. NIJHOFF (eds)
256 Aspects of Galois theory, HELMUT VÖLKLEIN *et al*
257 An introduction to noncommutative differential geometry and its physical applications 2ed, J. MADORE
258 Sets and proofs, S.B. COOPER & J. TRUSS (eds)
259 Models and computability, S.B. COOPER & J. TRUSS (eds)
260 Groups St Andrews 1997 in Bath, I, C.M. CAMPBELL *et al*
261 Groups St Andrews 1997 in Bath, II, C.M. CAMPBELL *et al*
262 Analysis and logic, C.W. HENSON, J. IOVINO, A.S. KECHRIS & E. ODELL
263 Singularity theory, BILL BRUCE & DAVID MOND (eds)
264 New trends in algebraic geometry, K. HULEK, F. CATANESE, C. PETERS & M. REID (eds)
265 Elliptic curves in cryptography, I. BLAKE, G. SEROUSSI & N. SMART
267 Surveys in combinatorics, 1999, J.D. LAMB & D.A. PREECE (eds)
268 Spectral asymptotics in the semi-classical limit, M. DIMASSI & J. SJÖSTRAND
269 Ergodic theory and topological dynamics, M.B. BEKKA & M. MAYER
270 Analysis on Lie groups, N.T. VAROPOULOS & S. MUSTAPHA
271 Singular perturbations of differential operators, S. ALBEVERIO & P. KURASOV
272 Character theory for the odd order theorem, T. PETERFALVI
273 Spectral theory and geometry, E.B. DAVIES & Y. SAFAROV (eds)
274 The Mandlebrot set, theme and variations, TAN LEI (ed)
275 Descriptive set theory and dynamical systems, M. FOREMAN *et al*
276 Singularities of plane curves, E. CASAS-ALVERO
277 Computational and geometric aspects of modern algebra, M.D. ATKINSON *et al*
278 Global attractors in abstract parabolic problems, J.W. CHOLEWA & T. DLOTKO
279 Topics in symbolic dynamics and applications, F. BLANCHARD, A. MAASS & A. NOGUEIRA (eds)
280 Characters and automorphism groups of compact Riemann surfaces, THOMAS BREUER
281 Explicit birational geometry of 3-folds, ALESSIO CORTI & MILES REID (eds)
282 Auslander-Buchweitz approximations of equivariant modules, M. HASHIMOTO
283 Nonlinear elasticity, Y. FU & R.W. OGDEN (eds)
284 Foundations of computational mathematics, R. DEVORE, A. ISERLES & E. SÜLI (eds)
285 Rational points on curves over finite fields, H. NIEDERREITER & C. XING
286 Clifford algebras and spinors 2ed, P. LOUNESTO
287 Topics on Riemann surfaces and Fuchsian groups, E. BUJALANCE *et al*
288 Surveys in combinatorics, 2001, J. HIRSCHFELD (ed)
289 Aspects of Sobolev-type inequalities, L. SALOFF-COSTE
290 Quantum groups and Lie theory, A. PRESSLEY (ed)
291 Tits buildings and the model theory of groups, K. TENT (ed)
292 A quantum groups primer, S. MAJID
293 Second order partial differential equations in Hilbert spaces, G. DA PRATO & J. ZABCZYK
294 Introduction to the theory of operator spaces, G. PISIER
296 Lectures on invariant theory, IGOR DOLGACHEV
297 The homotopy category of simply connected 4-manifolds, H.-J. BAUES
300 Introduction to Möbius differential geometry, UDO HERTRICH-JEROMIN

London Mathematical Society Lecture Note Series. 303

# Number Theory and Algebraic Geometry

## to Peter Swinnerton-Dyer on his 75th Birthday

Edited by

Miles Reid
*University of Warwick*

Alexei Skorobogatov
*Imperial College, London*

**CAMBRIDGE**
**UNIVERSITY PRESS**

32 Avenue of the Americas, New York NY 10013-2473, USA

Cambridge University Press is part of the University of Cambridge.

It furthers the University's mission by disseminating knowledge in the pursuit of education, learning and research at the highest international levels of excellence.

www.cambridge.org
Information on this title: www.cambridge.org/9780521545181



First published 2003

*A catalogue record for this publication is available from the British Library*

*Library of Congress Cataloguing in Publication data*

Number theory and algebraic geometry / edited by Miles Reid, Alexei Skorobogatov.
p. cm. – (London Mathematical Society lecture note series; 303)
Includes bibliographical references.
ISBN 0-521-54518-8 (pb.)
1. Number theory. 2. Geometry, Algebraic. I. Reid, Miles (Miles A.)
II. Skorobogatov, Alexei, 1961 - III. Series.
QA241.N8663 2004
512.7 – dc22 2003055797

ISBN 978-0-521-54518-1 Paperback

# Contents

# In lieu of Birthday Greetings

B. J. Birch Jean-Louis Colliot-Thélène
G. K. Sankaran Miles Reid Alexei Skorobogatov

This is a volume of papers in honour of Peter Swinnerton-Dyer's 75th birthday; we very much regret that it appears a few months late owing to the usual kind of publication delays. This preface contains four sections of reminiscences, attempting the impossible task of outlining Peter's many-sided contributions to human culture. Section 5 is the editor's summary of the 12 papers making up the book, and the preface ends with a bibliographical section of Peter's papers to date.

## 1 Peter's first sixty years in Mathematics by Bryan Birch

Peter Swinnerton-Dyer wrote his first paper [1] as a young schoolboy just 60 years ago, under the abbreviated name P. S. Dyer; in it, he gave a new parametric solution for $x^4 + y^4 = z^4 + t^4$. It is very appropriate that his first paper was on the arithmetic of surfaces, the theme that recurs most often in his mathematical work; indeed, for several years he was almost the only person writing substantial papers on the subject; and he is still writing papers about the arithmetic of surfaces sixty years later. Peter went straight from school to Trinity College (National Service had not quite been introduced); after his BA, he began research as an analyst, advised by J E Littlewood. At the time, Littlewood's lectures were fairly abstract, heading towards functional analysis; in contrast, Peter was advised to work on the very combinatorial, down-to-earth, theory of the van der Pol equation (the subject of Littlewood's wartime collaboration with Mary Cartwright), where a surprising sequence of stable periodic orbits arise completely unexpectedly from a simple-looking but non-linear ordinary differential equation. Lurking in the background was the three body problem, together with ambitions to prove the stability of the solar system, compare [20].

After a couple of years, Peter was elected to a Trinity Junior Research Fellowship, and became a full member of the mathematical community (he never needed to submit a doctoral thesis). In 1954, he was selected for a Commonwealth Fund Fellowship, and went to Chicago intending to work with Zygmund; but when he reached Chicago, he met Weil, who converted him to geometry; I believe that Weil was the person who most influenced Peter's mathematics. Ever since his year in Chicago, Peter has been an arithmetic geometer, with unexpected expertise in classical analysis.

Peter returned to Cambridge in 1955. In the 1950s, mathematical life in Cambridge was vigorous and sociable; everyone collaborated with everyone else. It was the heyday of the Geometry of Numbers (it was sad that so much excellent mathematical work was poured into such an unworthy subject!) and Peter joined in. In particular, he and Eric Barnes (later Professor at Adelaide) wrote a massive series of papers [5] on the inhomogeneous minima of binary quadratic forms, which completely settled the problem of which real quadratic fields are norm-Euclidean; like the van der Pol equation, this is a case where a 'discrete' phenomenon arises from a 'continuous' question. He went on to collaborate with Ian Cassels [8], trying to obtain a similar theory for products of three linear forms; their work was highly interesting, but only partially successful, and to this day there has been (I believe) no further progress on the problem.

I first came into contact with Peter in 1953, when he read my Rouse Ball essay on the Theory of Games (one of Peter's lesser interests, that does not show up in his list of publications), and I got to know him well after he returned from Chicago. Over the next couple of years, we talked a lot and he taught me to enjoy opera and we wrote two or three pretty but unimportant papers together; but at that stage, he wanted to be a geometer, and I was turning towards analytic number theory, under the influence of Harold Davenport. In my turn, I went to the States with a Commonwealth Fund Fellowship, and while I was away Peter took a post in the fledgling Computer laboratory. When I returned, I was excited by the Tamagawa numbers of linear algebraic groups, one of us (probably Peter) wondered about algebraic groups that aren't affine, and we set to, computing elliptic curves.

Those four years, from 1958–62, were probably the best of my life; they were the most productive, and I married Gina (who had a desk in Peter's office in the Computer laboratory). We were under no pressure to publish: we both had Fellowships, and knew we could get another job whenever we needed one; and we didn't have to worry about anyone else anticipating our work. In the first phase, we made a frontal assault; for the curves $E(a, b) : y^2 = x^3 + ax + b$ with $|a| \leq 20$ and $|b| \leq 30$ we computed the Mordell–Weil rank, the 2-part of the Tate–Shafarevich group, and a substitute $T(E, P)$ for a Tamagawa number $\tau(E)$, namely the product of $p$-adic densities taken over primes $p \leq P$ where $P$ was as high as the market would stand. Peter did the programming, which he made feasible by dealing with many curves simultaneously; for good primes the $p$-adic density was of course $N_p/p$ where $N_p$ was the number of points mod $p$, and the crude methods of computing $N_p$ for medium-sized $p$ were nearly as fast for a batch of curves as for a single curve; there was an even better batch-processing gain in the rank computations. (For the finitely many 'bad' primes one needed so-called fudge factors, which I seem to remember were part of my job). To our delight, the numbers $T(E, P)$ increased roughly as $c(E) \log^r P$, where $r$ was the Mordell–Weil rank of $E$; so we prepared [17] for publication, and proceeded to the second phase. Here, Davenport and Cassels were very helpful; urged by their prodding, we realised that, rather than considering the product $T(E, P)$ as $P$ got large, one should be considering $L(E, s)^{-1}$ as $s$ tends to 1 (so that $L(E, s)$ should have a zero of order $r$ at $s = 1$). (As Weil remarked to a colleague in Chicago, 'it was time for them to learn some mathematics'.) Hecke had tamed this Dirichlet series for elliptic curves with complex multiplication, giving an explicit formula that actually converged at $s = 1$. So we approximated to the Dirichlet series $L(E, 1)$, in case $E$ had complex multiplication and Mordell–Weil rank 1; and we got numbers that really seemed to mean something: after the junk factors had been scraped off, they seemed to be the order of the Tate–Shafarevich group divided by the torsion squared. Next, Davenport showed us how to evaluate $L(E, 1)$ explicitly in terms of the Weierstrass $\wp$-function; we computed some more, and [18], containing the main B–S-D conjectures, was the result.

In 1962 I left Cambridge to take a job in Manchester, and our collaboration became less close; we had expected to write further Notes in the series 'On Elliptic Curves', but they didn't happen. Note III might have been a plan of Peter's, to test the conjecture for abelian varieties by starting with products of elliptic curves; this turned into the thesis of Damerell, which essentially computed critical values of $L(E^{(3)}, s)$, where $E^{(3)}$ is the cube of a curve; the numbers were interesting but he was not able to interpret them. The intended Note IV was more important; Nelson Stephens was able to compute the higher derivatives $L^{(r)}(E, 1)$, where $r$ is the Mordell–Weil rank; he was the first to obtain exact evidence for the conjectured formula, for elliptic curves of higher rank over the rationals, and indeed his thesis [93] is where it is first precisely stated. In July 1965, Peter received a letter from Weil [94] which set the tone for further progress in the area. Weil reminded us that our conjectures make sense only if the relevant functions $L(E, s)$ have functional equations, and this is likely to be true only if the elliptic curve $E/\mathbf{Q}$ is parametrised by modular functions invariant by some $\Gamma_0(N)$. So we had better be looking at modular curves! I was in Cambridge on sabbatical for the next term, so we set to work. Indeed, we worked very hard; on one occasion we were so engrossed talking mathematics after dinner, on Trinity Backs, that an unobservant porter locked us out; fortunately, we were able to regain entry by successfully charging the New Court gate.

Weil's letter led to three developments; first, modular symbols: I am pretty certain that Peter had the first idea [89], but he was very busy, so I and my students had to make them work, and Manin [91] formalised the concept. Next, the tabulation of elliptic curves of small conductor (Table I of [43]); this involved many people, starting with Peter and then me, as described in the introduction to the table. Finally, a few years later, Heegner points came on the scene.

I was most excited in our work on elliptic curves; but indeed Peter's interests in this period were exceedingly diverse. He did seminal work on his earliest love, the arithmetic of surfaces: in [15] he found the first counterexample to the Hasse principle for cubic surfaces (I think he found this example in 1959, as I reported on it in Boulder). A little later, he improved a result of Mordell, that the Hasse principle is valid for the intersection of two quadric hypersurfaces in $\mathbf{P}^n$ so long as the dimension $n$ is large enough — this paper [19] is of interest as a very early example of Peter's technique of working out what one can prove if one assumes various useful but unprovable 'facts'; with luck, one may remove such unwanted hypotheses later. His 1969 paper [34] at the Stony Brook conference reviewed what was known, and contained new material. At last, in 1970, Peter ceased to be a lone voice crying in the wilderness, when Manin introduced the so-called Manin obstruction in his lecture at Nice [90], and went on to write his book on cubic surfaces [92]. Also in 1969–70, Colliot-Thélène went to Cambridge to work with Peter; since then the theory has flourished, as this volume amply testifies.

Meanwhile, Peter remained an analyst; in particular, Noel Lloyd was his research student between 1969 and 1972. He also became interested in modular forms for their own sake; with Atkin, he investigated modular forms on non-congruence subgroups [32]. Surprisingly, their results suggested that the power series of such modular forms should have good $p$-adic properties (their conjecture was proved long afterwards by Scholl). Peter corresponded with Serre, and published the basic paper on the structure of (ordinary) modular forms modulo $p$ in the third volume of the Antwerp Proceedings [43]; this volume was of course the beginning of the theory of $p$-adic modular forms. Peter made yet another important contribution in the Computing Laboratory, where he was responsible for implementing Autocode for Titan.

He worried about the inefficiencies of university governance, and took an increasing interest in administrative matters. In 1973 he was elected Master of St Catherine's College, from 1979–81 he was Vice Chancellor, and from 1983–89 he was Chairman of the University Grants Committee. All this involved an immense amount of committee work, but miraculously (and with the help of Harriet, and of Jean-Louis Colliot-Thélène) he remained in touch with mathematics. When he returned to Cambridge in 1989 he resumed full-time research, principally on the arithmetic theory of surfaces, but also on analysis.

Oxford, 10th Dec 2002

# 2 Peter Swinnerton-Dyer's work on the arithmetic of higher dimensional varieties by Jean-Louis Colliot-Thélène

In parallel to his well-known contributions to elliptic curves, modular forms, $L$-functions, differential equations, bridge, chess and other respectable topics, Peter has a lifelong interest in the arithmetic geometry of some – at first sight – rather special varieties: cubic surfaces and hypersurfaces, complete intersections of two quadrics defining a variety of dimension $\geq 2$, and quartic surfaces.

I happened to spend a year in Cambridge when I started research, and Peter passed on to me his keen interest in the corresponding diophantine questions. I am thus happy to report here on Peter's past and ongoing work on these problems. As will be clear from what follows, Peter, at age 75, is still doing entirely original innovative research.

Much of the progress achieved in arithmetic geometry during the twentieth century has been concerned with curves. For these, we now have a clear picture: for genus zero, the Hasse principle holds; for genus one, many problems remain, but we have the Birch and Swinnerton-Dyer conjecture, and we hope that the Tate–Shafarevich groups are finite; for genus at least two, Faltings proved the Mordell conjecture.

In higher dimension the situation is much less clear. For the three types of varieties mentioned above, one is still grappling with the basic diophantine questions: How can we decide whether there are rational points on such a variety? Is there a local-to-global principle, or at least some substitute for such a principle? What are the density properties of rational points on such varieties (in the sense of the Chinese remainder theorem)? Can one "parametrize" the rational points? Can one estimate the number of rational points of bounded height?

The time when varieties were classified according to their degree, as in Mordell's book, is long gone, and one may view the varieties just mentioned as belonging to some general classes of varieties. One general class of interest is that of rational varieties (varieties birational to projective space after a finite extension of the ground field). A wider class, whose interest has been recognized only in the last ten years, is that of rationally connected varieties. These are now considered as the natural higher dimensional analogues of curves of genus zero. Nonsingular intersections of two quadrics (of dimension $\geq 2$) are rational varieties, hence rationally connected; so are nonsingular cubic surfaces. Higher dimensional cubic hypersurfaces are rationally connected. Nonsingular quartic surfaces are not rationally connected, but there are interesting density questions for rational points on them.

Until 1965, there were two kinds of general results on the arithmetic of rational varieties. One series of works, going back to the papers of H. Hasse in the twenties (local-to-global principle for the existence of rational points on quadrics), was concerned with homogeneous spaces of connected linear algebraic groups. A very different series of works, going back to the work of G. H. Hardy and J. E. Littlewood, proved very precise estimates on the number of points of bounded height (hence in particular proved existence of rational points) on complete intersections when the number of variables is considerably larger than the multidegree.

There had also been isolated papers by F. Enriques, Th. A. Skolem, B. Segre, L. J. Mordell, E. S. Selmer, F. Châtelet, J. W. S. Cassels and M. J. T. Guy. Peter himself made various contributions to the topic in his early work: he produced the first counterexamples to the Hasse principle and to weak approximation for cubic surfaces [15], he extended results of Mordell on the existence of rational points on complete intersections of two quadrics in higher dimensional projective space [19], and he proved the Hasse principle for cubic surfaces with special rationality properties of the lines.

Over the years 1965–1970, after some prodding by I. R. Shafarevich, Yu. I. Manin and V. A. Iskovskikh looked at this field of research in the light of Grothendieck's algebraic geometry. They did not solve all the diophantine problems, but they put some order on them. A typical illustration was Manin's appeal [90] to Grothendieck's Brauer group to reinterpret most known counterexamples to the Hasse principle, including Peter's.

I spent the academic year 1969/1970 in Cambridge – I was hoping to learn more about concrete diophantine problems, not the kind of arithmetic geometry I was exposed to in France. Professor Cassels advised me to take Peter as a research supervisor. I was first taken aback, because, ignorant as I was, the only thing I knew about Peter was that he had written a paper entitled "An application of computing to number theory", and I was not too keen on computing. I wanted concrete diophantine equations, but with abstract theory. I nevertheless asked Peter, and this was certainly one of the most important moves in my mathematical career.

In those days, Peter was neither a Sir nor a Professor. He was known to Trinity students as "The Dean", whose function I understand was to preserve moral order among the students. To this he contributed by serving sherry ("Sweet, medium or dry?") each evening in his small flat in New Court. Sherry time was the ideal time to ask him for advice, mathematical or other – I do not remember Peter as a great addict of long sessions in the Mathematics Department. Well, at least one could enjoy his beautifully prepared lectures (the young Frenchman enjoyed the very clear, classical English as much as the mathematics). Peter was well known for his wit, and Swinnerton-Dyer quotations and stories abounded. His students enjoyed his avuncular behaviour – he was not a thesis adviser in the classical sense – and at the same time one vaguely feared him as the possible mastermind of many things going on in Cambridge. (His masterminding was later to extend to a wider scene – I remember Spencer Bloch being rather impressed by a 1982 newspaper representation of Peter Swinnerton-Dyer portrayed as King Kong climbing up one of London University's main buildings.)

One day in April 1970, on Burrell's walk, I asked Peter for a research topic. He mentioned the question of understanding and generalizing some work of François Châtelet, who had performed for cubic surfaces of the shape $y^2 - az^2 = f(x)$ (with $f(x)$ a polynomial of the third degree) something which looked like descent for elliptic curves – Peter also had handwritten lists of questions on a similar process for diagonal cubic surfaces.

In July 1970 I went back to France, and learned "French algebraic geometry" with J.-J. Sansuc. He and I discussed étale cohomology and Grothendieck's papers on the Brauer group, but I kept on thinking about Châtelet surfaces and Peter's questions. In 1976–77, Sansuc and I laid out the general mechanism of descent, which appeals to principal homogeneous spaces (so-called torsors) with structure group a torus (as opposed to the finite commutative group schemes used in the study of curves of genus one). One aim was to find the right descent varieties on Châtelet surfaces (and to answer a question of Peter, whether descent here was a one-shot process, as opposed to what happens for elliptic curves). The theory was first applied to more amenable varieties, namely to smooth compactification of tori. As far as Châtelet surfaces are concerned, there were two advances: In 1978, Sansuc and I realized that Schinzel's hypothesis (a wild generalization of the twin prime conjectures) – also considered much earlier by Bouniakowsky, Dickson, and Hardy and Littlewood – would imply statements of the type: the Brauer–Manin obstruction is the only obstruction to the Hasse principle for generalized Châtelet surfaces, namely for surfaces of the shape $y^2 - az^2 = f(x)$ with $f(x)$ a polynomial of arbitrary degree (over the rationals). The second advance took place in 1979: following a rather devious route, D. Coray, J.-J. Sansuc and I found a class of generalized Châtelet surfaces for which the Brauer–Manin obstruction entirely accounts for the defect of the Hasse principle.

During the period 1970–1982, Peter was busy with any number of different projects: the Antwerp tables on elliptic curves [45], understanding Ramanujan congruences for coefficients of modular forms [43], [51], writing, jointly with B. Mazur, an influential paper [37] on the arithmetic of Weil curves and on $p$-adic $L$-functions, proving (jointly with M. Artin [46]) the Tate conjecture for K3 surfaces with a pencil of curves of genus one (a function-theoretic analogue of the finiteness of the Tate–Shafarevich group), and also writing a number of papers on differential equations. He also wrote a note on the number of lattice points on a convex curve [33], which was followed by papers of other writers (W. M. Schmidt, E. Bombieri and J. Pila). The ideas in those papers now play a rôle in the search for unconditional upper bounds for the number of rational points of bounded height (work of D. R. Heath-Brown).

During that period, Peter also contributed papers on rational varieties: he gave a proof of Enriques' claim that del Pezzo surfaces of degree 5 always have a rational point [42], he wrote a paper with B. Birch producing further counterexamples to the Hasse principle [47] and he wrote a paper on $R$-equivalence on cubic surfaces over finite fields and local fields [56]. This last paper used techniques specific to cubic surfaces to prove results which have just been generalized to all rationally connected varieties by J. Kollár and E. Szabó, who use modern deformation techniques. That paper and a later one [87] on a related topic exemplify how Peter is not deterred by inspection of a very high number of special cases.

Indeed it is Peter's general attitude that a combination of cleverness and brute force is just as powerful as modern cohomological machineries. As the development of many of his ideas has shown, cohomology often follows, and sometimes helps. As we say in France, "l'intendance suit".

Let me here include a parenthesis on Peter's ideal working set-up. Sitting at a conference and not listening to a lecture on a rather abstruse topic seems to be an ideal situation for him to conceive and write mathematical papers. The outcome, written without a slip of the pen, is then imposed upon the lesser mortal who will definitely take much more time to digest the contents than it took Peter to write them.

In 1982, I spent another six months in Cambridge. I did not see Peter too often, as I was rather actively working on algebraic K-theory, not a field which attracts his attention. However, shortly before I left Cambridge, in June 1982, Peter invited me for lunch at high table in Trinity, and while reminding me how to behave in this respectable environment, he inadvertently mentioned that he could say something new on descent varieties attached to Châtelet surfaces – the topic he had offered to me as a research topic 12 years earlier. If my memory is correct, what he did was to sketch how to prove the Hasse principle on the specific intersections of two quadrics appearing in the descent process on Châtelet surfaces, the method being a reduction by clever hyperplane sections to some very special intersections of two quadrics in 4-dimensional projective space. Sansuc and I quickly saw how the descent mechanism we had developed in 1976–77 could combine with this new result. This was to develop into a Comptes Rendus note of Sansuc, Swinnerton-Dyer and myself [57] in 1984, then into a 170 page paper of the three of us in Crelle three years later [61]. Among other results, we obtained a characterisation of rational numbers that are sums of two squares and a fourth power, and we proved that over a totally imaginary number field two quadratic forms in at least 9 variables have a nontrivial common zero (this is the analogue of Meyer's result for one form in 5 variables). An outcome of the algebraic geometry in our work was a negative answer (joint work of the three of us with A. Beauville [59]) to a 1949 problem of Zariski: some varieties are stably rational but not rational.

Around 1992, the idea to use Schinzel's hypothesis to explore the validity of the Hasse principle (or of its Brauer–Manin substitute) was revived independently by J-P. Serre and by Peter [67]. In that paper, conceived during a lengthy coach trip in Anatolia, Peter simultaneously started developing something he calls the Legendre obstruction. In many cases, this obstruction can be shown to be equivalent to the Brauer–Manin obstruction, but Peter tells me there are cases where this yields information not reachable by means of the Brauer–Manin obstruction. In 1988, P. Salberger had obtained a remarkable result on zero-cycles on conic bundles over the projective line. The paper involved a mixture of algebraic K-theory and approximation of polynomials. Peter saw how to get rid of the K-theory and how to isolate the essence of Salberger's trick, which turned out to be an unconditional analogue of Schinzel's hypothesis. This was developed in papers of Peter, in a paper with me [66] and in a paper with A. N. Skorobogatov and me [73]. The motto here is: it is worth exploring results conditional on Schinzel's hypothesis for rational points, because if one succeeds, then one may hope to replace Schinzel's hypothesis by Salberger's trick and prove unconditional results for zero-cycles.

Up until about ten years ago, work in this area was concerned with the total space of one-parameter families of varieties which were close to being rational. In 1993 Peter invented a very intricate new method, which enables one to attack pencils of curves of genus one. In its general form, the method builds upon two well-known but very hard conjectures, already mentioned: Schinzel's hypothesis and finiteness of Tate–Shafarevich groups of elliptic curves. The original paper [69], in Peter's own words, looks like a series of lucky coincidences and "rather uninspiring" explicit computations (not many of us have the good fortune to come across such series). It already had striking applications to surfaces which are complete intersections of two quadrics.

It took several years for Skorobogatov and me to get rid of as many lucky coincidences as possible (one instance being a brute force computation which turned out to be Peter's rediscovery of Tate's duality theorem for abelian varieties over local fields). The outcome was a long joint paper of the three of us [74] in 1998. In that paper Peter's original method is extended beyond rational surfaces: the method can predict a substitute of the Hasse principle and density results for rational points on some elliptic surfaces (surfaces with a pencil of curves of genus one). This came as quite a surprise.

Since 1998, Peter has been developing subtle variants of the method, with application to some of the simplest unsolved diophantine equations: systems of two quadratic forms in as low as 5 variables [69], [74], [84], diagonal quartics [80] (hence some K3 surfaces, whose geometry is known to be far more complicated than that of rational surfaces); diagonal cubic surfaces and hypersurfaces over the rationals [85]. The first two applications assume Schinzel's hypothesis and finiteness of Tate–Shafarevich groups, but [85] (on diagonal cubic surfaces) only assumes the latter finiteness: this theorem of Peter's on diagonal cubic surfaces, both by the result and by the subtlety of the proof, is certainly the most spectacular one obtained in the area in the last ten years. For instance, under the finiteness assumption on Tate–Shafarevich groups, the local-to-global principle holds for diagonal cubic forms in at least 5 variables over the rationals.

In 1996, rather wild guesses were made on two different topics: For which varieties do we expect potential density of rational points? For varieties over the rationals with a Zariski-dense set of rational points, what should we expect about the closure of the set of rational points in the set of real points (question of B. Mazur)? Peter had the idea to call in bielliptic surfaces to produce unexpected answers to the second question. Skorobogatov and I elaborated, and applied the mechanism to get rid of preliminary guesses for the first question. This led to a joint work between the three of us [70]. There has been recent (conjectural) progress on an answer to the first question (work of complex algebraic geometers). The same bielliptic surfaces were later used by Skorobogatov (1999) to produce the first ever example of a surface for which the Brauer–Manin obstruction is not the only obstruction to the Hasse principle. This has led to further developments by D. Harari and Skorobogatov (descent under noncommutative groups).

Peter also contributed two papers [65], [77] to a topic which has seen quite some activity over the last ten years: the behaviour of the counting function for points of bounded height on Fano varieties. He pointed out the way to the correct guess for the constant in the standard conjecture (later important work in this area was done by E. Peyre and others). The lower bound he obtained (jointly with J. B. Slater [77]) for cubic surfaces is still one of the best results in this area.

The line of investigation Peter started in 1994 with the paper [69] is very delicate, and while his 2001 paper on diagonal cubic surfaces [85] is quite a feat, I am sure that Peter will produce much more in this exciting new direction. I am confident that he will keep on being as generous with his ideas as he has always been and that he will allow some of us to accompany him along the way.

Orsay, the 13th of February, 2003

# 3 Peter Swinnerton-Dyer: Geometer and politician by G.K. Sankaran

Peter Swinnerton-Dyer's interest in algebraic geometry derives arguably from its relation to number theory, and from the formative period he spent with André Weil in Chicago in the 1950s, but he has also made important contributions to geometry over algebraically closed fields. Probably his most notable technical result of a purely geometric nature is the proof (described elsewhere in this preface by Jean-Louis Colliot-Thélène) that stable rationality does not imply rationality [59]. This was, probably, contrary to the expectations of the majority of algebraic geometers at the time; though, as often happens, it is hard with hindsight to imagine why anybody ever thought the opposite was true. The result, published in French in a joint paper with Beauville, Sansuc and Colliot-Thélène, uses a wide range of techniques from different parts of algebraic geometry: torsors, linear systems with base points, Prym varieties and singularities of the theta divisor. It arose, however, out of arithmetic work with Sansuc and Colliot-Thélène. Many of Peter's arithmetic results have a geometric flavour, especially his work with Bombieri and with Artin; and it is now appreciated among geometers that arithmetic information can be made to yield geometrical or topological information (in addition to the well-known consequences of the Weil Conjectures). Rational and abelian varieties particularly feature in his work: these topics are represented in this volume by the papers of Reid and Suzuki and of Sankaran respectively.

Within algebraic geometry, however, Peter's chief influence has been as teacher, expositor, supplier of encouragement and enthusiasm, and éminence grise. He recognised, at a time when few in Britain were more than dimly aware of it, the power of the French school of algebraic geometry of Weil, Serre and Grothendieck. In the 1970s he encouraged his then student Miles Reid to visit Paris and learn directly from Deligne. The flourishing state of

British algebraic geometry at the present day owes much to this development, and to Peter's encouragement and direction of later students. His Cambridge Part III courses have been a source of inspiration to many, and his book on abelian varieties and his account of the basic facts of Hodge theory have been of great service to even more.

Many of Peter's multifarious activities are completely unrepresented in this book. The purpose of the rest of this note is to allude to some of them. I am not the best person to write such a note (that would be Peter Swinnerton-Dyer): I have drawn on my memories of conversations with many people, among them Carl Baron, Arnaud Beauville, Bryan Birch, Béla Bollobás, Jean-Louis Colliot-Thélène, James Davenport, Nicholas Handy, Richard Pinch, Colin Sparrow, Miles Reid, Pelham Wilson, Rachel Wroth and, above all, Peter Swinnerton-Dyer.

Mathematically the most obvious of Peter's other activities is his substantial contribution to the theory of differential equations, including a paper with Dame Mary Cartwright published only in Russian [55]. He is still active in differential equations. Readers of the present volume will have no difficulty in finding more information about this part of Peter's work. Slightly further afield, Peter was a member of the computing group in Cambridge in the 1960s, in the days of the Cambridge University computer TITAN. The original operating system for this famous machine, known as the Temporary Supervisor, was written by Peter single-handed, and it worked. He wrote the computer language Autocode for the same machine, and most Cambridge mathematicians of the 1960s had their first programming instruction in this language. Who could ask for anything more?

Peter, then Dean of Trinity College, was elected Master of St Catharine's College in 1973 and remained there for ten years. Littlewood is said to have greeted the news with Clemenceau's remark on hearing that the pianist Paderewski was to be Prime Minister of Poland: 'Ah, quelle chute!'. But St Catharine's afforded Peter considerable scope, and by all the numerous accounts I heard, as a later Fellow of St Catharine's, he was highly successful. The head of a Cambridge College (of Oxford I cannot speak) is commonly all but invisible to the students, and in some cases even to the Fellows. Peter was not: he has never been averse to the company of students and he was even willing to do College teaching. As he could and would teach almost any course in the Mathematical Tripos, the task of the Director of Studies (who is responsible for arranging for the students to be taught) was occasionally much simplified.

While at St Catharine's he served as Vice Chancellor of the University. This is now a full-time post held for a long period, but at the time the Vice Chancellor was chosen from among the heads of the various colleges and served for two years only. The role of the Chancellor (then, as now, the Duke of Edinburgh) is purely ceremonial, and the Vice Chancellor is in effect at the head of the University. It is a job for a skilled diplomatist. Cambridge University is a highly visible organisation, under constant and occasionally hostile scrutiny by newspapers and television. Internal matters can lead to very acrimonious public debate, and in extreme cases, which are quite common, the Vice Chancellor is expected to reconcile the factions. During Peter's term of office there was one especially well-publicised dispute about whether a tenured post should be awarded to a particular person. It was clearly impossible to satisfy all parties, but Peter nevertheless managed to bring the matter to a conclusion without offending anybody further. Who could ask for anything more?

Peter left St Catharine's to take up a post as Chairman of the University Grants Commission, a semi-independent Government body which was charged with deciding how Government funding ought to be apportioned among different universities. He had already written an influential, and in some quarters unpopular, report on the structure of the University of London, and was thus well known to be of a reforming cast of mind. He was also widely assumed to be in general political sympathy with the government of the time (otherwise, the reasoning ran, why did they appoint him?); but this was far from the case. He was nevertheless able to use his position to defend the reputation of the universities for financial responsibility, and in particular to establish the principle that research is a core activity for any university and therefore merits funding on its own account, independently of teaching. The price to be paid was investigation by government of the research activities of universities. Peter is thus often held responsible for, or credited with, the Research Assessment Exercise, which attempts to grade British university departments (not individuals) roughly according to the quality of research that they produce, and then hopes that they will be funded accordingly. The system is agreed to be imperfect, but it is easier to think of worse alternatives than better ones.

Peter's first involvement in politics dates from early in his tenure as Master of St Catharine's. The Member of Parliament for Cambridge resigned his seat and a by-election had to be held. Among the candidates was a representative of the Science Fiction Loony Party, whose aim in standing was to have some fun, and if possible to do better than the extreme right-wing candidate. Candidates in British parliamentary elections are required to pay a deposit of a few hundred pounds, returnable if they receive a certain proportion (then one-eighth) of the votes cast. In this case there was no prospect of that, so the deposit was, in effect, a fee: Peter, a wealthy man, paid it. He explained that the candidate "deserved every possible support, short of actually voting for him". Later his own name was mentioned as a possible parliamentary candidate, on behalf of the more serious but probably less entertaining Social

Democratic Party formed by Roy Jenkins and other disaffected members of the Labour Party in 1981. Nothing came of the plan, if it ever existed. The SDP seems, understandably, to have been unable to believe that all Peter's activities were the work of one man, and on occasion sent him two copies of the same letter, one for Swinnerton and one for Dyer.

Peter is a strong Chess player. Even when Vice Chancellor, he used to put in occasional appearances at the Cambridge University Chess Club, playing five-minute against undergraduates. The story is that when appointed to a Trinity research fellowship, he was strongly advised to cut down the time he spent on Chess; and that his interest in Bridge dates from this time. He was to become a very strong Bridge player. He was a member of the team that won the British Gold Cup in 1963, and he acted as non-playing captain of the Great Britain Ladies' Bridge team.

On leaving UGC (by then renamed UFC) Peter resumed work as a mathematician as if nothing had happened. He also continued his life of public service, working on behalf of such diverse institutions as the World Bank and the Isaac Newton Institute: he is still frequently to be found at the latter, at least.

Peter's work at UGC/UFC was recognised by the award of a knighthood (a KBE, to be precise). The editors of this volume tell me that "how did Swinnerton-Dyer get his title?" is a frequently asked question after seminars in places such as Buenos Aires and Vladivostok: at the risk of spoiling the fun, here is an explanation.

Peter is a baronet: he is also a knight. A baronet is entitled to call himself "Sir", and when he dies his eldest son, or some other male relative if he has none, inherits the title. It is only a title: it does not give him a seat in the House of Lords, and never has. Baronetcies were invented by King James I, early in the seventeenth century, as a way of raising money: they were simply sold. Later baronetcies were awarded for actual achievement, but the oldest ones are purely mercenary affairs. Since no baronetcies have been created for many years, all current baronets have inherited their titles rather than earning or buying them. A knight is also entitled to call himself "Sir", but the title dies with him. Knighthoods, which are still awarded in quite large numbers, are for specific personal achievements: they are given by the Queen on the recommendation of the Prime Minister. By the time he was knighted, Peter was already a baronet, so already entitled to call himself "Sir Peter" (not "Sir Swinnerton-Dyer"). For this reason he is sometimes referred to as $(\text{Sir})^2$ Peter, although strictly speaking "Sir" is idempotent: he is technically Professor Sir Henry Peter Francis Swinnerton-Dyer, FRS, Bt., KBE.

Bath, 12th Feb 2003

# 4 Peter Swinnerton-Dyer, man and legend by Miles Reid

I was supervised by Peter as a second year Trinity undergraduate. From then on, I was among the many Cambridge students who were occasionally invited for sherry at 7:30 pm (before Hall at 8:00 pm). For me and many other middle-class students of my generation, this provided an education into hitherto unsuspected areas of culture, such as good quality sherry, opera, college politics, famous math visitors, the workings of the British upper classes, etc. Peter is 16th Baronet Swinnerton-Dyer, and his family was an illustration that the feudal system was still alive and well, in Shropshire, at least in 1949: he had an elegant clock on the mantlepiece of his Trinity New Court apartment, with an inscription

> "Presented to Henry Peter Francis Swinnerton-Dyer Esq by the tenants, cottagers and employees of the Westhope estate on the occasion of his coming of age".

Peter's legendary status was already well established – as a sample of the stories in circulation, when Galois theory was introduced as a Part II course lectured by Cassels, Peter claimed that the whole course could be given in 4 hours, and made good his claim one evening between 10 pm and 2 am. Another story about bridge, that I heard from Peter himself: At a tournament, Peter called over the referee, told him formally that he was not making an error or oversight, then bid 8 clubs. Although this bid is impossible, he had calculated that he would lose less going down in it than allowing his opponents to make their grand slam. He knew the fine wording of the rules of bridge, and the match referee was forced to accept the impossible bid, since it was not made by error or oversight; the rules were subsequently changed to block this obscure loophole.

At that time Peter was Dean of Trinity; the position included disciplinary control of students. Those caught walking on the grass in College would be sent to Peter, and would in theory be fined in multiples of 6/8 (that is, 6 shillings and 8 pence, a third of a pound). In my case, for a particularly unpleasant misdemeanour, my sentence was to wash Peter's car.

Peter had an affinity with math students, and would drop in on friends in the evening to see if there was a conversation going on; I can well believe that student company was more fun than that of the senior combination room. He would often join in conversations, or dominate them – his predilection for that well-turned phrase certainly had a lasting effect on my literary pretensions. (For example: Would he send his son to Eton? "Certainly, it has advantages both in this life, and in the life that is to come.") Or, he would sometimes simply be comfortable among student friends and nod off to sleep (presumably this mainly happened after wine in the Combination room following High Table dinner). On one occasion, we played the board game Diplomacy from after dinner until breakfast the following morning – with great cunning and skill, Peter unexpectedly murdered me treacherously at about 6:00 am. Outside board games, Peter was extremely generous with friends and colleagues – many of us were invited to accompany him on a trip to the opera in London, or on a car trip to Norway, Paris or Italy, with appropriate stops to appreciate the great cathedrals and the starred restaurants of the Michelin guide.

As a PhD student I started to get more specific mathematical benefit from Peter's advice. He helped Jean-Louis Colliot-Thélène and me set up a seminar to study Mumford's little red book, and was always in a position to illustrate our questions with some example from his own research experience, although his background in Weil foundations meant that there was always the added challenge of a language barrier. The subject of my thesis (the cohomology of the intersection of two quadrics), given to me by Pierre Deligne, turned out to be closely related to Peter's work with Bombieri on the cubic 3-fold [22]. Peter was also in the thick of the action surrounding modular forms at the time of the 1972 Antwerp conference [43]–[45].

From 1978, when I got married and left Cambridge for Warwick, my contact with Peter became less frequent. A few years later, Peter married the distinguished archeologist Harriet Crawford (reader at UCL and author of 3 books in the current Amazon catalogue). Together with everyone else in British academia, I was frequently aware, often through the media, of his activities as Vice Chancellor of Cambridge, as Chairman of the University Grants Committee, as the person who persuaded the conservative government of Mrs Thatcher ("We shall not see her like again!") to accept research as the main criterion for judging the quality of universities, and in numerous other capacities. As a member of the British Great and Good, he chaired any number of committees or public enquiries, investigating anything from parochial malpractice at British universities (see

http://www.freedomtocare.org/page37.htm),

to the disastrous storm of 16th October 1987 (this on behalf of the Secretary of State, see Meteorological Magazine 117, 141–144). I met him, for example, in Japan on a mission to investigate the state of university libraries.

At about the time Peter retired from the UGC, Warwick University had the foresight to offer him the position of Honorary Professor. He has visited us on many occasions in this capacity, both on Vice Chancellor's business and for mathematical visits, on each occasion giving us the full benefit of his wit and wisdom (for example, his scathing comments on teaching assessment in universities: "The Teaching Quality Assessment was an extremely tedious farce, bloody silly"). On several occasions he has given two mathematical lectures on the same day, one in Diophantine geometry and another in differential equations, before taking us all out to a very good dinner.

Peter is more active in research than ever at age 75, and in closer contact with us at Warwick: he has repeated his lecture series "New methods for Diophantine equations" (first given in Arizona in December 2002) as a Warwick M.Sc. course, driving over each week and meeting us for lunch in a Kenilworth pub, at which Peter takes two pints of cider to put him in good voice for the afternoon lectures.

I close with some Swinnerton-Dyer quotes:

- To have a computer job rejected by the EDSAC 2 Priorities Committee, "You had to be both stupid and arrogant – neither alone would do it."
- On meeting Colin Sparrow in King's Parade "I have been made Chairman of UGC. Waste of a knighthood!"
- "They aren't true, of course, but one believes them at least as much as one believes the Thirty-Nine Articles of the Church of England."
- In Trinity College parlour with Alexei Skorobogatov (in connection with the dogma of the Orthodox Church): "In order to become a clergyman in the Church of England you need to believe only one thing – that it is better to be wealthy than poor."

Warwick, 21st Feb 2002

## 5 Editor's preface to the volume by Alexei Skorobogatov

The papers in this volume offer a representative slice of the delicately intertwined tissue of analytic, geometric and cohomological methods used to attack the fundamental questions on rational solutions of Diophantine equations. A unique feature of the study of rational points is the enormous variety of methods that interact and contribute to our understanding of their behaviour: to name but a few, the Hardy–Littlewood circle method, the geometry of the underlying complex algebraic varieties, arithmetic and geometry over finite and $p$-adic fields, harmonic analysis, Manin's use of the Brauer–Grothendieck group to define a systematic obstruction to the Hasse principle, the theory of universal torsors of Colliot-Thélène–Sansuc, and the analysis of Shafarevich–Tate groups. It is no exaggeration to say that pioneering work of Peter

Swinnerton-Dyer was an early example of many of these techniques, and a source of inspiration for others. The contents of this volume, that we now describe, reflect this vast influence.

## Analytic number theory

The paper by **Enrico Bombieri** and **Paula B. Cohen** "*An elementary approach to effective diophantine approximation on* $\mathbf{G}_m$" concerns approximations of high order roots of algebraic numbers, with applications to Diophantine approximation in a number field by a finitely generated multiplicative subgroup. Such results can be obtained from the theory of linear forms in logarithms, whereas Bombieri's new approach is based on the Thue–Siegel–Roth theorem. The main improvement comes from a new zero lemma that is simpler than the lemma of Dyson employed up to now. The results sharpen Liouville's inequality for $r$th roots of an algebraic number $a$. More precisely, the authors obtain a lower bound for the distance $|a^{1/r} - \gamma|$, where $\gamma$ is an algebraic number, and $|\cdot|$ a non-Archimedean absolute value.

**Roger Heath-Brown's** paper "*Linear relations amongst sums of two squares*" is an inspiring example of what analytic methods can do for the study of rational points. The main result of the paper is an asymptotic formula for the number of integral points of prescribed height on a class of intersections of two quadratic forms in six variables. This formula accounts for possible failures of weak approximation. The result is a significant advance in the state of knowledge on density of rational points, for existing methods (such as the circle method) provide asymptotic formulas given by the product of local densities. Heath-Brown determines the additional factor that reflects the failure of weak approximation — a conclusion that was hitherto inaccessible. Such a result should provide a stimulus to establish analogous conclusions for a broader range of examples. The proof involves descent to an intersection of quadratic forms, to which analytic methods can be applied. The analysis here is delicate, and motivated by earlier work of Hooley and Daniel.

## Diophantine equations

**Andrew Bremner's** short note "*A Diophantine system*" finds infinitely many nontrivial $\mathbf{Q}$-rational points on the complete intersection surface given by

$$x_1^k + x_2^k + x_3^k = y_1^k + y_2^k + y_3^k \quad \text{for } k = 2, 3, 4.$$

Trivial solutions to this system, with the second triple a permutation of the first, are of no interest, but only one nontrivial rational solution was previously known. The proof is the observation that the hyperplane section $x_1 + x_2 + y_1 + y_2 = 0$ gives an elliptic curve of rank 1.

In "*Valeurs d'un polynôme à une variable représentées par une norme*", **Jean-Louis Colliot-Thélène**, **David Harari** and **Alexei Skorobogatov** consider the Diophantine equation $P(t) = N_{K/k}(z)$, where $P(t)$ is a polynomial and $N_{K/k}(z)$ the norm form defined by a finite field extension $K/k$. The paper builds on previous work by Heath-Brown and Skorobogatov, who combined the circle method and descent to prove results on rational solutions of this equation for $P(t)$ a product of two linear factors and $k = \mathbf{Q}$. It studies in detail the Brauer group of a smooth and proper model of the variety given by $P(t) = N_{K/k}(z)$, with $k$ an arbitrary field, and calculates it explicitly under some additional assumptions. On the other hand, when $k = \mathbf{Q}$ and $P(t)$ is a product of arbitrary powers of two linear factors, the Brauer–Manin obstruction is proved to be the only obstruction to the Hasse principle and to weak approximation. This leads to some new cases of the Hasse principle.

The consensus among experts seems to be that the failure of the Hasse principle for rational surfaces can be characterised in terms of the Brauer–Manin obstruction (this is far from being settled; possibly the closely related problem for zero-cycles of degree 1 has more chances of success). Recent work of Skorobogatov shows that this fails for some bielliptic surfaces; the paper of **Laura Basile and Alexei Skorobogatov** "*On the Hasse principle for bielliptic surfaces*" explores this area, providing positive and negative results as testing ground for a future overall conjecture.

In his contribution "*On the obstructions to the Hasse principle*", **Per Salberger** gives a new proof of the main theorem of the descent theory of Colliot-Thélène and Sansuc. Surprisingly, this new approach avoids an explicit computation of the Poitou–Tate pairing at the crucial point of the proof, relying instead on standard functoriality properties of étale cohomology. One of the results was obtained independently by Colliot-Thélène and Swinnerton-Dyer, following Salberger's innovative 1988 paper. It is interesting to note that whereas Colliot-Thélène and Swinnerton-Dyer extended Salberger's original method, in the present paper Salberger uses for the first time Colliot-Thélène and Sansuc's universal torsors to prove results about zero-cycles. This demonstrates in a striking way that universal torsors are well adapted not only for rational points, but also for zero-cycles. This approach may eventually advance our understanding of the following question of Colliot-Thélène: is the Brauer–Manin obstruction to the existence of a zero-cycle of degree 1 the only obstruction, if we assume the existence of such cycles everywhere locally?

## Shafarevich–Tate groups

**Neil Dummigan**, **William Stein** and **Mark Watkins'** paper "*Constructing elements in Shafarevich–Tate groups of modular motives*" gives a criterion for the existence of nontrivial elements of certain Shafarevich–Tate groups. Their methods build upon Cremona and Mazur's notion of "visibility", but in the context of motives rather than abelian varieties. The motives considered are attached to modular forms on $\Gamma_0(N)$ of weight $> 2$. Examples are found in which the Beilinson–Bloch conjectures imply the existence of nontrivial elements of these Shafarevich–Tate groups. Modular symbols and Tamagawa numbers are used to compute nontrivial conjectural lower bounds for the orders of the Shafarevich–Tate groups of modular motives of low level and weight $\leq 12$.

**Tom Fisher's** paper "*A counterexample to a conjecture of Selmer*" answers the following question. Let $K$ be a number field containing a primitive cube root of unity, and $E$ an elliptic curve over $K$ having complex multiplication by $\sqrt{-3}$. Is the kernel of this complex multiplication on the Shafarevich–Tate group of $E$ over $K$ of square order? The answer is positive if $E$ is defined over a subfield $k \subset K$ such that $[K : k] = 2$, $K = k(\sqrt{-3})$, assuming that the Shafarevich–Tate group of $E$ over $k$ is finite. Examples show that without this assumption the answer can be negative. These results play an important rôle in the new method for proving the Hasse principle for pencils of curves of genus 1, first used by Heath-Brown and then artfully employed by Swinnerton-Dyer in his recent paper on the Hasse principle for diagonal cubic forms.

In "*On Shafarevich–Tate groups of Fermat jacobians*", **William McCallum** and **Pavlos Tzermias** find all the points on the Fermat curve of degree 19 with quadratic residue field; these turn out to be the points previously described by Gross and Rohrlich. The result about rational points is an application of the following result about the Shafarevich–Tate groups. For an odd prime $p$, let $F$ be a quotient of the $p$th Fermat curve by $\mu_p$, and let $J$ be the jacobian of $F$. Then $J$ has complex multiplication by the ring of integers of the cyclotomic field $K = \mathbf{Q}(\zeta_p)$. The authors prove that in certain cases there are nontrivial elements of order exactly $(1 - \zeta_p)^3$ in the Shafarevich–Tate group of $J$ over $K$.

## Zagier's conjectures

In his paper "*Kronecker double series and the dilogarithm*", **Andrey Levin** gives an explicit expression for the value of a certain Kronecker double series at a point of complex multiplication as a sum of dilogarithms whose arguments are values of some modular unit of higher level. This result can be interpreted in the spirit of Zagier's conjecture. The special value of the Kronecker double series is equal to the value of the partial zeta function of an ideal class for an order in an imaginary quadratic field. The values of the modular unit mentioned above belong to the ray class field corresponding to this order. This gives an explicit formula for the value of a partial zeta function at $s = 2$ as a combination of dilogarithms of algebraic numbers.

## Complex algebraic geometry

In "*Cascades of projections from log del Pezzo surfaces*", **Miles Reid** and **Kaori Suzuki** weave a fantasy around the fascinating old algebraic geometric construction (del Pezzo, 1890) of the blowup of $\mathbf{P}^2$ in $d \leq 8$ general points and its anticanonical embedding. Some natural families of del Pezzo surfaces with quotient singularities are organized in 'cascades' of projections, similar to the way that the classic nonsingular del Pezzo surfaces are obtained by successive projections from the del Pezzo surface of degree 9 in $\mathbf{P}^9$ (in other words, $\mathbf{P}^2$ in its anticanonical embedding). Apart from their geometric beauty, these examples illustrate the technique of 'unprojection', a good working substitute for an as yet missing structure theory of Gorenstein rings of small codimension, and a possible tool to eventually construct one. The authors also sketch a program for the study of singular Fano 3-folds of index $\geq 2$ according to their Hilbert series, modelled on the 2-dimensional case.

**Gregory Sankaran** studies the bilevel structures on abelian surfaces first introduced by Mukai. Given a $(1,t)$-polarized abelian surface $A$, a bilevel structure on $A$ consists of a (canonical) level structure on $A$ and a (canonical) level structure on the dual variety $\widehat{A}$, which also carries a natural $(1,t)$-polarization. The corresponding moduli problem gives rise to a Siegel modular threefold $\mathcal{A}_t^{\mathrm{bil}}$. Mukai proved the rationality of these moduli spaces for $t = 2, 3$ and 5. He also related them to the symmetry groups of the Platonic solids and to projective threefolds with many nodes. In "*Abelian surfaces with odd bilevel structure*" Sankaran proves that $\mathcal{A}_t^{\mathrm{bil}}$ is of general type for odd $t \geq 17$. A result of Borisov says that $\mathcal{A}_t^{\mathrm{bil}}$ is of general type for all but finitely many $t$. Borisov's method, however, gives no explicit bound.

Imperial College, Mon 24th Feb 2003

Bryan Birch,
Mathematical Institute,
24-29 St Giles',
Oxford, OX1 3LB, UK
e-mail: birch@maths.ox.ac.uk

Jean-Louis Colliot-Thélène,
C.N.R.S., U.M.R. 8628,
Mathématiques, Bâtiment 425,
Université de Paris-Sud,
F-91405 Orsay, France
e-mail: colliot@math.u-psud.fr

Miles Reid,
Math Inst., Univ. of Warwick,
Coventry CV4 7AL, England
e-mail: miles@maths.warwick.ac.uk
web: www.maths.warwick.ac.uk/~miles

G.K. Sankaran,
Department of Mathematical Sciences,
University of Bath,
Bath BA2 7AY, England
e-mail: gks@maths.bath.ac.uk

Alexei Skorobogatov,
Department of Mathematics,
Imperial College London,
South Kensington Campus,
London SW7 2AZ, England
e-mail: a.skorobogatov@ic.ac.uk

# Peter Swinnerton-Dyer's mathematical papers to date


[1] P. S. Dyer, A solution of $A^4 + B^4 = C^4 + D^4$, J. London Math. Soc. **18** (1943) 2–4

[2] H. P. F. Swinnerton-Dyer, On a conjecture of Hardy and Littlewood, J. London Math. Soc. **27** (1952) 16–21

[3] H. P. F. Swinnerton-Dyer, A solution of $A^5 + B^5 + C^5 = D^5 + E^5 + F^5$, Proc. Cambridge Phil. Soc. **48** (1952) 516–518

[4] H. P. F. Swinnerton-Dyer, Extremal lattices of convex bodies, Proc. Cambridge Phil. Soc. **49** (1953) 161–162

[5] E. S. Barnes and H. P. F. Swinnerton-Dyer, The inhomogeneous minima of binary quadratic forms. I, Acta Math. **87** (1952) 259–323. II, same J. **88** (1952) 279–316. III, same J. **92** (1954) 199–234

[6] H. P. F. Swinnerton-Dyer, Inhomogeneous lattices, Proc. Cambridge Phil. Soc. **50** (1954) 20–25

[7] A. O. L. Atkin and P. Swinnerton-Dyer, Some properties of partitions, Proc. London Math. Soc. (3) **4** (1954) 84–106

[8] J. W. S. Cassels and H. P. F. Swinnerton-Dyer, On the product of three homogeneous linear forms and the indefinite ternary quadratic forms, Phil. Trans. Roy. Soc. London. Ser. A. **248** (1955) 73–96

[9] H. Davenport and H. P. F. Swinnerton-Dyer, Products of inhomogeneous linear forms, Proc. London Math. Soc. (3) **5** (1955) 474–499

[10] B. J. Birch and H. P. F. Swinnerton-Dyer, On the inhomogeneous minimum of the product of $n$ linear forms, Mathematika **3** (1956) 25–39

[11] H. P. F. Swinnerton-Dyer, On an extremal problem, Proc. London Math. Soc. (3) **7** (1957) 568–583

[12] K. Rogers and H. P. F. Swinnerton-Dyer, The geometry of numbers over algebraic number fields, Trans. Amer. Math. Soc. **88** (1958) 227–242

[13] B. J. Birch and H. P. F. Swinnerton-Dyer, Note on a problem of Chowla, Acta Arith. **5** (1959) 417–423

[14] D. W. Barron and H. P. F. Swinnerton-Dyer, Solution of simultaneous linear equations using a magnetic-tape store, Comput. J. **3** (1960/1961) 28–33

[15] H. P. F. Swinnerton-Dyer, Two special cubic surfaces, Mathematika **9** (1962) 54–56

[16] H. T. Croft and H. P. F. Swinnerton-Dyer, On the Steinhaus billiard table problem, Proc. Cambridge Phil. Soc. **59** (1963) 37–41

[17] B. J. Birch and H. P. F. Swinnerton-Dyer, Notes on elliptic curves. I, J. reine angew. Math. **212** (1963) 7–25

[18] B. J. Birch and H. P. F. Swinnerton-Dyer, Notes on elliptic curves. II, J. reine angew. Math. **218** (1965) 79–108

[19] H. P. F. Swinnerton-Dyer, Rational zeros of two quadratic forms, Acta Arith. **9** (1964) 261–270

[20] H. P. F. Swinnerton-Dyer, On the formal stability of the solar system, Proc. London Math. Soc. (3) **14a** (1965) 265–287

[21] H. P. F. Swinnerton-Dyer, The zeta function of a cubic surface over a finite field, Proc. Cambridge Phil. Soc. **63** (1967) 55–71

[22] E. Bombieri and H. P. F. Swinnerton-Dyer, On the local zeta function of a cubic threefold, Ann. Scuola Norm. Sup. Pisa (3) **21** (1967) 1–29

[23] H. P. F. Swinnerton-Dyer, An application of computing to class field theory, in Algebraic Number Theory (Brighton 1965), Thompson, Washington, D.C. (1967), pp. 280–291

[24] H. P. F. Swinnerton-Dyer, $A^4+B^4=C^4+D^4$ revisited, J. London Math. Soc. **43** (1968) 149–151

[25] P. Swinnerton-Dyer, The conjectures of Birch and Swinnerton-Dyer, and of Tate, in Proc. Conf. Local Fields (Driebergen 1966), Springer, Berlin (1967), pp. 132–157

[26] P. Swinnerton-Dyer, The use of computers in the theory of numbers, in Proc. Sympos. Appl. Math., Vol. XIX, Amer. Math. Soc., Providence, R.I. (1967), pp. 111–116

[27] F. K. C. Rankin and H. P. F. Swinnerton-Dyer, On the zeros of Eisenstein series, Bull. London Math. Soc. **2** (1970) 169–170

[28] H. P. F. Swinnerton-Dyer, On a problem of Littlewood concerning Riccati's equation, Proc. Cambridge Phil. Soc. **65** (1969) 651–662

[29] H. P. F. Swinnerton-Dyer, The birationality of cubic surfaces over a given field, Michigan Math. J. **17** (1970) 289–295

[30] H. P. F. Swinnerton-Dyer, On the product of three homogeneous linear forms, Acta Arith. **18** (1971) 371–385

[31] H. P. F. Swinnerton-Dyer, The products of three and of four linear forms, in Computers in number theory (Oxford 1969), A. O. L. Atkin and B. J. Birch (eds.), Academic Press, London-New York (1971), pp. 231–236

[32] A. O. L. Atkin and H. P. F. Swinnerton-Dyer, Modular forms on non-congruence subgroups, in Combinatorics (UCLA 1968), Proc. Sympos. Pure Math., Vol. XIX, Amer. Math. Soc., Providence, R.I. (1971), pp. 1–25

[33] H. P. F. Swinnerton-Dyer, The number of lattice points on a convex curve, J. Number Theory **6** (1974) 128–135

[34] H. P. F. Swinnerton-Dyer, Applications of algebraic geometry to number theory, in Number Theory (Stony Brook 1969), Proc. Sympos. Pure Math., Vol. XX, Amer. Math. Soc., Providence, R.I. (1971), pp. 1–52

[35] H. P. F. Swinnerton-Dyer, Applications of computers to the geometry of numbers, in Computers in algebra and number theory (New York 1970), Proc. Sympos. Appl. Math., SIAM-AMS Proc., Vol. IV, Amer. Math. Soc., Providence, R.I. (1971), pp. 55–62

[36] H. P. F. Swinnerton-Dyer, An enumeration of all varieties of degree 4, Amer. J. Math. **95** (1973) 403–418

[37] B. Mazur and P. Swinnerton-Dyer, Arithmetic of Weil curves, Invent. Math. **25** (1974) 1–61

[38] Mary L. Cartwright and H. P. F. Swinnerton-Dyer, Boundedness theorems for some second order differential equations. I, Collection of articles dedicated to the memory of Tadeusz Ważewski, III, Ann. Polon. Math. **29** (1974) 233–258

[39] H. P. F. Swinnerton-Dyer, Almost-conservative second-order differential equations, Math. Proc. Cambridge Phil. Soc. **77** (1975) 159–169

[40] H. P. F. Swinnerton-Dyer, Analytic theory of abelian varieties, London Mathematical Society Lecture Note Series, No. 14. Cambridge University Press, London-New York, 1974. viii+90 pp.

[41] H. P. F. Swinnerton-Dyer, An outline of Hodge theory, in Algebraic geometry (Oslo 1970), Wolters-Noordhoff, Groningen (1972), pp. 277–286

[42] H. P. F. Swinnerton-Dyer, Rational points on del Pezzo surfaces of degree 5, in Algebraic geometry (Oslo 1970), Wolters-Noordhoff, Groningen (1972), pp. 287–290

[43] H. P. F. Swinnerton-Dyer, On $l$-adic representations and congruences for coefficients of modular forms, in Modular functions of one variable, III (Antwerp 1972), Lecture Notes in Math., Vol. 350, Springer, Berlin, 1973, pp. 1–55. Correction in [44], p. 149

[44] H. P. F. Swinnerton-Dyer and B. J. Birch, Elliptic curves and modular functions, in Modular functions of one variable, IV (Antwerp 1972), Lecture Notes in Math., Vol. 476, Springer, Berlin (1975), pp. 2–32

[45] H. P. F. Swinnerton-Dyer, N. M. Stephens, James Davenport, J. Vélu, F. B. Coghlan, A. O. L. Atkin and D. J. Tingley, Numerical tables on elliptic curves, in Modular functions of one variable, IV (Antwerp 1972), Lecture Notes in Math., Vol. 476, Springer, Berlin (1975), pp. 74–144

[46] M. Artin and H. P. F. Swinnerton-Dyer, The Shafarevich–Tate conjecture for pencils of elliptic curves on K3 surfaces, Invent. Math. **20** (1973) 249–266

[47] B. J. Birch and H. P. F. Swinnerton-Dyer, The Hasse problem for rational surfaces, in Collection of articles dedicated to Helmut Hasse on his seventy-fifth birthday. III, J. reine angew. Math. **274/275** (1975) 164–174

[48] Peter Swinnerton-Dyer, The Hopf bifurcation theorem in three dimensions, Math. Proc. Cambridge Phil. Soc. **82** (1977) 469–483

[49] M. L. Cartwright and H. P. F. Swinnerton-Dyer, The boundedness of solutions of systems of differential equations, in Differential equations (Keszthely 1974), Colloq. Math. Soc. János Bolyai, Vol. 15, North-Holland, Amsterdam (1977), pp. 121–130

[50] H. P. F. Swinnerton-Dyer, Arithmetic groups, in Discrete groups and automorphic functions (Cambridge, 1975), Academic Press, London (1977), pp. 377–401

[51] H. P. F. Swinnerton-Dyer, On $l$-adic representations and congruences for coefficients of modular forms. II, in Modular functions of one variable, V (Bonn 1976), Lecture Notes in Math., Vol. 601, Springer, Berlin (1977), pp. 63–90

[52] Peter Swinnerton-Dyer, Small parameter theory: the method of averaging, Proc. London Math. Soc. (3) **34** (1977) 385–420

[53] Peter Swinnerton-Dyer, The Royal Society and its impact on the intellectual and cultural life of Britain, Jbuch. Heidelberger Akad. Wiss. 1979 (1980) 136–143

[54] H. P. F. Swinnerton-Dyer, The method of averaging for some almost-conservative differential equations, J. London Math. Soc. (2) **22** (1980) 534–542

[55] M. L. Kartraĭt and H. P. F. Svinnerton-Daĭer, Boundedness theorems for some second-order differential equations. IV, Differentsial'nye Uravneniya **14** (1978) 1941–1979 and 2106 = Differ. Equations **14** (1979) 1378–1406

[56] H. P. F. Swinnerton-Dyer, Universal equivalence for cubic surfaces over finite and local fields, in Severi centenary symposium (Rome 1979), Symposia Mathematica, Vol. XXIV, Academic Press, London-New York (1981), pp. 111–143

[57] Jean-Louis Colliot-Thélène, Jean-Jacques Sansuc and Peter Swinnerton-Dyer, Intersections de deux quadriques et surfaces de Châtelet, C. R. Acad. Sci. Paris Sér. I Math. **298** (1984) 377–380

[58] H. P. F. Swinnerton-Dyer, The basic Lorenz list and Sparrow's conjecture A, J. London Math. Soc. (2) **29** (1984) 509–520

[59] Arnaud Beauville, Jean-Louis Colliot-Thélène, Jean-Jacques Sansuc et Peter Swinnerton-Dyer, Variétés stablement rationnelles non rationnelles, Ann. of Math. (2) **121** (1985) 283–318

[60] H. P. F. Swinnerton-Dyer, The field of definition of the Néron-Severi group, in Studies in pure mathematics in memory of Paul Turán, Birkhäuser, Basel (1983), pp. 719–731

[61] Jean-Louis Colliot-Thélène, Jean-Jacques Sansuc and Peter Swinnerton-Dyer, Intersections of two quadrics and Châtelet surfaces. I, J. reine angew. Math. **373** (1987) 37–107. II, same J., **374** (1987) 72–168

[62] H. P. F. Swinnerton-Dyer, Congruence properties of $\tau(n)$, in Ramanujan revisited (Urbana-Champaign 1987), Academic Press, Boston (1988), pp. 289–311

[63] R. G. E. Pinch and H. P. F. Swinnerton-Dyer, Arithmetic of diagonal quartic surfaces. I, in $L$-functions and arithmetic (Durham 1989), London Math. Soc. Lecture Note Ser., 153, Cambridge Univ. Press, Cambridge (1991), pp. 317–338

[64] Peter Swinnerton-Dyer, The Brauer group of cubic surfaces, Math. Proc. Cambridge Phil. Soc. **113** (1993) 449–460

[65] Peter Swinnerton-Dyer, Counting rational points on cubic surfaces, in Classification of algebraic varieties (L'Aquila 1992), Contemp. Math., 162, Amer. Math. Soc., Providence, RI (1994), pp. 371–379

[66] Jean-Louis Colliot-Thélène and Peter Swinnerton-Dyer, Hasse principle and weak approximation for pencils of Severi-Brauer and similar varieties, J. reine angew. Math. **453** (1994) 49–112

[67] Peter Swinnerton-Dyer, Rational points on pencils of conics and on pencils of quadrics, J. London Math. Soc. (2) **50** (1994) 231–242

[68] H. P. F. Swinnerton-Dyer and C. T. Sparrow, The Falkner–Skan equation. I, The creation of strange invariant sets, J. Differential Equations **119** (1995) 336–394

[69] Peter Swinnerton-Dyer, Rational points on certain intersections of two quadrics, in Abelian varieties (Egloffstein 1993), de Gruyter, Berlin (1995), pp. 273–292

[70] J.-L. Colliot-Thélène, A. N. Skorobogatov and Peter Swinnerton-Dyer, Double fibres and double covers: paucity of rational points, Acta Arith. **79** (1997) 113–135

[71] Peter Swinnerton-Dyer, Diophantine equations: the geometric approach, Jahresber. deutsch. Math.-Verein. **98** (1996) 146–164

[72] Peter Swinnerton-Dyer, Brauer–Manin obstructions on some Del Pezzo surfaces, Math. Proc. Cambridge Phil. Soc. **125** (1999) 193–198

[73] J.-L. Colliot-Thélène, A. N. Skorobogatov and Peter Swinnerton-Dyer, Rational points and zero-cycles on fibred varieties: Schinzel's hypothesis and Salberger's device, J. reine angew. Math. **495** (1998) 1–28

[74] J.-L. Colliot-Thélène, A. N. Skorobogatov and Peter Swinnerton-Dyer, Hasse principle for pencils of curves of genus one whose Jacobians have rational 2-division points, Invent. Math. **134** (1998) 579–650

[75] Peter Swinnerton-Dyer, A stability theorem for unsymmetric Liénard equations, Dynam. Stability Systems **14** (1999) 93–94

[76] Peter Swinnerton-Dyer, Some applications of Schinzel's hypothesis to Diophantine equations, in Number theory in progress, Vol. 1 (Zakopane-Kościelisko 1997), de Gruyter, Berlin (1999), pp. 503–530

[77] John B. Slater and Peter Swinnerton-Dyer, Counting points on cubic surfaces. I, in Nombre et répartition de points de hauteur bornée (Paris 1996), Astérisque No. 251 (1998), pp. 1–12

[78] D. F. Coray, D. J. Lewis, N. I. Shepherd-Barron and Peter Swinnerton-Dyer, Cubic threefolds with six double points, in Number theory in progress, Vol. 1 (Zakopane-Kościelisko 1997), de Gruyter, Berlin (1999), pp. 63–74

[79] Peter Swinnerton-Dyer, Rational points on some pencils of conics with 6 singular fibres, Ann. Fac. Sci. Toulouse Math. (6) **8** (1999) 331–341

[80] Peter Swinnerton-Dyer, Arithmetic of diagonal quartic surfaces. II, Proc. London Math. Soc. (3) **80** (2000) 513–544, and Corrigenda, same J. **85** (2002) 564

[81] Peter Swinnerton-Dyer, A note on Liapunov's method, Dyn. Stab. Syst. **15** (2000) 3–10

[82] H. P. F. Swinnerton-Dyer, A brief guide to algebraic number theory, London Mathematical Society Student Texts, 50. Cambridge University Press, Cambridge, 2001

[83] Peter Swinnerton-Dyer, Bounds for trajectories of the Lorenz equations: an illustration of how to choose Liapunov functions, Phys. Lett. A **281** (2001) 161–167

[84] A. O. Bender and Peter Swinnerton-Dyer, Solubility of certain pencils of curves of genus 1, and of the intersection of two quadrics in $\mathbf{P}^4$, Proc. London Math. Soc. (3) **83** (2001) 299–329

[85] Peter Swinnerton-Dyer, The solubility of diagonal cubic surfaces, Ann. Sci. École Norm. Sup. (4) **34** (2001) 891–912

[86] Peter Swinnerton-Dyer, The invariant algebraic surfaces of the Lorenz system, Math. Proc. Cambridge Phil. Soc. **132** (2002) 385–393

[87] Peter Swinnerton-Dyer, Weak approximation and $R$-equivalence on cubic surfaces, in Rational points on algebraic varieties, Progr. Math., 199, Birkhäuser, Basel (2001), pp. 357–404

[88] C. Sparrow and H. P. F. Swinnerton-Dyer, The Falkner–Skan equation. II, Dynamics and the bifurcations of $P$- and $Q$-orbits, J. Differential Equations **183** (2002) 1–55

## Other references

[89] B. J. Birch, Elliptic curves over **Q**: A progress report, in Number Theory (Stony Brook 1969), Proc. Sympos. Pure Math., Vol. XX, Amer. Math. Soc., Providence, R.I. (1971), pp. 396–400

[90] Yu. I. Manin, Le groupe de Brauer–Grothendieck en géométrie diophantienne, in Actes du Congrès International des Mathématiciens (Nice 1970), Gauthier-Villars, Paris (1971), Tome 1, pp. 401–411

[91] Yu. I. Manin, Parabolic points and zeta functions of modular curves, Izv. Akad. Nauk SSSR Ser. Mat. **36** (1972) 19–66

[92] Yu. I. Manin, Cubic forms: algebra, geometry, arithmetic, North-Holland 1974

[93] N. M. Stephens, The diophantine equation $X^3 + Y^3 = DZ^3$ and the conjectures of Birch and Swinnerton-Dyer, J. reine angew. Math. **231** (1968) 121–162

[94] André Weil, Letter to Peter Swinnerton-Dyer dated 24/7/1965

# On the Hasse principle for bielliptic surfaces

Carmen Laura Basile and Alexei Skorobogatov

*To Sir Peter Swinnerton-Dyer*

From the geometer's point of view, bielliptic surfaces can be described as quotients of abelian surfaces by freely acting finite groups, that are not abelian surfaces themselves. Together with abelian, K3 and Enriques surfaces they exhaust the class of smooth and projective minimal surfaces of Kodaira dimension 0. Because of their close relation to abelian surfaces, bielliptic surfaces are particularly amenable to computation. At the same time they display phenomena not encountered for rational, abelian or K3 surfaces, for example, torsion in the Néron–Severi group, finite geometric Brauer group, non-abelian fundamental group. This curious geometry is reflected in amusing arithmetical properties of these surfaces over number fields.

The behaviour of rational points on bielliptic surfaces was first studied by Colliot-Thélène, Swinnerton-Dyer and the second author [CSS] in relation with Mazur's conjectures on the connected components of the real closure of $\mathbb{Q}$-points. The second author then constructed a bielliptic surface over $\mathbb{Q}$ that has points everywhere locally but not globally; moreover, this counterexample to the Hasse principle cannot be explained by the Manin obstruction [S1] (see also [S2], Ch. 8). D. Harari [H] showed that bielliptic surfaces give examples of varieties with a Zariski dense set of rational points that do not satisfy weak approximation; moreover this failure cannot be explained by the Brauer–Manin obstruction.

A discrete invariant of a bielliptic surface is the order $n$ of the canonical class in the Picard group. The possible values of $n$ are 2, 3, 4 and 6. The surface contructed in [S1] has $n = 2$. Until now this was the only known counterexample to the Hasse principle that cannot be explained by the Manin obstruction. In this note we construct a similar example in the case $n = 3$. The difference is that we now need to consider elliptic curves with complex multiplication. The actual construction turns out to be somewhat simpler than in [S1]. In contrast, for the bielliptic surfaces with $n = 6$ we prove that the Manin obstruction to the Hasse principle is the only one (under the assumption that the Tate–Shafarevich group of its Albanese variety is finite).

# 1 Bielliptic surfaces

Let $k$ be a field of $\operatorname{char} k = 0$, and $\overline{k}$ be an algebraic closure of $k$. For a $k$-variety $X$ we write $\overline{X} = X \times_k \overline{k}$.

**Definition 1** A *bielliptic surface* $X$ over $k$ is a smooth projective surface such that $\overline{X}$ is a minimal surface of Kodaira dimension 0, and is not a K3, abelian or Enriques surface.

Bielliptic surfaces over $\overline{k}$ were classified by Bagnera and de Franchis (see [B], VI.20). Their theorem says that any bielliptic surface over $\overline{k}$ can be obtained as the quotient of the product of two elliptic curves $E \times F$ by a freely acting finite abelian group. The geometric genus of any bielliptic surface is 0. For a bielliptic surface $X$ let $n$ be the order of $K_{\overline{X}}$ in $\operatorname{Pic} \overline{X}$. It follows from the Bagnera–de Franchis classification that $n$ can be $2, 3, 4$ or $6$ (*loc.cit.*).

**Proposition 1** *Let $X$ be a bielliptic surface over $k$. There exists an abelian surface $A$, a principal homogeneous space $Y$ of $A$, and a finite étale morphism $f \colon Y \to X$ of degree $n$, that is a torsor under the group scheme $\mu_n$.*

**Proof** The natural map $\operatorname{Pic} X \to \operatorname{Pic} \overline{X}$ is injective, hence $nK_X$ is a principal divisor. We write $nK_X = (\phi)$, where $\phi \in k(X)^*$. Let $Y$ be the normalization of the covering of $X$ given by $t^n = \phi$. Then the natural map $f \colon Y \to X$ is unramified, and is a torsor under $\mu_n$ (cf. [CS], 2.3.1, 2.4.1). This implies that $K_Y = f^* K_X = 0$. By the classification of surfaces, $\overline{Y}$ is an abelian surface. (It is not K3 as the only unramified quotients of K3 surfaces are Enriques surfaces.) Let $A$ be the Albanese variety of $Y$, defined over $k$ (see [L], II.3). Then $\overline{A}$ is the Albanese variety of $\overline{Y}$. The choice of a base point makes $\overline{Y}$ an abelian variety isomorphic to $\overline{A}$, so that $\overline{Y}$ is naturally a principal homogeneous space of $\overline{A}$. Choose $\overline{y}_0 \in Y(\overline{k})$, then we have an isomorphism $\overline{Y} \to \overline{A}$ that sends $\overline{y}$ to $\overline{y} - \overline{y}_0$. Then $\rho(g) = {}^g\overline{y}_0 - \overline{y}_0$ is a continuous 1-cocycle of $\operatorname{Gal}(\overline{k}/k)$ with coefficients in $A(\overline{k})$. Let $A^\rho$ be the principal homogeneous space of $A$ defined by $\rho$; it corresponds to the twisted Galois action $(g, \overline{a}) \mapsto {}^g\overline{a} + \rho(g)$, where $\overline{a} \in A(\overline{k})$ (see [S], III.1, or [S2], 2.1). Then the above $\overline{k}$-isomorphism $\overline{Y} \to \overline{A}$ descends to a $k$-isomorphism $Y \to A^\rho$. □

Note that the analogue of the proposition fails in higher dimension because there are many more possibilities for $Y$.

# 2 Group action on principal homogeneous spaces of abelian varieties

Let $A$ be an abelian variety over $k$, and $Z$ a principal homogeneous space of $A$. Suppose that a $k$-group scheme $\Gamma$ acts on $Z$. This gives rise to a Galois-equivariant action of the group $\Gamma(\overline{k})$ on the set $Z(\overline{k})$. The action of $\Gamma$ on $Z$ naturally defines an action of $\Gamma$ on $A$, the Albanese variety of $Z$. Then the action of the group $A(\overline{k})$ on the set $Z(\overline{k})$ is both Galois and $\Gamma$-equivariant. Let $A^\Gamma$ be the $\Gamma$-invariant group subscheme of $A$. Similarly, let $Z^\Gamma \subset Z$ be the closed subscheme consisting of points fixed by $\Gamma$.

**Proposition 2** *Suppose that a $k$-group scheme $\Gamma$ acts on $Z$ in such a way that $Z^\Gamma$ is a nonempty scheme (i.e., some $\overline{k}$-point of $Z$ is fixed by $\Gamma(\overline{k})$). Then $[Z] \in \mathrm{Im}[H^1(k, A^\Gamma) \to H^1(k, A)]$.*

**Proof** Take $\overline{x} \in Z(\overline{k})$, fixed by $\Gamma(\overline{k})$. Then a 1-cocycle of $\mathrm{Gal}(\overline{k}/k)$ sending $g \in \mathrm{Gal}(\overline{k}/k)$ to ${}^g\overline{x} - \overline{x} \in A(\overline{k})$ represents the class $[Z] \in H^1(k, A)$. For any $\gamma \in \Gamma(\overline{k})$ we have

$$\gamma({}^g\overline{x} - \overline{x}) = \gamma \cdot {}^g\overline{x} - \gamma \cdot \overline{x} = {}^g({}^{g^{-1}}\gamma \cdot \overline{x}) - \overline{x} = {}^g\overline{x} - \overline{x}.$$

Therefore, ${}^g\overline{x} - \overline{x} \in A^\Gamma(\overline{k})$. □

It is easy to see that $Z^\Gamma$ is a principal homogeneous space of $A^\Gamma$. The $A$-torsor $Z$ is the push-forward of the $A^\Gamma$-torsor $Z^\Gamma$ with respect to the natural injection of group schemes $A^\Gamma \to A$. This gives an alternative proof of the proposition.

**Corollary 1** *Let $A_1 = A/A^\Gamma$, and $\alpha\colon A \to A_1$ the natural surjection. Then $[Z] \in H^1(k, A)[\alpha_*]$, where*

$$H^1(k, A)[\alpha_*] = \ker[\alpha_*\colon H^1(k, A) \to H^1(k, A_1)].$$

We now consider the case when $Z = C$ is a curve of genus 1 equipped with a faithful action of $\Gamma$ that has a fixed point. Then $A = E$ is the Jacobian of $C$. We shall write $\mathrm{Aut}_0(\overline{E})$ for the automorphism group of $\overline{E}$ as an elliptic curve. Now $\Gamma(\overline{k}) \subset \mathrm{Aut}_0(\overline{E})$, hence $\Gamma(\overline{k})$ is a cyclic group of order $n$, where $n$ can be 1, 2, 3, 4 or 6. A straightforward calculation shows that, excluding the trivial case $n = 1$, we have one of the following possibilities:

| $n$ | $\#E^\Gamma(\overline{k})$ |
|---|---|
| 6 | 1 |
| 4 | 2 |
| 3 | 3 |
| 2 | 4 |

By Corollary 1 the first line of this table shows that if a cyclic group scheme of order 6 acts on a curve of genus 1, then this curve has a $k$-point. As a consequence of this fact we obtain in the next section a simple description of bielliptic surfaces with $n = 6$.

# 3 A case when the Manin obstruction to the Hasse principle is the only one

**Proposition 3** *Let $X$ be a bielliptic surface over $k$ such that the order of $K_{\overline{X}}$ in $\operatorname{Pic}\overline{X}$ is 6. There exist an elliptic curve $E$ and a curve $D$ of genus 1 such that the group scheme $\mu_6$ acts on $E$ by automorphisms of an elliptic curve (in particular, preserving the origin), and acts on $D$ by translations, in such a way that $X = (E \times D)/\mu_6$.*

**Proof** The Bagnera–de Franchis classification ([B], VI.20) says that for any bielliptic surface $\overline{X}$ with $K_{\overline{X}}$ of order 6 in $\operatorname{Pic}\overline{X}$ there exist elliptic curves $C_1$ and $C_2$ over $\overline{k}$ such that:

(1) $\mu_6$ acts on $C_1$ by automorphisms of an elliptic curve (in particular, preserving the origin);

(2) the group scheme $\mu_6$ is a subgroup of $C_2$;

(3) $X = (C_1 \times C_2)/\mu_6$.

The free action of $\mu_6$ on $C_1 \times C_2$ makes the finite étale map $C_1 \times C_2 \to \overline{X}$ a torsor under $\mu_6$. Let us compare it with the torsor $\overline{Y} \to \overline{X}$ constructed in Proposition 1.

Recall that the type of a $Z$-torsor under a group of multiplicative type $S$ is a certain functorial map $\widehat{S} \to \operatorname{Pic}\overline{Z}$, where $\widehat{S}$ is the module of characters of $S$ (see [S2], Definition 2.3.2). A torsor under a group of multiplicative type over an integral projective $\overline{k}$-variety is uniquely determined up to isomorphism by its type (this follows from the fundamental exact sequence of Colliot-Thélène and Sansuc, see [CS], [S2], (2.22)). Therefore it is enough to compare the respective types. There is an exact sequence

$$0 \to \operatorname{Hom}(\mu_6, \overline{k}^*) = \mathbb{Z}/6 \to \operatorname{Pic}\overline{X} \to \operatorname{Pic}\overline{Y},$$

where the second arrow is the type of the torsor $\overline{Y} \to \overline{X}$, and a similar sequence for $C_1 \times C_2 \to \overline{X}$ ([S2], (2.4) and Lemma 2.3.1). Since the canonical class of an abelian surface is trivial, $K_{\overline{X}}$ is in the image of $\mathbb{Z}/6$ in $\operatorname{Pic}\overline{X}$, and hence it is a generator of that image. Thus the types of both torsors are the same (up to sign). Hence the pair ($\overline{Y}$, the action of $\mu_6$) can be identified with the pair ($C_1 \times C_2$, the action of $\mu_6$).

Let $A$ be the Albanese variety of $Y$. This is an abelian surface defined over $k$. Let $s$ be the $k$-endomorphism of $A$ given by $s = \sum_{\sigma \in \mu_6} \sigma$. Let $A_1$ (respectively $A_2$) be the connected component of 0 in $\ker(s)$ (respectively in

$A^{\mu_6}$). Note that $s$ acts as 0 on $J_1 = \mathrm{Jac}(\overline{C}_1) \subset \overline{A}$, and as multiplication by 6 on $J_2 = \mathrm{Jac}(\overline{C}_2) \subset \overline{A}$. Therefore, $\overline{A}_1 = J_1$, $\overline{A}_2 = J_2$. Now the map

$$A_1 \times A_2 \to A, \quad (x, y) \mapsto x + y,$$

is an isomorphism, since over $\overline{k}$ it is the natural isomorphism $J_1 \times J_2 \to \overline{A}$. This proves that $A$ is a product of two elliptic curves over $k$. Hence $Y$, which is a principal homogeneous space of $A$, is a product of two curves of genus 1 over $k$: $Y = E \times D$, where $C_1 \simeq \overline{E}$, $C_2 \simeq \overline{D}$.

By the Bagnera–de Franchis theorem the group scheme $\mu_6$ acts on $E$ with a fixed point. By the remark preceding the statement of the proposition, this point is unique, and hence is $k$-rational. Hence $E$ is an elliptic curve (isomorphic to $A_1$). □

See the beginning of the next section (or, in more generality, [S2], 5.2) for the definition of the Manin obstruction.

**Corollary 2** *Let $k$ be a number field. The Manin obstruction is the only obstruction to the Hasse principle on the bielliptic surfaces $X$ over $k$ such that the order of $K_{\overline{X}}$ in $\mathrm{Pic}\,\overline{X}$ is 6, and the Tate–Shafarevich group of the Albanese variety of $X$ is finite.*

**Proof** By the previous proposition we have $X = (E \times D)/\mu_6$. Consider the curve $D' = D/\mu_6$ of genus 1, and let $p \colon X \to D'$ be the natural surjective map. Let $J'$ be the Jacobian of $D'$. It is known ([B], VI) that the Albanese variety of any bielliptic surface has dimension 1. Using the universal property of the Albanese variety (see [L], II.3) and the connectedness of the fibres of $p$ one easily checks that $J'$ is the Albanese variety of $X$.

Let $\{Q_v\}$ be a collection of local points on $X$, for all places $v$ of $k$, that satisfies the Brauer–Manin conditions. Then $\{p(Q_v)\}$ satisfies the Brauer–Manin conditions on $D'$. If $\text{Ш}(J')$ is finite, then $D'$ has a $k$-point by a theorem of Manin (see [S2], Theorem 6.2.3). Call this point $Q$. The inverse image of $Q$ in $D$ defines a class $\rho \in H^1(k, \mu_6) = k^*/k^{*6}$. Consider the twisted torsor $E^\rho \times D^\rho \to X$. Now $D^\rho$ has a $k$-point over $Q$. But the action of $\mu_6$ on $E$ preserves the origin, hence the twisted curve $E^\rho$ has a $k$-point. Therefore, we obtain a $k$-point on $E^\rho \times D^\rho$, and hence on $X$. □

Note that for the bielliptic surfaces of Corollary 2 the quotient of $\mathrm{Br}\,X$ by the image of $\mathrm{Br}\,k$ is infinite, but in the proof we only used the Brauer–Manin conditions given by the elements of the conjecturally finite group $\text{Ш}(J')$.

Corollary 2 is a particular case of a more general situation. Let $\Gamma$ be an algebraic group acting on varieties $V$ and $W$ such that the action on $W$ is free. Suppose that $V$ has a $k$-point fixed by $\Gamma$. If the Manin obstruction to the Hasse principle is the only one on $W/\Gamma$, then the same is true for $(V \times W)/\Gamma$.

# 4 Main construction and example

Now assume $k = \mathbb{Q}$, and let $\mathbf{A}_\mathbb{Q}$ be the ring of adèles of $\mathbb{Q}$. For a projective variety $X$ we have $X(\mathbf{A}_\mathbb{Q}) = \prod_v X(\mathbb{Q}_v)$, where $v$ ranges over all places of $\mathbb{Q}$ including the real place. Let $X(\mathbf{A}_\mathbb{Q})^{\mathrm{Br}}$ be the subset of $X(\mathbf{A}_\mathbb{Q})$ consisting of the families of local points $\{P_v\}$ satisfying all the Brauer–Manin conditions. These conditions, one for each $A \in \mathrm{Br}\, X$, are

$$\sum_{\text{all } v} \mathrm{inv}_v\, A(P_v) = 0,$$

where $\mathrm{inv}_v$ is the local invariant at the place $v$, which is a canonical map $\mathrm{Br}\, \mathbb{Q}_v \to \mathbb{Q}/\mathbb{Z}$ provided by local class field theory. The Brauer–Manin conditions are satisfied for any $\mathbb{Q}$-point of $X$ by the Hasse reciprocity law, so that we have $X(\mathbb{Q}) \subset X(\mathbf{A}_\mathbb{Q})^{\mathrm{Br}}$. If the last set is empty, this is an obstruction to the existence of a $\mathbb{Q}$-point on $X$; it is called the Manin obstruction.

We now give a construction of bielliptic surfaces $X$ for which $X(\mathbf{A}_\mathbb{Q})^{\mathrm{Br}} \neq \emptyset$, but $X(\mathbb{Q}) = \emptyset$. Then $X$ is a counterexample to the Hasse principle that is not explained by the Manin obstruction.

**Theorem 1** *Let $E$ be an elliptic curve over $\mathbb{Q}$ with a nontrivial action of the group scheme $\mu_3$. Let $\alpha\colon E \to E_1$ be the degree 3 isogeny with kernel $E^{\mu_3}$. Let $D$ be an elliptic curve with a group subscheme isomorphic to $\mu_3$. Assume that:*

(i) $\mathrm{Gal}(\overline{\mathbb{Q}}/\mathbb{Q})$ *acts nontrivially on* $E^{\mu_3}$*;*

(ii) $\#\text{Ш}(E)[\alpha_*] = 3$*;*

(iii) *$C$ is a principal homogeneous space of $E$ representing a nontrivial element of $\text{Ш}(E)[\alpha_*]$;*

(iv) $\mathrm{Sel}(D, \mu_3) = 0$*, that is, for any principal homogeneous space of $D$ obtained from a nontrivial class in $H^1(\mathbb{Q}, \mu_3) = \mathbb{Q}^*/\mathbb{Q}^{*3}$, there exists a place $v$ where it has no $\mathbb{Q}_v$-point.*

*Then $X = (C \times D)/\mu_3$ is a counterexample to the Hasse principal not explained by the Manin obstruction.*

Let us give an example of curves $C$ and $D$ satisfying the conditions of the theorem. Let $\zeta$ be a primitive cubic root of unity.

Let $C$ be the plane cubic curve $x^3 + 11y^3 + 43z^3 = 0$, where the root of unity $\zeta$ acts by $(x : y : z) \mapsto (x : y : \zeta z)$. The Jacobian $E$ of $C$ is the plane curve $x^3 + y^3 + 473z^3 = 0$, with the action of $\mu_3$ given by the same formula. One easily checks that Condition (i) is satisfied. Condition (ii) is verified in Example 4.3 of [F2]. The elements of $H^1(\mathbb{Q}, E)[\alpha_*]$ are given by the curves $mx^3 + m^2y^3 + 473z^3 = 0$ with $m$ a cube-free integer. The curve $C$ corresponds to $m = 11$. It has been known for some time [Se] that $C$ has points everywhere locally but not globally. This gives Condition (iii). (See also [Ba], VI.)

Let $D$ be the elliptic curve $u^3+v^3+w^3 = 0$, with $(1 : -1 : 0)$ as the origin. The group subscheme of $D$ generated by $(1 : -\zeta : 0)$ is isomorphic to $\mu_3$. The translation by this element is $(u, v, w) \mapsto (u : \zeta v : \zeta^2 w)$. The elements of the Selmer group $\mathrm{Sel}(D, \mu_3)$ are represented by the principal homogeneous spaces $D_a$ defined by $u^3 + av^3 + a^2u^3 = 0$, where $a$ is a cube free integer. Let $p$ be a prime factor of $a$. Then $D_a$ has no $\mathbb{Q}_p$-point. Therefore, the only curve $D_a$ with points everywhere locally is $D$ itself, so that $\mathrm{Sel}(D, \mu_3) = 0$, which is our Condition (iv).

**Remark** On changing some of the conditions of the theorem one obtains bielliptic surfaces for which the Manin obstruction to the Hasse principle is the only one. We replace Condition (ii) by the condition $Ш(E)[\alpha_*] = 0$, and instead of Condition (iii) we require that $C$ is any principal homogeneous space of $E$ whose class is in $H^1(\mathbb{Q}, E)[\alpha_*]$. We drop Condition (i) and keep Condition (iv). Then the Manin obstruction is the only obstruction to the Hasse principle for the surfaces $(C \times D)/\mu_3$. For the proof, consider the torsor $C \times D \to (C \times D)/\mu_3$ under $\mu_3$. Under our assumptions the class of twists $C^\rho \times D^\rho$, $\rho \in \mathbb{Q}^*/\mathbb{Q}^{*3}$, satisfies the Hasse principle. By descent theory ([S2], Corollary 6.1.3 (2)) this implies our statement.

# 5 Proof of the theorem

Consider the alternating Cassels pairing $Ш(E) \times Ш(E) \to \mathbb{Q}/\mathbb{Z}$. Its restriction to $Ш(E)[\alpha_*]$ gives an alternating pairing

$$Ш(E)[\alpha_*] \times Ш(E)[\alpha_*] \to \mathbb{Q}/\mathbb{Z}. \qquad (1)$$

The kernel of the last pairing is the image of $\alpha_*^t \colon Ш(E_1) \to Ш(E)$, where $\alpha^t \colon E_1 \to E$ is the dual isogeny. (This seems to be part of the folklore; see [F1] for a proof.) Since $Ш(E)[\alpha_*] \cong \mathbb{Z}/3\mathbb{Z}$ by Condition (ii), the pairing (1) must be zero. Therefore, there exists a principal homogeneous space $C_1$ of $E_1$ with points everywhere locally, that lifts $C$. This means that the map $\alpha_*^t \colon H^1(\mathbb{Q}, E_1) \to H^1(\mathbb{Q}, E)$ sends $[C_1]$ to $[C]$. There is a finite étale morphism $C_1 \to C$ that represents $C$ as the quotient of $C_1$ by the action of $\ker(\alpha^t)$. Let $Y = C \times D$, $Y_1 = C_1 \times D$. This gives rise to a finite étale morphism $Y_1 \to Y$ which is the identity on $D$. Let $f_1$ be the composition of the finite étale maps $Y_1 \to Y \to X$, and let $\pi \colon Y_1 \to D$ be the projection to the second factor. In this notation we have the following key property analogous to ([S1], Theorem 1):

$$f_1^*(\mathrm{Br}\, X) \subset \pi^*(\mathrm{Br}\, D). \tag{2}$$

To prove this we note that for any smooth and projective surface $X$ with $p_g = 0$, in particular, for a bielliptic surface, we have an isomorphism of Galois modules $\mathrm{Br}\, \overline{X} = \mathrm{Hom}(\mathrm{NS}(\overline{X})_{\mathrm{tors}}, \mathbb{Q}/\mathbb{Z})$ (see [G], II, Corollary 3.4, III, (8.12)). As in the proof of Corollary 2 one shows that the Albanese variety of $X$ is $D/\mu_3$. The same argument as in ([S1], pp. 403–404) works in our situation, and we obtain $\mathrm{NS}(\overline{X})_{\mathrm{tors}} = E^{\mu_3}$. Then (i) implies that $(\mathrm{Br}\, \overline{X})^{\mathrm{Gal}(\overline{\mathbb{Q}}/\mathbb{Q})} = 0$. Therefore, $\mathrm{Br}\, X = \ker[\mathrm{Br}\, X \to \mathrm{Br}\, \overline{X}]$. A well known Leray spectral sequence shows that the quotient of this group by the image of $\mathrm{Br}\, \mathbb{Q}$ is naturally isomorphic to $H^1(\mathbb{Q}, \mathrm{Pic}\, \overline{X})$ ([S2], (2.23); here we use the fact that $H^3(\mathbb{Q}, \overline{\mathbb{Q}}^*) = 0$). The analysis of the morphism of Galois modules $f_1^* \colon \mathrm{Pic}\, \overline{X} \to \mathrm{Pic}\, \overline{Y}_1$ is carried out in the same way as in the proof of Lemma 2 of [S1], where the multiplication by 2 on $E$ has now to be replaced by the isogeny $\alpha \colon E \to E_1$. The result is that the image $f_1^*(H^1(\mathbb{Q}, \mathrm{Pic}\, \overline{X}))$ in $H^1(\mathbb{Q}, \mathrm{Pic}\, \overline{Y}_1)$ is contained in $\pi^*(H^1(\mathbb{Q}, \mathrm{Pic}\, \overline{D}))$. Formula (2) now follows from the functoriality of the Leray spectral sequence.

Let us construct an adelic point on $X$ satisfying all the Brauer–Manin conditions. Take a rational point $R \in D(\mathbb{Q})$, and a collection $\{P_v\} \in C_1(\mathbf{A}_\mathbb{Q})$. Then $f_1(\{(P_v, R)\}) \in X(\mathbf{A}_\mathbb{Q})^{\mathrm{Br}}$, as follows from (2) and the Hasse reciprocity law.

It remains to show that there are no $\mathbb{Q}$-points on $X$. Indeed, rational points on $X$ come from twists of $Y$ given by $a \in H^1(\mathbb{Q}, \mu_3) = \mathbb{Q}^*/\mathbb{Q}^{*3}$. Any such twist of $Y$ is the product $C_a \times D_a$, where $C_a$ and $D_a$ are curves of genus 1. Moreover, $D_a$ is a principal homogeneous space of $D$ of the kind described in Condition (iv) of the theorem. By that condition, if $D_a$ has points everywhere locally, then $a$ is trivial, so that $D_a = D$. Thus there are no $\mathbb{Q}$-points on the nontrivial twists of $Y$. On the other hand, $Y$ has no $\mathbb{Q}$-points since by Condition (iii) there are no $\mathbb{Q}$-points on $C$. Therefore, $X(\mathbb{Q}) = \emptyset$. This completes the proof.

More details can be found in the thesis of the first author [Ba]. The preparation of this paper was speeded up by John Voight's notes of the conference "Rational and Integral Points on Higher Dimensional Varieties" at the American Institute of Mathematics in December, 2002. We thank him for the notes, and the organizers for stimulating atmosphere. We are very grateful to Tom Fisher for telling us about the curve $x^3 + 11y^3 + 43z^3 = 0$. We thank Ekaterina Amerik for useful discussions.

# References


[Ba] C.L. Basile, *On the Hasse principle for certain surfaces fibred into curves of genus one*, Thesis, Imperial College London, March 2003, 92 pp.

[B] A. Beauville, *Surfaces algébriques complexes*, Astérisque **54** (1978)

[CS] J.-L. Colliot-Thélène et J.-J. Sansuc, La descente sur les variétés rationnelles. II, *Duke Math. J.* **54** (1987) 375–492

[CSS] J.-L. Colliot-Thélène, A.N. Skorobogatov and Sir Peter Swinnerton-Dyer, Double fibres and double covers: paucity of rational points, *Acta Arithm.* **79** (1997) 113–135

[F1] T. Fisher, The Cassels–Tate pairing and the Platonic solids, *J. Number Theory* **98** (2003) 105–155

[F2] T. Fisher, A counterexample to a conjecture of Selmer, *This volume*, 121–133

[H] D. Harari, Weak approximation and non-abelian fundamental groups, *Ann. Sci. École Norm. Sup.* **33** (2000) 467–484

[L] S. Lang, *Abelian varieties*, Springer-Verlag, 1983

[Se] E.S. Selmer, The diophantine equation $ax^3+by^3+cz^3=0$. Completion of the tables, *Acta Math.* **92** (1954) 191–197

[S] J.-P. Serre, *Cohomologie galoisienne.* 5ème éd., Lecture Notes Math. **5**, Springer-Verlag, 1994

[S1] A.N. Skorobogatov, Beyond the Manin obstruction, *Invent. Math.* **113** (1999) 399–424

[S2] A.N. Skorobogatov, *Torsors and rational points*, Cambridge Tracts in Mathematics **144**, Cambridge University Press, 2001

Carmen Laura Basile,
Department of Mathematics,
Imperial College London,
South Kensington Campus,
London SW7 2AZ, England
e-mail: laura.basile@ic.ac.uk

Alexei Skorobogatov,
Department of Mathematics,
Imperial College London,
South Kensington Campus,
London SW7 2AZ, England
e-mail: a.skorobogatov@ic.ac.uk

# An elementary approach to effective diophantine approximation on $\mathbb{G}_m$

Enrico Bombieri and Paula B. Cohen

*To Sir Peter Swinnerton-Dyer, on his 75th birthday*

## 1 Introduction

Effective results in the diophantine approximation of algebraic numbers are difficult to obtain, and for a long time the only general method available was Baker's theory of linear forms in logarithms. An alternative, more algebraic, method was later proposed in Bombieri [2] and Bombieri and Cohen [3]. This new method is quite different from the classical approach through the theory of linear forms in logarithms.

In this paper, we improve on results derived in [3]. These results concern effective approximations to roots of high order of algebraic numbers and their application to diophantine approximation in a number field by a finitely generated multiplicative subgroup. We restrict our attention to the non-archimedean case, although our results and methods should go over *mutatis mutandis* to the archimedean setting.

We do not claim that our theorems are the best that are known in this direction. Linear forms in two logarithms (which are easier to treat than the general case) suffice to prove somewhat better results than our Theorem 5.1, see Bugeaud [6] and Bugeaud and Laurent [7]; we give an explicit comparison in §5, Remark 5.1.

Theorem 5.2, which is useful for general applications, follows from Theorem 5.1 by means of a trick introduced for the first time in [2] and improved in [3]. Thus any improved form of Theorem 5.1 carries automatically an improvement of Theorem 5.2. Note however that Theorem 5.2, in the form given here, is still far from what one can obtain directly from Baker's theory of logarithmic forms in many variables, as in Baker and Wüstholz [1] in the archimedean case and Kunrui Yu [10, 11] in the $p$-adic case.

Notwithstanding the comparison with Baker's theory, we feel that there is some untapped potential here. For example, one treats with equal ease the archimedean and the $p$-adic case, while this is not so in Baker's theory because of the bad analytic behaviour of the $p$-adic exponential.

The auxiliary construction involves a universal family of two-variable polynomials invariant under an action of roots of unity of a certain order. The main new feature in the current paper is the use of an elementary Wronskian argument, involving differentiation only in a single variable, to derive a zero estimate which bypasses former appeals to a more sophisticated two-variable Dyson's Lemma. This was initially inspired by private communication between the first author and David Masser in 1984. We reproduce part of that communication in §6.

Although the method of the current paper is more elementary, the results obtained are sharper than those of [3]. The main results are stated in §5, Theorem 5.1 and Theorem 5.2. Theorem 5.1 represents an improvement over the corresponding result of [3] both in the absolute constants and in the lower bound for $r$ in (H1), where $(\log \frac{1}{\kappa})^7$ is replaced by $(\log \frac{1}{\kappa})^5$, as well as in the lower bound for $h(\alpha')$ in (H2) of [3], which is no longer required. These improvements automatically carry over to Theorem 5.2, which we restate for convenience here in the Main Theorem. We follow the notations of [3], §2. In particular $H(\ )$ denotes the absolute Weil height, $h(\ ) = \log H(\ )$ the absolute logarithmic Weil height and $|\ |_v$ is the absolute value associated to a place $v \in M_K$, normalized so that $h(x) = \sum_{v \in M_K} \max(0, \log |x|_v)$.

We define $\rho(x)$ to be the solution $\rho(x) > e^5$ of $\rho/(\log \rho)^5 = x$ if $x > e^5 5^{-5}$, and $\rho(x) = e^5$ otherwise; for large $x$ we have $\rho(x) \sim x(\log x)^5$.

**Main Theorem** *Let $K$ be a number field of degree $d$ and $v$ a place of $K$ dividing a rational prime $p$. We denote by $f_v$ the residue class degree of the extension $K_v/\mathbb{Q}_p$ and set $D_v^* = \max(1, \frac{d}{f_v \log p})$. Define a modified logarithmic height of $x \in K$ by $h'(x) = \max\big(h(x), \frac{1}{D_v^*}\big)$, and let $H'(x) = \exp h'(x)$.*

*Let $\Gamma$ be a finitely generated subgroup of the multiplicative group $K^*$, and write $\xi_1, \ldots, \xi_t$ for generators of $\Gamma/\mathrm{tors}$. Let $\xi \in \Gamma$, $A \in K^*$ and $\kappa > 0$ be such that*

$$0 < |1 - A\xi|_v < H'(A\xi)^{-\kappa}.$$

*Define*

$$C = 66 p^{f_v} (D_v^*)^6 \quad \textit{and} \quad Q = \big(2t\rho(C/\kappa)\big)^t \prod_{i=1}^{t} h'(\xi_i).$$

*Then we have*

$$h'(A\xi) \ \leq \ 16 p^{f_v} \rho(C/\kappa)\, Q \max\Big(h'(A), 4p^{f_v} Q\Big).$$

It is an interesting problem to try to refine the auxiliary construction of §2 to the point where the nonvanishing of $P(x, y)$ at the point $(\alpha, \alpha')$ is immediate, that is $P(\alpha, \alpha') \neq 0$. In Cohen and van der Poorten [8] it is shown that this would lead to a result comparable with the best known consequences of Baker's method.

**Acknowledgements** The second author thanks the Institute for Advanced Study, Princeton, where much of this research was carried out with the support of The Ellentuck Fund. We also thank David Masser for permission to include in the Appendix the text of his letter to the first author.

# 2 Equivariant polynomials

Let $r \geq 2$ be a positive integer and $l$ an integer with $(l, r) = 1$. For $0 \leq j < r$ we define $e_j$ to be the integer with $0 \leq e_j < r$ such that

$$le_j \equiv -j \pmod{r}.$$

Let $0 \leq s < r$ and consider the polynomial

$$P(x, y) = \sum_{j=0}^{s} A_j(x^r) x^{e_j} y^j,$$

where the $A_j(x) \in \mathbb{Q}[x]$ are polynomials in $x$ of degree at most $n$, not all identically 0. This polynomial is invariant under the action $(x, y) \mapsto (\varepsilon^l x, \varepsilon y)$ of $r$th roots of unity, in the sense that

$$P(\varepsilon^l x, \varepsilon y) = P(x, y) \quad \text{whenever } \varepsilon^r = 1.$$

We define the *index* $i(P; \xi, \eta)$ of $P(x, y)$ at a point $(\xi, \eta)$ to be the order of zero of $P(x, \eta)$ at $x = \xi$, namely

$$i(P; \xi, \eta) = \operatorname{ord}_\xi P(x, \eta).$$

In what follows, for a real number $t$ we abbreviate $t^+ = \max(t, 0)$.

**Lemma 2.1** *We have*

$$\sum_{\xi \in \mathbb{C}^*/\{\varepsilon^r = 1\}} \max_\eta \big(i(P; \xi, \eta) - s\big)^+ \leq (s + 1)n.$$

**Proof** We use the classical Wronskian argument. Let $I \subset \{0, 1, \ldots, s\}$ be the set of indices $j$ such that $A_j(x)$ is not identically 0 and $t+1$ its cardinality. Clearly $t \leq s$. We calculate the $(t+1) \times (t+1)$ Wronskian determinant

$$\det\left[\frac{\partial^{h+k}}{\partial x^h \partial y^k} P(x,y)\right]_{0\leq h,k\leq t} = \det\left[\sum_{j\in I} \frac{\partial^h}{\partial x^h}\left(A_j(x^r)x^{e_j}\right) \cdot \frac{\partial^k}{\partial y^k}\left(y^j\right)\right]_{0\leq h,k\leq t}$$

$$= \det\left[\frac{\partial^h}{\partial x^h}\left(A_j(x^r)x^{e_j}\right)\right]_{0\leq h\leq t,\, j\in I} \cdot \det\left[\frac{\partial^k}{\partial y^k}\left(y^j\right)\right]_{j\in I,\, 0\leq k\leq t}.$$

Thus this Wronskian is a polynomial $W(x)y^\Delta$, with

$$\Delta = \sum_{j\in I} j - \frac{t(t+1)}{2}.$$

Moreover, $W(x)$ is not identically 0, because the polynomials $A_j(x^r)x^{e_j}$ are linearly independent over $\mathbb{Q}$ and the monomials $y^j$, $j \in I$, are also linearly independent over $\mathbb{Q}$ (the $A_j(x)$ for $j \in I$ are not identically 0 by hypothesis and the exponents in the monomials in $A_j(x^r)x^{e_j}$ belong to different arithmetic progressions as $j$ varies). By looking at the determinant of the matrix $\left[(\mathrm{d}/\mathrm{d}x)^h A_j(x^r)x^{e_j}\right]$ we verify that

$$\operatorname{ord}_0 W(x) \geq \sum_{j\in I} e_j - \frac{t(t+1)}{2} \quad \text{and}$$

$$\operatorname{ord}_\infty W(x) \leq r(t+1)n + \sum_{j\in I} e_j - \frac{t(t+1)}{2}.$$

Now, if we specialize $y$ to any $\eta \neq 0$ (which does not affect the vanishing of $W(x)$) and look at the first column of the Wronskian, we see that

$$\operatorname{ord}_\xi W(x) \geq \left(i(P;\xi,\eta) - t\right)^+;$$

therefore we have

$$\operatorname{ord}_\xi W(x) \geq \max_{\eta\neq 0}\left(i(P;\xi,\eta) - t\right)^+.$$

Since $P(x,y)$ is invariant we have $i(P;\varepsilon^l\xi,\varepsilon\eta) = i(P;\xi,\eta)$. Hence

$$r \sum_{\xi\in\mathbb{C}^*/\{\varepsilon^r=1\}} \max_\eta\left(i(P;\xi,\eta) - t\right)^+ = \sum_{\xi\in\mathbb{C}^*} \max_\eta\left(i(P;\xi,\eta) - t\right)^+$$

$$\leq \sum_{\xi\in\mathbb{C}^*} \operatorname{ord}_\xi W(x) = \operatorname{ord}_\infty W(x) - \operatorname{ord}_0 W(x) \leq r(t+1)n,$$

concluding the proof. □

Consider now the $\mathbb{Q}$-vector spaces $V_0 \supseteq V_1 \supseteq \cdots \supseteq V_k \supseteq \cdots$ defined by

$$V_0 = \{P : P = \sum_{j=0}^{s} A_j(x^r)x^{e_j}y^j\},$$
$$V_k = \{P : P \in V_0 \quad and \quad i(P;1,1) \geq k\}.$$

**Lemma 2.2** *The vector space $V_k$ has dimension*

$$\dim V_k = (s+1)(n+1) - k.$$

**Proof** We abbreviate $\partial_k$ for $(\partial/\partial x)^k$.

It is clear that $\dim V_0 = (s+1)(n+1)$. Also, we have $\dim V_k/V_{k+1} \leq 1$, because

$$V_{k+1} = \{P \in V_k : (\partial_k P)(1,1) = 0\}.$$

Thus the lemma follows from the statement that

$$\dim V_{(s+1)(n+1)} = 0.$$

Suppose this is not the case. Then there is a polynomial $P$, not identically 0, with $i(P;1,1) \geq (s+1)(n+1)$. By Lemma 2.1 we get $(s+1)(n+1) - s \leq (s+1)n$, a contradiction. This completes the proof. □

Our next result gives us a small basis of the vector space $V_k$.

**Lemma 2.3** *There is a basis $\{P_l\}$ of $V_k$ such that*

$$\sum_{l=1}^{(s+1)(n+1)-k} h(P_l) \leq \frac{1}{2}k^2 \log\left(\frac{r(n+1)}{4k}\right) + \frac{3}{4}k^2.$$

**Proof** Consider

$$P(x,1) = \sum_{j=0}^{s} A_j(x^r)x^{e_j} = \sum_{j=0}^{s}\sum_{h=0}^{n} a_{jh}x^{rh+e_j}.$$

Then $V_k$ can be identified with the subspace of $\{a_{jh}\} \in \mathbb{Q}^{(s+1)(n+1)}$ defined by the linear equations

$$\sum_{j=0}^{s}\sum_{h=0}^{n} a_{jh}\binom{rh+e_j}{i} = 0, \quad \text{for} \quad i = 0, 1, \ldots, k-1,$$

which has codimension $k$ by Lemma 2.2. Let $\mathcal{A}$ be the associated matrix

$$\mathcal{A} = \left[\binom{rh+e_j}{i}\right]_{\substack{i=0,1,\ldots,k-1\\(j,h)\in\{0,\ldots,s\}\times\{0,\ldots,n\}}}.$$

By Lemma 2.2, $\mathcal{A}$ has maximal rank $k$, therefore it is a submatrix of maximal rank of the matrix

$$\mathcal{B} = \left[\binom{l}{i}\right]_{\substack{i=0,1,\ldots,k-1\\l=0,\ldots,rn+r-1}}.$$

It follows that $H(\mathcal{A}) \le H(\mathcal{B})$ where $H(\ )$ is the height. In our case, where everything is over $\mathbb{Z}$, the height of $\mathcal{B}$ is given by

$$H(\mathcal{B}) = \sqrt{\sum \det\left[\binom{n_j}{i}\right]^2_{\substack{i=0,\ldots,k-1\\j=1,\ldots,k}}}$$

where the sum ranges over all $k$-tuples $0 \le n_1 < n_2 < \cdots < n_k < r(n+1)$ (note that the greatest common divisor of the determinants of all maximal minors of $\mathcal{B}$ is 1).

We have

$$\det\left[\binom{n_j}{i}\right]_{\substack{i=0,\ldots,k-1\\j=1,\ldots,k}} = \frac{\prod_{h<j}(n_j-n_h)}{1!2!\cdots(k-1)!},$$

as one sees by transforming the determinant into $\det(n_j^i/i!)$ and computing the Vandermonde determinant $\det(n_j^i)$. For the logarithmic height, this gives

$$h(\mathcal{B}) = \frac{1}{2}\log\Big(\frac{1}{1!2!\cdots(k-1)!}\sum_{0\le n_1<\cdots<n_k<r(n+1)}\prod_{h<j}(n_j-n_h)^2\Big).$$

An exact calculation based on the theory of orthogonal polynomials can be found in Bombieri and Vaaler [5]. Writing for simplicity $N = r(n+1)$, we have

$$h(\mathcal{B}) = \frac{1}{2}\sum_{m=-k+1}^{k-1}(k-|m|)\log\left(\frac{N+m}{k+m}\right) \le N^2 u\left(\frac{k}{N}\right),$$

where[1]

$$\begin{aligned} u(\theta) &= \frac{1}{4}\theta^2\log\frac{1-\theta^2}{16\theta^2} + \frac{1}{2}\theta\log\frac{1+\theta}{1-\theta} + \frac{1}{4}\log(1-\theta^2) \\ &= \frac{1}{2}\theta^2\log\frac{1}{4\theta} + \frac{3}{4}\theta^2 - \sum_{h=2}^{\infty}\frac{\theta^{2h}}{4(h-1)h(2h-1)} \\ &\le \frac{1}{2}\theta^2\log\frac{1}{4\theta} + \frac{3}{4}\theta^2. \end{aligned}$$

[1] The series expansion is given incorrectly in [5], p.57 with $h^2-2h+2$ in place of $(h-1)h$.

To conclude the proof of Lemma 2.3 we simply apply the main theorem in Bombieri and Vaaler [4] which, in our case over the rational field $\mathbb{Q}$, gives the existence of a basis $\{P_l\}$ for $V_k$ such that

$$\sum_{l=0}^{(s+1)(n+1)-k} h(P_l) \leq h(\mathcal{A}) \leq h(\mathcal{B}). \quad \square$$

Let $a \in K$ and suppose that $a$ is neither 0 nor a root of unity. We fix an $r$th root $\alpha = a^{1/r}$ and set $\alpha' = \gamma^{-1}\alpha$ with $\gamma \in K$ and $\gamma \neq 0$.

**Lemma 2.4** *Let $M \geq 1$. There is an invariant polynomial $P \in V_k$ with rational integral coefficients such that*

$$\sum_{\substack{m=-M \\ m\neq 0}}^{M} \left(i(P;\alpha^{lm},(\alpha')^m) - s\right)^+ \leq (s+1)(n+1) - k - 1$$

*and*

$$h(P) \leq \frac{1}{2}\frac{k^2}{(s+1)(n+1)-k}\left[\log\left(\frac{r(n+1)}{4k}\right) + \frac{3}{2}\right].$$

**Proof** By Lemma 2.3, there is an invariant polynomial $P(x, y) \in V_k$ with rational integral coefficients such that

$$h(P) \leq \frac{1}{2}\frac{k^2}{(s+1)(n+1)-k}\left[\log\left(\frac{r(n+1)}{4k}\right) + \frac{3}{2}\right].$$

Lemma 2.1 gives

$$\sum_{m=-M}^{M} \left(i(P;\alpha^{lm},(\alpha')^m) - s\right)^+ \leq (s+1)n,$$

while on the other hand $i(P; 1, 1) \geq k$ because $P \in V_k$. It follows that

$$\sum_{\substack{m=-M \\ m\neq 0}}^{M} \left(i(P;\alpha^{lm},(\alpha')^m) - s\right)^+ \leq (s+1)n - k + s,$$

completing the proof. $\square$

# 3 The Thue–Siegel method

In this section we prove

**Lemma 3.1** *Let $v \in M_K$ be a finite place for which*

$$\alpha \in K_v, \quad |\alpha^l - 1|_v < 1 \quad \textit{and} \quad |\alpha' - 1|_v < 1.$$

*Let also $k$, $s$, $n$ be positive integers such that $s < k < (s+1)(n+1)$, $s < r$, and $\Lambda$ a positive real number such that*

$$\log \frac{1}{|1-\alpha^l|_v} \geq \Lambda \quad \textit{and} \quad \log \frac{1}{|1-\alpha'|_v} \geq (k-s)\Lambda. \tag{A1}$$

*Define $D = \frac{1}{2M}\big((s+1)(n+1) - k - 1\big) + s + 1$. Then we have*

$$\begin{aligned}(k - D + 1)\Lambda \;\leq\; & M + \frac{1}{2}[(n+1)|l|\, h(a) + s\, h(\gamma)] \\ & + \frac{1}{2}\frac{k^2}{(s+1)(n+1)-k}\left[\log\left(\frac{r(n+1)}{4k}\right) + \frac{3}{2}\right] \\ & + D\left[\log\left(\frac{r(n+1)}{D}\right) + 1\right].\end{aligned}$$

**Proof** Fix $m$ with $-M \leq m \leq M$, $m \neq 0$. We write for simplicity $(\xi, \eta) = (\alpha^{lm}, (\alpha')^m)$, $i_m = i(P; \xi, \eta)$.

Let $P(x,y) = \sum a_{hj} x^{rh+e_j} y^j$ and $Q(x,y) = x^{i_m}(\partial_{i_m} P)(x,y)$. We have

$$Q(x,y) = \sum_{hj} a_{hj} \binom{rh+e_j}{i_m} x^{rh+e_j} y^j$$

whence, setting $\beta = Q(\xi, \eta)$, we have

$$\beta = \sum_{hj} a_{hj} \binom{rh+e_j}{i_m} a^{lhm + [(le_j+j)/r]m} \gamma^{-mj} \in K.$$

The fact that $\beta \in K$ rather than an extension of $K$ is essential for our next argument.

By definition of $i_m$ we have

$$(\partial_{i_m} P)(\xi, \eta) \neq 0,$$

therefore $\beta \neq 0$ and the product formula in $K$ yields

$$\sum_{w \in M_K} \log |\beta|_w = 0.$$

Now we estimate each term $\log|\beta|_w$ as follows.

We have

$$|le_j + j| \le |l|(r-1) + (r-1) = (|l|+1)(r-1) < (|l|+1)r,$$

and since the left-hand side of this inequality is divisible by $r$ we find $|le_j+j| \le |l|\,r$. This gives $|lm|h + [(le_j+j)/r]\,|m| \le |lm|(n+1)$. Hence for every $w \ne v$ we have

$$\begin{aligned}\log|\beta|_w \;\le\; & |lm|(n+1)\log^+|a|_w + |m|s\log^+|1/\gamma|_w \\ & + \varepsilon_w \max_{hj}\log|a_{hj}|_w + \varepsilon_w \log\left|\binom{r(n+1)}{i_m+1}\right|_w\end{aligned}$$

where as usual $\varepsilon_w = [K_w : \mathbb{Q}_w]/[K : \mathbb{Q}]$ if $w \mid \infty$ and $\varepsilon_w = 0$ otherwise. In the proof of the above estimate, we have used the obvious majorization

$$\sum_{hj}\binom{rh+e_j}{i_m} \le \sum_{k=0}^{rn+r-1}\binom{k}{i_m} = \binom{r(n+1)}{i_m+1}.$$

If instead $w = v$, we note that since

$$\alpha \in K_v, \quad |\alpha^l - 1|_v < 1 \quad \text{and} \quad |\alpha' - 1|_v < 1,$$

we also have $|\xi|_v = 1$, $|\eta|_v = 1$ and

$$|\xi - 1|_v \le |\alpha^l - 1|_v < 1, \quad |\eta - 1|_v \le |\alpha' - 1|_v < 1.$$

The Taylor series of $Q(x,y)$ with center at $(1,1)$ has rational integral coefficients because $Q(x,y) \in \mathbb{Z}[x,y]$. Moreover, by construction, the polynomial $Q(x,1) = x^{i_m}(\partial_{i_m}P)(x,1)$ has a zero of order $\ge (k-i_m)^+$ at $x = 1$, therefore

$$\begin{aligned}|\beta|_v &= |\xi^{i_m}(\partial_{i_m}P)(\xi,\eta)|_v \\ &\le \max\left(|\xi-1|_v^{(k-i_m)^+}, |\eta-1|_v\right) \\ &\le \max\left(|\alpha^l-1|_v^{(k-i_m)^+}, |\alpha'-1|_v\right).\end{aligned}$$

If we combine these estimates with the product formula we find

$$\begin{aligned}&\min\Big((k-i_m)^+\log\frac{1}{|\alpha^l-1|_v}, \log\frac{1}{|\alpha'-1|_v}\Big) \\ &\qquad \le |lm|(n+1)h(a) + |m|s\,h(\gamma) + h(P) + \log\binom{r(n+1)}{i_m+1}.\end{aligned}$$

Using $(x-y)^+ \geq (x-z)^+ - (y-z)^+$ and (A1), this implies the new inequality

$$(k-s)^+\Lambda \leq |lm|(n+1)\,h(a) + |m|s\,h(\gamma) + h(P) + \log\left(\frac{r(n+1)}{(i_m-s)^+ + s+1}\right) + (i_m - s)^+\Lambda.$$

We take the average of this inequality for $-M \leq m \leq M$, $m \neq 0$. In view of the easy estimate

$$\log\binom{p}{q} \leq q\log\frac{p}{q} + q$$

we obtain

$$\begin{aligned}(k-s)^+\Lambda \leq\ & \frac{M+1}{2}\left[(n+1)|l|\,h(a) + s\,h(\gamma)\right] + h(P)\\ & + \frac{1}{2M}\sum_{\substack{m=-M\\ m\neq 0}}^{M}\left((i_m-s)^+ + s+1\right)\left[\log\left(\frac{r(n+1)}{(i_m-s)^+ + s+1}\right) + 1\right]\\ & + \frac{1}{2M}\sum_{\substack{m=-M\\ m\neq 0}}^{M}(i_m-s)^+\Lambda.\end{aligned}$$

In order to bound the right-hand side of this inequality we replace $(i_m - s)^+$ by a positive continuous variable $z_m$ subject to $\sum z_m \leq (s+1)(n+1) - k - 1$ and estimate the maximum using Lagrange multipliers. The maximum is achieved if $z_m$ is constant, hence $z_m + s + 1 = D$ with $D$ as in the statement of the lemma. Since $(k-s)^+ - (D-s-1) \geq k - D + 1$, this completes the proof of Lemma 3.1. □

# 4 Simplification of the main inequality

In order to apply Lemma 3.1 we make some further assumptions and introduce new variables, with the aim of tidying up the inequality stated in the conclusion of the lemma.

First, we remove the condition that $k$ be a positive integer. To this end, it suffices to note that the right-hand side of our inequality increases in $k$ and $D$ for $4k \leq r(n+1)$; thus we may drop the integrality condition on $k$, replacing $D$ by $D' = \left((s+1)(n+1) - k\right)/(2M) + s + 1$ throughout and $k+1$ by $k$ in the left-hand side of the inequality. Note also that dropping the integrality condition on $k$ makes condition (A1) even more stringent. This gives

$$\begin{aligned}(k-D')\Lambda \;\le\; & M+\frac{1}{2}\Big[|l|(n+1)h(a)+s\,h(\gamma)\Big]\\ &+\frac{1}{2}\frac{k^2}{(s+1)(n+1)-k}\left[\log\left(\frac{r(n+1)}{4k}\right)+\frac{3}{2}\right]\\ &+D'\left[\log\left(\frac{r(n+1)}{D'}\right)+1\right].\end{aligned} \tag{4.1}$$

Next, we choose $k=\lambda(s+1)(n+1)$. Then

$$D'=(1-\lambda)(s+1)(n+1)/(2M)+s+1.$$

After dividing by $(s+1)(n+1)$ the resulting inequality becomes

$$\begin{aligned}\left(\lambda-\frac{1-\lambda}{2M}-\frac{1}{n+1}\right)\Lambda \;\le\; & \frac{M+1}{2}\left[|l|\frac{h(a)}{s+1}+\frac{s}{s+1}\frac{h(\gamma)}{n+1}\right]\\ &+\frac{1}{2}\frac{\lambda^2}{1-\lambda}\left[\log\left(\frac{r}{4\lambda(s+1)}\right)+\frac{3}{2}\right]\\ &+\frac{1}{2M}\left[\log\left(\frac{2Mr}{s+1}\right)+1\right].\end{aligned} \tag{4.2}$$

Now let

$$r>A\ge h(a),\;\; G\ge h(\gamma),\;\; r=\rho A,\;\; s+1=\sigma A\le r,\;\; n+1=\nu G, \tag{A2}$$

and set

$$\lambda\log\rho=\Lambda;$$

decreasing $\lambda$ if needed, we also assume that

$$\lambda<1,\qquad 0<\Lambda\le 1. \tag{A3}$$

If we suppose

$$G>\frac{2M}{\nu\lambda}, \tag{A4}$$

which implies $\lambda/(2M)\ge 1/(n+1)$, the above inequality simplifies to

$$\begin{aligned}\left(\lambda-\frac{1}{2M}\right)\lambda\log\rho \;\le\; & \frac{M+1}{2}\left(\frac{|l|}{\sigma}+\frac{1}{\nu}\right)\\ &+\frac{1}{2}\frac{\lambda^2}{1-\lambda}\left[\log\left(\frac{\rho}{4\sigma\lambda}\right)+\frac{3}{2}\right]+\frac{1}{2M}\left[\log\left(\frac{2M\rho}{\sigma}\right)+1\right].\end{aligned} \tag{4.3}$$

This inequality is obtained under assumptions (A1), (A2), (A3), (A4) and the further assumption, implicitly made along the way, that $M$, $\sigma A$ and $\nu G$ are integers. We choose[2]

$$M = \lceil 2\lambda^{-2} \rceil. \tag{4.4}$$

With this choice, (4.3) can be replaced by

$$\begin{aligned} \left(\lambda - \frac{1}{4}\lambda^2\right)\lambda \log\rho \le & \left(\frac{1}{\lambda^2}+1\right)\left(\frac{|l|}{\sigma}+\frac{1}{\nu}\right) \\ & + \frac{1}{2}\frac{\lambda^2}{1-\lambda}\left[\log\left(\frac{\rho}{4\sigma\lambda}\right)+\frac{3}{2}\right] + \frac{\lambda^2}{4}\left[\log\left(\frac{\rho}{\sigma}\frac{2+2\lambda^2}{\lambda^2}\right)+1\right]. \end{aligned} \tag{4.5}$$

We now choose

$$\nu = \frac{1}{G}\left\lceil \frac{G\sigma}{|l|} \right\rceil, \qquad \sigma = A^{-1}\left\lceil \frac{8|l|A}{\lambda^4 \log\rho} \right\rceil; \tag{4.6}$$

note that $M$, $\sigma A$ and $\nu G$ are integers, hence our implicit assumption is verified. An easy majorization of the right-hand side of (4.5) shows that

$$\begin{aligned} \left(\lambda - \frac{1}{4}\lambda^2\right)\lambda \log\rho \le & \left(\frac{1}{\lambda^2}+1\right)\frac{1}{4}\lambda^4 \log\rho + \frac{\lambda^2(3-\lambda)}{4(1-\lambda)}\log\rho \\ & + \frac{\lambda^2}{4(1-\lambda)}\left[\log\left(\frac{1+\lambda^2}{8\sigma^3\lambda^4}\right)+4\right]. \end{aligned} \tag{4.7}$$

Since $\sigma \ge 8\lambda^{-4}(\log\rho)^{-1}$, we see that (4.7) implies

$$\begin{aligned} \left(\lambda - \frac{1}{4}\lambda^2\right)\lambda \log\rho \le & \left(\frac{1}{\lambda^2}+1\right)\frac{1}{4}\lambda^4 \log\rho + \frac{\lambda^2(3-\lambda)}{4(1-\lambda)}\log\rho \\ & + \frac{\lambda^2}{4(1-\lambda)}\left[\log\left(8^{-4}(1+\lambda^2)\lambda^8(\log\rho)^3\right)+4\right]. \end{aligned} \tag{4.8}$$

Since $\lambda \log\rho \le 1$, inequality (4.8) yields

$$\begin{aligned} \left(\lambda - \frac{1}{4}\lambda^2\right)\lambda \log\rho \le & \left(\frac{1}{\lambda^2}+1\right)\frac{1}{4}\lambda^4 \log\rho + \frac{\lambda^2(3-\lambda)}{4(1-\lambda)}\log\rho \\ & + \frac{\lambda^2}{4(1-\lambda)}\left[\log(\lambda^5+\lambda^7) - 4.317\right]. \end{aligned} \tag{4.9}$$

Note that $\log(\lambda^5+\lambda^7) - 4.317 < \log 2 - 4.317 < -3.623 < 0$. Dividing both sides of (4.9) by $\lambda^2 \log\rho$ and using the lower bound $1/\log\rho \ge \lambda$ gives

$$1 - \frac{1}{4}\lambda \le (1+\lambda^2)\frac{1}{4} + \frac{3-\lambda}{4(1-\lambda)} - 3.623\frac{\lambda}{4(1-\lambda)}$$

[2] We use here the ceiling function $\lceil x \rceil = \min_{n\in\mathbb{Z}}\{n : n \ge x\}$.

and after multiplication by $4(1-\lambda)$ and an easy simplification we find

$$0 \leq -0.623\lambda - \lambda^3 < 0.$$

This is a contradiction, and shows that one of the hypotheses (A1) to (A4), together with the choices (4.4) and (4.6), is untenable. Therefore, (A1) does not hold if we assume (A2), (A3), (A4) and (4.4) and (4.6). Our choice of parameters in (4.4) and (4.6) guarantees that (A2), (A3), (A4) are verified, except possibly for the condition $s+1 \leq r$ in (A2) that must be compatible with our choice of $\sigma$ in (4.6). Let us assume for the time being that this is the case. Then if we assume the first half of (A1), namely $\log|1-\alpha^l|_v \leq -\Lambda$, we conclude that the second half of (A1) does not hold. Note also that by (4.6) we have

$$\sigma \geq 8|l|\lambda^{-4}(\log\rho)^{-1} \quad \text{and} \quad \nu \geq \sigma/|l| \geq 8\lambda^{-4}(\log\rho)^{-1}; \tag{4.10}$$

therefore, $2\lceil 2\lambda^{-2}\rceil/(\nu\lambda) \geq \frac{1}{2}\lambda\log\rho$ and *a fortiori* (A4) can be replaced by $G \geq \lambda\log\rho$.

If we recall that we had chosen $k = \lambda(s+1)(n+1)$, we conclude that

**Proposition 4.1** *Let $K$, $v$, $r$, $a$, $\alpha = a^{1/r}$, $\gamma$ be as before. Assume that $A$, $\rho$, $G$, $\lambda$ satisfy $r > A \geq h(a)$, $\rho = r/A$, $G \geq \max(h(\gamma), \lambda\log\rho)$ and $0 < \lambda < \min(1, 1/\log\rho)$. Suppose further that*

$$\log\left|1-\alpha^l\right|_v \leq -\lambda\log\rho.$$

*Let*

$$\sigma = A^{-1}\left\lceil \frac{8|l|A}{\lambda^4\log\rho} \right\rceil.$$

*Then if $\sigma \leq \rho$ we have*

$$\log|1-\gamma^{-1}\alpha|_v > -\lambda^2 \left\lceil \frac{G\sigma}{|l|} \right\rceil \left\lceil \frac{8|l|A}{\lambda^4\log\rho} \right\rceil \log\rho.$$

# 5 Applications to diophantine approximation in a number field by a finitely generated multiplicative group

As a corollary of Proposition 4.1, we derive in this section improvements of Theorem 1 and Theorem 2 of [3]. As in that paper, we let $K(v)$ be the residue field of $K_v$ and $f_v$, $e_v$ the residue class degree and ramification index of the extension $K_v/\mathbb{Q}_v$. We abbreviate

$$d_v^* = \frac{d}{f_v \log p}, \qquad D_v^* = \max(1, d_v^*).$$

We assume that $|a|_v = 1$, so that if we choose $l = p^{f_v} - 1$, then $|a^l - 1|_v < 1$. From Lemma 1 of [3] we may suppose that

$$\log \frac{1}{|1-\alpha^l|_v} = \log \frac{1}{|1-a^l|_v} \geq \frac{f_v \log p}{d} = \frac{1}{d_v^*} \geq \frac{1}{D_v^*}. \tag{5.1}$$

Continuing with the notations of §4, we suppose that $r > 2A$ and choose

$$\lambda = (D_v^* \log \rho)^{-1}. \tag{5.2}$$

Then we can apply Proposition 4.1 and deduce that

$$\log |1 - \gamma^{-1}\alpha|_v > -\lambda^2 \left\lceil \frac{G\sigma}{|l|} \right\rceil \left\lceil \frac{8|l|A}{\lambda^4 \log \rho} \right\rceil \log \rho \tag{5.3}$$

provided that $G \geq \max(h(\gamma), 1/D_v^*)$ and also $\sigma \leq \rho$.

With the modified height $h'(x)$ defined in the statement of the Main Theorem, the condition on $G$ becomes $G \geq h'(\gamma)$. Our choice for $A$ will be $A = h'(a)$.

For the application we have in mind, $r$ must be relatively large compared to $h'(a)$ if we want a nontrivial conclusion for our final result. Thus to begin with we assume that

$$r > e^4 D_v^* h'(a). \tag{5.4}$$

In particular, $\log \rho \geq 4$.

The next step in simplifying (5.3) consists in removing the brackets in the ceiling function. By (4.10), (5.4), $A \geq 1/D_v^*$ and our choice of $\lambda$ we have

$$\left\lceil \frac{8|l|A}{\lambda^4 \log \rho} \right\rceil = A\sigma \geq 8|l|(D_v^*)^3(\log \rho)^3 \geq 512,$$

hence we may remove the brackets at the cost of multiplying by $1 + 1/512$, at most. In a similar way, we have

$$\left\lceil \frac{G\sigma}{|l|} \right\rceil \geq 8G(D_v^*)^3(\log \rho)^3 \geq 512,$$

because $G \geq 1/D_v^*$. Therefore, the cost of removing the brackets is at most a factor of $1 + 1/512$. Again, removing the brackets from $\sigma$ will not cost us more than a further factor $1+1/512$. Thus the total cost in this simplification is at most a factor $(1+1/512)^3 < 1.006$.

We can replace $h'(\gamma)$ by $h'(\gamma^{-1}\alpha)$, at a small cost. Indeed, $h(\gamma) \leq h(\gamma^{-1}\alpha) + h(\alpha)$, hence using $h' \geq 1/D_v^*$ we find

$$h'(\gamma) \leq h'(\gamma^{-1}\alpha) + \frac{h(a)}{r} < h'(\gamma^{-1}\alpha) + \frac{1}{e^4 D_v^*} < (1+e^{-4})h'(\gamma^{-1}\alpha). \quad (5.5)$$

Thus the total cost of these simplifications is a factor of at most $1.006 \times (1+e^{-4}) < 1.03$. Therefore, after removing the brackets, taking into account this small correction and making a further rounding off of constants, (5.3) becomes the simpler

$$\log|1-\gamma^{-1}\alpha|_v > -66p^{f_v}(D_v^*)^6 h'(a)\left(\log\frac{r}{h'(a)}\right)^5 h'(\gamma^{-1}\alpha). \quad (5.6)$$

This inequality has been obtained under the assumption that $s+1 \leq r$. If however $s+1 \geq r+1$, we must have

$$\left\lceil \frac{8|l|A}{\lambda^4 \log\rho} \right\rceil = A\sigma = s+1 \geq r+1 = \rho A + 1,$$

hence $8|l| \geq \lambda^4 \rho \log\rho$. With our choice of $\lambda$ and $l$, this means that if

$$\rho(\log\rho)^{-3} \geq 8p^{f_v}(D_v^*)^4 \quad (A5)$$

then the condition $\sigma \leq \rho$ in Proposition 1 is verified.

We now summarize our results as follows.

**Theorem 5.1** *Let $K$ be a number field of degree $d$ and $v$ an absolute value of $K$ dividing a rational prime $p$. Let $a \in K$ with $a$ not 0 or a root of unity, and suppose that $a$ satisfies $|a|_v = 1$.*

*Let $r$ be a positive integer coprime to $p$. Then $a$ has an $r$th root $\alpha \in K_v$ satisfying $0 < |1-\alpha^{p^{f_v}-1}|_v < 1$. Let $\alpha' = \alpha\gamma^{-1}$ with $\gamma \in K$, $\gamma \neq 0$. Let $C = 66\, p^{f_v}\,(D_v^*)^6$ and $0 < \kappa$, and suppose that*

$$r \geq \rho\left(\frac{C}{\kappa}\right) h'(a). \quad (H1)$$

*Then*

$$|\alpha' - 1|_v \geq H'(\alpha')^{-r\kappa}.$$

*Moreover, if $|a-1|_v < 1$ then $a$ has an $r$th root $\alpha \in K_v$ satisfying*

$$0 < |\alpha - 1|_v < 1,$$

*and (H1) can be further improved by replacing $C$ with the smaller constant $C' = 66(D_v^*)^6$.*

**Remark 5.1** Before completing the proof of Theorem 5.1, a comparison with the explicit result in [6] is in order. To avoid undue complications, we only consider asymptotic bounds as $h(\gamma) \to \infty$ and $r/h'(a) \to \infty$. Then, with the optimal choice of $\kappa$, the bound given by our Theorem 5.1 is

$$\log \frac{1}{|\alpha' - 1|_v} \leq (66 + o(1))\, p^{f_v}\, (D_v^*)^6 h'(a) \left(\log \frac{r}{h'(a)}\right)^5 h(\gamma).$$

On the other hand, from [6] we may show that

$$\log \frac{1}{|\alpha' - 1|_v} \leq (24 + o(1))\, p^{f_v}\, (d_v^*)^4 h'(a) \left(\log \frac{r}{h'(a)}\right)^2 h(\gamma),$$

which is better than Theorem 5.1. Thus the interest of Theorem 5.1 is more in the method of proof than in the result itself.

**Proof** By (5.6) it suffices that $r$ be so big that

$$\kappa r \geq C h'(a) \left(\log \frac{r}{h'(a)}\right)^5,$$

that is $\rho(\log \rho)^{-5} \geq C/\kappa$. Note that, with our value for $C$, this condition takes care of (A5) as soon as $\kappa \leq 8(D_v^*)^2$.

On the other hand, we have the Liouville lower bound

$$|\alpha' - 1|_v \geq (2H'(\alpha'))^{-r},$$

while $H'(\alpha')^{D_v^*} \geq e > 2$, hence in any case we have $|\alpha' - 1|_v \geq H'(\alpha')^{-2D_v^* r}$. This shows that the conclusion of Theorem 5.1 is trivial as soon as $\kappa > 2D_v^*$. Thus condition (A5) is of no consequence for the verification of Theorem 5.1, completing the proof. □

In applications, condition (H1) is the most important. A direct comparison with Theorem 1 of [3] shows a big improvement in the absolute constant of (H1) and a reduction in the power of the logarithmic term from 7 to 5. The condition (H2) of [3] is now eliminated.

**Theorem 5.2** *Let $K$ be a number field of degree $d$ and $v$ a place of $K$ dividing a rational prime $p$.*

*Let $\Gamma$ be a finitely generated subgroup of $K^*$ and let $\xi_1, \ldots, \xi_t$ be generators of $\Gamma/\mathrm{tors}$. Let $\xi \in \Gamma$, $A \in K^*$ and $\kappa > 0$ be such that*

$$0 < |1 - A\xi|_v < H'(A\xi)^{-\kappa}.$$

*Define*

$$C = 66p^{f_v}(D_v^*)^6 \quad \text{and} \quad Q = \left(2t\rho(C/\kappa)\right)^t \prod_{i=1}^{t} h'(\xi_i).$$

*Then we have*

$$h'(A\xi) \leq 16p^{f_v}\rho(C/\kappa)\, Q \max\left(h'(A), 4p^{f_v}Q\right).$$

**Proof** The main idea is to find $r$ coprime to $p$, and $a \in K$ not a root of unity and $\gamma \in \Gamma$ such that $A\xi = a\gamma^{-r} = (\alpha')^r$, without $h(a)$ being too large and with some control on the range of $r$. In [3], Lemma 6.2 uses a geometry of numbers argument to show that if $|1 - A\xi|_v < H(A\xi)^{-\kappa}$ we can do this with

$$h'(a) \leq h'(A) + rt\left(Q^{-1}\prod_{i=1}^{t} h'(\xi_i)\right)^{1/t} + \frac{4}{r}h(\xi) \tag{5.7}$$

and $r$ in a range[3]

$$\frac{N}{2(p^{f_v}-1)Q} - 1 \leq r \leq N + 3, \tag{5.8}$$

for any $Q \geq (tD_v^*)^t \prod h'(\xi_i)$ and $N \geq 8p^{f_v}D_v^*h'(A)Q$. By (5.8), this lower bound for $N$ implies $r \geq 4$.

Since $r$ and $p$ are coprime, we have $|1 - A\xi|_v = |1 - \alpha'|_v$ for some choice of the $r$th root $\alpha$; note also that Lemma 6.2 of [3] also guarantees that $\alpha$ is not a root of unity.

In what follows, we abbreviate $\rho$ for $\rho(C/\kappa)$; note that we must have $\kappa < 2D_v^*$ (see the end of the proof of Theorem 5.1), hence $\rho \geq 33D_v^*$.

Suppose now that

$$r \geq \rho h'(a). \tag{5.9}$$

Then Theorem 5.1 yields

$$|1 - \alpha'|_v \geq H'(\alpha')^{-r\kappa}.$$

This contradicts

$$|1 - \alpha'|_v = |1 - A\xi|_v < H(A\xi)^{-\kappa} = H(\alpha')^{-r\kappa}$$

[3] Lemma 6.2 in [3] has $2\sqrt{2}$ in place of 2, but a more accurate evaluation of constants appearing in Lemma 6.1 of [3] yields the cleaner bound stated here.

unless $H(\alpha') < H'(\alpha')$, *i.e.* $h(\alpha') < 1/D_v^*$ or equivalently

$$h(A\xi) < r/D_v^*. \tag{5.10}$$

We have shown that (5.9) implies the bound (5.10) for $h(A\xi)$. It remains to localize $r$ by choosing $Q$ and $N$ appropriately so as to satisfy the hypothesis $r \geq \rho h'(a)$ of Theorem 5.1.

We begin by choosing $Q$ as

$$Q = (2\rho t)^t \prod_{i=1}^{t} h'(\xi_i), \tag{5.11}$$

which we may because $2\rho t > tD_v^*$.

We need to bound $h'(a)$ and for this we use (5.7). In view of (5.11), $r \geq 4$ and $h(\xi) \leq h(A\xi) + h'(A)$, we have

$$h'(a) \leq h'(A) + \frac{1}{2\rho} r + \frac{4}{r} h(\xi) \leq 2\,h'(A) + \frac{1}{2\rho} r + \frac{4}{r} h(A\xi). \tag{5.12}$$

Now we choose $N$ to be

$$N = \left\lceil 2(p^{f_v} - 1)Q\left(1 + \max\left(8\rho h'(A), \sqrt{16\rho h(A\xi)}\right)\right)\right\rceil.$$

Then (5.8) implies that

$$r \geq \max\left(8\rho h'(A), \sqrt{16\rho h(A\xi)}\right)$$

hence (5.12) yields

$$h'(a) \leq \frac{1}{4\rho} r + \frac{1}{2\rho} r + \frac{1}{4\rho} r = \frac{1}{\rho} r,$$

hence (5.9), and *a fortiori* (5.10), holds with this choice of $N$.

On the other hand, $r \leq N + 3$ and finally from (5.10) we have

$$h(A\xi) \leq (D_v^*)^{-1} \left\lceil 2(p^{f_v} - 1)Q\left(1 + \max\left(8\rho h'(A), \sqrt{16\rho h(A\xi)}\right)\right)\right\rceil + 3.$$

The first alternative for the maximum easily yields

$$h(A\xi) \leq 16 p^{f_v} \rho h'(A) Q,$$

because $\rho h'(A)Q$ is fairly large (use $\rho \geq 33D_v^*$ to get $\rho h'(A)Q \geq (66t)^t$), hence the small corrections in going from $1 + \max$ to $\max$ and in removing the ceiling brackets and the constant 3 are easily absorbed in replacing $p^{f_v} - 1$ by $p^{f_v}$.

The second alternative for the maximum yields

$$h(A\xi) \leq 2p^{f_v} Q \sqrt{16\rho h(A\xi)}$$

and finally

$$h(A\xi) \leq 64 p^{2f_v} \rho Q^2,$$

completing the proof of Theorem 5.2. □

# 6 Appendix: from a private communication by David Masser

In this appendix, we reproduce material from a letter of David Masser to the first author dated 8th January 1984. These ideas of Masser inspired our §2 and are reproduced here with his permission.

" ... My own method was based on zero estimates rather than heights, using a 'dividing out' trick from transcendence. It gives the following general result.

**Theorem** *Suppose $\theta$ is algebraic of degree $d \geq 2$ and of absolute height $H \geq 1$. Fix an integer $e$ with*

$$1 \leq e < d$$

*and real $\varepsilon$ with*

$$0 < \varepsilon < \frac{1}{e+1}.$$

*Put*

$$\delta = \frac{d}{e+1} + \varepsilon, \qquad \alpha = \frac{d\delta}{(e+1)\varepsilon},$$
$$\beta = d\delta + \alpha, \qquad \gamma = 1 - (e+1)\varepsilon.$$

*Suppose the integers $p_0, q_0 \geq 1$ satisfy*

$$\Lambda = (4H)^{-\beta} q_0^{-\delta} \left| \theta - \frac{p_0}{q_0} \right|^{-\gamma} > 1.$$

*Then the effective strict type of $\theta$ is at most*

$$\frac{-e\gamma \log |\theta - \frac{p_0}{q_0}|}{\log \Lambda}.$$

I didn't try to improve the constant 4, although this could certainly be done by using asymptotics for binomial coefficients.

The proof can be expressed in three lemmas, where $c_1, c_2, \ldots$ denote constants depending only on $d$, $H$, $\varepsilon$. For $P(x,y)$ in $\mathbb{C}[x,y]$ write as in Siegel's set-up

$$P_l(x,y) = \frac{1}{l!} \left( \frac{\partial}{\partial x} \right)^l P(x,y).$$

**Lemma 1** *For each* $k \geq 1$ *there exists a nonzero polynomial* $P(x, y)$ *in* $\mathbb{Z}[x, y]$, *of degree at most* $\delta k$ *in* $x$ *and at most* $e$ *in* $y$, *with coefficients of absolute value at most* $c_1(4H)^{\alpha k}$, *such that*

$$P_l(\theta, \theta) = 0, \qquad (0 \leq l < k).$$

*Furthermore* $P(x, y)$ *is not divisible by any nonconstant element of* $\mathbb{C}[y]$.

Without the last sentence this is routine (I myself like to use the version of Siegel's Lemma proved as the Proposition (p. 32) of the enclosed offprint[4]). Then one simply divides $P(x, y)$ by its greatest monic factor in $\mathbb{C}[y]$. It is not hard to see that the resulting quotient also satisfies the conditions of the lemma.

**Lemma 2** *Suppose* $k \geq e$, *and let* $\xi$, $\eta$ *be arbitrary numbers with* $\xi$ *not a conjugate of* $\theta$. *Then there exists* $l$ *with*

$$0 \leq l \leq (e + 1)\varepsilon k + ed$$

*such that*

$$P_l(\xi, \eta) \neq 0.$$

Again the proof is essentially routine, on taking a minimal representation

$$P(x, y) = A_0(x)B_0(y) + \cdots + A_f(x)B_f(y).$$

The point is that

$$B_0(\eta) = \cdots = B_f(\eta) = 0$$

is impossible by the last sentence of Lemma 1. This is the step usually done by Gauss's Lemma. It is interesting that the Dyson Lemma appears to give only $\sqrt{2e\delta - d}$ in place of $(e + 1)\varepsilon$ multiplying $k$.

**Lemma 3** *Suppose* $k \geq ed/\gamma$ *and let* $p_0, q_0, p, q$ *be integers with* $q_0 \geq 1$, $q \geq 1$. *Then we have*

$$q_0^{-\delta k} q^{-e} \leq c_2(4H)^{\beta k} \left( \left| \theta - \frac{p_0}{q_0} \right|^{\gamma k - ed} + \left| \theta - \frac{p}{q} \right| \right).$$

[4] M. Anderson, D. W. Masser, *Lower bounds for heights on elliptic curves*, Math. Z. **174** (1980) 23–34

This follows from a straightforward comparison of estimates for $P_l(\frac{p_0}{q_0}, \frac{p}{q})$ with $l$ chosen as in Lemma 2.

The Theorem now follows by taking $k$ asymptotic to $e \log q / \log \Lambda$. The usual ineffective arguments give any exponent

$$\lambda > \frac{d}{e+1} + e$$

as in Siegel. The optimal choice $e = 10$, $\varepsilon = \frac{\sqrt{2}-1}{11}$ gives any exponent

$$\lambda > \frac{55}{14}\left(4 + \sqrt{2}\right) = 21.270\ldots$$

for the real root $\theta(m, d)$ of $x^d - mx^{d-1} + 1 = 0$ provided $d \geq d_0(\lambda)$ and $m \geq m_0(d)$.

I briefly looked at a similar approach in the Gelfond–Dyson set-up, with a fixed integer $t$ and derivatives $(\partial/\partial x)^l (\partial/\partial y)^s P(x, y)$ for

$$\frac{l}{k} + \frac{s}{t} < 1$$

But even if the analogous zero estimate could be made to work, it seems as if $t = 1$ (i.e. Siegel) gives the best results for $\theta(m, d)$. So I didn't try too hard with this."

# References


[1] A. Baker and G. Wüstholz, *Logarithmic forms and group varieties*, J. reine angew. Math. **442** (1993) 19–62

[2] E. Bombieri, *Effective Diophantine approximation on* $\mathbb{G}_m$, Ann. Scuola Norm. Sup. Pisa, Cl. Sci. **XX** S.IV (1993) 61–89

[3] E. Bombieri and P. B. Cohen, *Effective Diophantine approximation on* $\mathbb{G}_m$. *II*, Ann. Scuola Norm. Sup. Pisa, Cl. Sci. **XXIV** S.IV (1997) 205–225

[4] E. Bombieri, J. Vaaler, *On Siegel's Lemma*, Invent. Math. **73** (1983) 11–32. Addendum, *ibid* **66** (1984) 177

[5] E. Bombieri, J. Vaaler, *Polynomials with low height and prescribed vanishing*, in Analytic number theory and Diophantine problems (Stillwater, OK, 1984), Progress in Math. 70, Birkhäuser Boston, 1990, 53–73

[6] Y. Bugeaud, *Bornes effectives pour les solutions des équations en S-unités et des équations de Thue–Mahler*, J. Number Theory **71** (1998) 227–244

[7] Y. Bugeaud, M. Laurent, *Minoration de la distance p-adique entre puissances de nombres algébriques*, J. Number Theory **61** (1996) 311–342

[8] P.B. Cohen, A. J. van der Poorten, *Ideal constructions and irrationality measures of roots of algebraic numbers*, Illinois J. Math. **46** (2002) 63–80

[9] M. Laurent, M. Mignotte, Y. Nesterenko, *Formes linéaires en deux logarithmes et déterminants d'interpolation*, J. Number Theory **55** (1995) 285–321

[10] K. Yu, *p-adic logarithmic forms and group varieties.* I, J. reine angew. Math. **502** (1998) 29–92

[11] K. Yu, *p-adic logarithmic forms and group varieties.* II, Acta Arith. **LXXXIX** (1999) 337–378

Enrico Bombieri,
School of Mathematics,
Institute for Advanced Study,
Princeton, N.J. 08540, USA
e-mail: eb@math.ias.edu

Paula B. Cohen,
Department of Mathematics,
Texas A&M University,
College Station, TX 77843, USA
e-mail: pcohen@math.tamu.edu

# A Diophantine system

Andrew Bremner

*Dedicated to Peter Swinnerton-Dyer on the occasion of his seventy-fifth birthday*

This note concerns the Diophantine system

$$\begin{aligned} x_1^2+x_2^2+x_3^2 &= y_1^2+y_2^2+y_3^2 \\ x_1^3+x_2^3+x_3^3 &= y_1^3+y_2^3+y_3^3 \\ x_1^4+x_2^4+x_3^4 &= y_1^4+y_2^4+y_3^4 \end{aligned} \tag{1}$$

which represents a surface of degree 24. The system is of interest in that Palamà [7] in 1951 showed that the only *real* points on (1) in the positive quadrant ($x_i > 0$, $y_i > 0$ for $i = 1, 2, 3$) are trivial points, that is, points where $(x_1, x_2, x_3)$ is a permutation of $(y_1, y_2, y_3)$. Geometrically, the only real points on (1) in the positive quadrant lie upon a finite number of planes. Choudhry [2] discovers the nontrivial rational point (which we also refer to as a nontrivial *solution*)

$$(x_1, x_2, x_3; y_1, y_2, y_3) = (358, -815, 1224; -776, 1233, -410),$$

and we observe that this point also satisfies $x_1 + x_2 + y_1 + y_2 = 0$. In this note, we investigate the section of the surface (1) cut by the plane

$$x_1 + x_2 = t(y_1 + y_2) \tag{2}$$

for $t = \pm 1$. There are only finitely many nontrivial points on the section with $t = 1$, but infinitely many nontrivial points on the section with $t = -1$.

First, make the substitution

$$\begin{aligned} x_1 &= a_1 m + b_1 n, & y_1 &= a_1 m - b_1 n \\ x_2 &= a_2 m + b_2 n, & y_2 &= a_2 m - b_2 n \\ x_3 &= a_3 m + b_3 n, & y_3 &= a_3 m - b_3 n \end{aligned} \tag{3}$$

where $mn \neq 0$, so that (2) above becomes

$$m(t-1)(a_1 + a_2) = n(t+1)(b_1 + b_2).$$

## Case I: $t = 1$

Then $b_2 = -b_1$, and substituting (3) into (1) gives:

$$\begin{aligned} a_1b_1 - a_2b_1 + a_3b_3 &= 0 \\ 3(a_1^2b_1 - a_2^2b_1 + a_3^2b_3)m^2 + b_3^3n^2 &= 0 \\ (a_1^3b_1 - a_2^3b_1 + a_3^3b_3)m^2 + (a_1b_1^3 - a_2b_1^3 + a_3b_3^3)n^2 &= 0. \end{aligned} \tag{4}$$

Nontrivial solutions demand $a_1 \neq a_2$, $a_3 \neq 0$. Eliminating $m$, $n$ in (4) and using $b_3 = (-a_1 + a_2)b_1/a_3$ gives

$$\begin{aligned} a_1^4 - a_1^3a_2 - a_1a_2^3 + a_2^4 - 3a_1^3a_3 + 3a_1^2a_2a_3 + 3a_1a_2^2a_3 \\ - 3a_2^3a_3 + 2a_1^2a_3^2 - 4a_1a_2a_3^2 + 2a_2^2a_3^2 + 3a_1a_3^3 + 3a_2a_3^3 - 3a_3^4 = 0. \end{aligned}$$

This quartic curve is singular at the point $(a_1, a_2, a_3) = (1, 1, 0)$, and has genus 2. Put

$$a_1 = u, \quad a_2 = u - v, \quad a_3 = w$$

to give

$$3u^2v^2 - 3(v^3 + 2v^2w - 2w^3)u + (v^2 - w^2)(v^2 + 3vw + 3w^2) = 0. \tag{5}$$

The discriminant (as function of $u$) being square implies

$$-3(v^6 - 4v^4w^2 + 12v^2w^4 - 12w^6) = \text{square}. \tag{6}$$

This latter curve of genus 2 has of course only finitely many rational points. Its Jacobian is isogenous to the product of the two elliptic curves

$$\begin{aligned} E_1: \quad -3(V^3 - 4V^2 + 12V - 12) &= S_1^2, \\ E_2: \quad -3(1 - 4W + 12W^2 - 12W^3) &= S_2^2, \end{aligned}$$

both of which are of rank 1 (with generators $(0, 6)$ and $(1, 3)$ respectively). Accordingly, Chabauty's method (see for example Coleman [3]) for determining the rational points on (6) does not apply. It is possible that methods of Flynn and Wetherell [6] may be effective, but we have not pursued the calculation; see also Bruin and Elkies [1]. In any event, there are only finitely many solutions of the system (1) satisfying $x_1 + x_2 = y_1 + y_2$, and their determination is afforded by finding all rational points on the curve (6). A modest computer search finds only the points $(u, v, w) = (1, 0, 0)$, $(1, 0, 2)$, $(0, 1, 1)$, $(1, 1, 1)$ on (5), corresponding to trivial solutions of (1).

## Case II: $t = -1$

Now $a_1 + a_2 = 0$, and

$$\begin{aligned} a_1b_1 - a_1b_2 + a_3b_3 &= 0, \\ 3(a_1^2b_1 + a_1^2b_2 + a_3^2b_3)m^2 + (b_1^3 + b_2^3 + b_3^3)n^2 &= 0, \\ (a_1^3b_1 - a_1^3b_2 + a_3^3b_3)m^2 + (a_1b_1^3 - a_1b_2^3 + a_3b_3^3)n^2 &= 0. \end{aligned} \tag{7}$$

Nontrivial solutions demand $b_1 \neq b_2$, $a_3 \neq 0$. Eliminating $m$, $n$ at (7) and using $b_3 = (-b_1 + b_2)a_1/a_3$ gives

$$a_1(b_1 - b_2)(a_1b_1 - a_3b_1 - a_1b_2 - a_3b_2)P(a_1, a_3, b_1, b_2) = 0,$$

where

$$\begin{aligned} P(a_1, a_3, b_1, b_2) = {} & a_1^4b_1^2 - 2a_1^3a_3b_1^2 + 2a_1a_3^3b_1^2 - a_3^4b_1^2 - 2a_1^4b_1b_2 \\ & - 8a_1^2a_3^2b_1b_2 + a_3^4b_1b_2 + a_1^4b_2^2 + 2a_1^3a_3b_2^2 - 2a_1a_3^3b_2^2 - a_3^4b_2^2. \end{aligned}$$

Consequently, either $a_1 = 0$, or $b_1 = b_2$, or $(a_1 - a_3)b_1 = (a_1 + a_3)b_2$, all of which lead to trivial solutions of the original system, or

$$(a_1 - a_3)^3(a_1 + a_3)b_1^2 - (2a_1^4 + 8a_1^2a_3^2 - a_3^4)b_1b_2 + (a_1 - a_3)(a_1 + a_3)^3b_2^2 = 0.$$

The discriminant of the latter is

$$3a_3^2(2a_1^2 + a_3^2)^2(4a_1^2 - a_3^2)$$

which accordingly is square precisely when

$$4a_1^2 - a_3^2 = 3 \times (\text{square}).$$

Put

$$a_1 = 3u^2 + v^2 \quad \text{and} \quad a_3 = 6u^2 - 2v^2,$$

so that (without loss of generality, on changing the sign of $v$ if necessary)

$$\frac{b_1}{b_2} = -\frac{(u - v)(3u + v)^3}{3(3u - v)(u + v)^3}$$

and

$$\frac{b_3}{b_2} = \frac{2(3u^2 + v^2)(3u^2 + 2uv + v^2)}{3(3u - v)(u + v)^3}.$$

Then from (7),

$$-9(3u - v)^2(u + v)^6m^2 + b_2^2(9u^4 - 24u^3v - 26u^2v^2 - 8uv^3 + v^4)n^2 = 0,$$

that is,

$$U^2 = V^4 - 8V^3 - 26V^2 - 24V + 9, \tag{8}$$

where

$$V = v/u, \qquad U = \frac{3(3u-v)(u+v)^3}{b_2 u^2} m/n.$$

A Weierstrass model for (8) is

$$y^2 = x^3 + x^2 - 4x + 32, \tag{9}$$

and using the APECS program [4] of Ian Connell, or the tables of Cremona ([5]), where (9) is numbered as the curve 552E1, we discover that (8) is of rank 1, with generator $P(V,U) = (-9/4, -111/16)$. Accordingly, we can construct infinitely many rational points (equivalently, integer points) on (1). Indeed, from

$$a_1 = 3u^2 + v^2, \qquad a_2 = -3u^2 - v^2, \qquad a_3 = 6u^2 - 2v^2,$$
$$b_1 = -(u-v)(3u+v)^3, \qquad b_2 = 3(3u-v)(u+v)^3$$
$$\text{and} \quad b_3 = 2(3u^2+v^2)(3u^2+2uv+v^2),$$

with

$$U^2 = V^4 - 8V^3 - 26V^2 - 24V + 9, \qquad V = \frac{v}{u}, \qquad U = \frac{1}{u^2}\, m/n,$$

we obtain

$$\begin{aligned}
x_1 &= (3+V^2)U - 27 + 18V^2 + 8V^3 + V^4,\\
x_2 &= -(3+V^2)U + 9 + 24V + 18V^2 - 3V^4,\\
x_3 &= 2(3-V^2)U + 18 + 12V + 12V^2 + 4V^3 + 2V^4,\\
y_1 &= (3+V^2)U + 27 - 18V^2 - 8V^3 - V^4,\\
y_2 &= -(3+V^2)U - 9 - 24V - 18V^2 + 3V^4,\\
y_3 &= 2(3-V^2)U - 18 - 12V - 12V^2 - 4V^3 - 2V^4.
\end{aligned} \tag{10}$$

The point $P(V,U) = (-9/4, -111/16)$ pulls back to the solution

$$(-815, 358, 1224; -776, 1233, -410),$$

and the point $2P(V,U) = (-148/33, 29219/1089)$ to the solution

$$(378382959, -931219912, -156845590; 357088490, 195748463, -932263416).$$

# References

[1] Nils Bruin and Noam D. Elkies, *Trinomials* $ax^7 + bx + c$ *and* $ax^8 + bx + c$ *with Galois groups of order* 168 *and* $8 \cdot 168$, Proceedings 5th International Symposium, ANTS-V, Sydney, 2002, Springer-Verlag, 172–188

[2] Ajai Choudhry, *Equal sums of like powers*, Rocky Mountain Journal of Mathematics **31** (2001) 115–129

[3] Robert F. Coleman, Effective Chabauty, Duke Math. J. **52** (1985) 765–770

[4] Ian Connell, APECS, www.math.mcgill.ca/connell/public/apecs

[5] John E. Cremona, Algorithms for modular elliptic curves, Cambridge University Press, 1997

[6] Victor Flynn and J.L. Wetherell, *Finding rational points on bielliptic genus 2 curves*, Manuscripta Math. **100** (1999) 519–533

[7] Giuseppe Palamà, *Sistemi indeterminati impossibili*, Boll. Un. Mat. Ital. (3) **6** (1951) 113–117

Andrew Bremner,
Department of Mathematics and Statistics,
Arizona State University,
Tempe, AZ 85287-1804, USA
e-mail: bremner@asu.edu

# Valeurs d'un polynôme à une variable représentées par une norme

J.-L. Colliot-Thélène, D. Harari et A. N. Skorobogatov

*à Peter Swinnerton-Dyer en signe d'admiration*

**Abstract**

The aim of the paper is twofold. On the one hand, we study the Brauer group of a smooth and proper model of the $k$-variety given by $P(t) = \mathrm{Norm}_{K/k}(z)$, where $P(t)$ is a polynomial, and $\mathrm{Norm}_{K/k}(z)$ is the norm form defined by a finite field extension $K/k$, with $k$ an essentially arbitrary field. Under some additional hypotheses we compute this group explicitly. On the other hand, when $k$ is the field of rational numbers, and $P(t)$ is a product of arbitrary powers of two linear factors, we prove that the Brauer–Manin obstruction to the Hasse principle and weak approximation is the only one.

# 1 Introduction

Cet article est consacré à l'étude des points rationnels de variétés définies, sur un corps de nombres $k$, par une équation

$$P(t) = N_{K/k}(\mathbf{z}), \tag{1}$$

où $P(t) \in k[t]$ est un polynôme à une variable, $K/k$ une extension finie de corps et $N_{K/k}(\mathbf{z})$, pour $\mathbf{z}$ variable dans $K$, est la forme normique associée à cette extension. Pour l'histoire de ce problème, on pourra consulter [CTPest].

Lorsque $k$ est le corps $\mathbb{Q}$ des nombres rationnels, Heath-Brown et l'un des auteurs [HBSk] ont démontré que pour le polynôme $P(t) = \alpha t^a(t-1)^b$, où $(a,b) = 1$ et $\alpha \in k^*$, l'obstruction de Brauer–Manin est la seule obstruction au principe de Hasse et à l'approximation faible pour tout modèle propre et lisse $X^c$ de (1). Une variante de leur démonstration est donnée dans [CTPest].

Le but de cet article est double. D'une part on étudie, sur un corps $k$ essentiellement arbitraire, et pour un polynôme $P(t)$ quelconque, le groupe de Brauer $\mathrm{Br}(X^c)$. D'autre part, sur $k = \mathbb{Q}$, on généralise le résultat de [HBSk] au cas $P(t) = \alpha t^a(t-1)^b \in \mathbb{Q}[t]$ avec $a$ et $b$ quelconques.

En ce qui concerne le groupe de Brauer, la difficulté réside dans le fait qu'il semble difficile, pour une extension $K/k$ arbitraire, d'écrire un modèle projectif et lisse $X^c$ de la $k$-variété définie par (1) ; nous calculons le groupe de Brauer $\mathrm{Br}(X)$ d'un modèle lisse $X$ non propre, mais assez gros, de (1). Ce calcul est fait au paragraphe 2, sur un corps $k$ de caractéristique zéro, et *pour* $K/k$ *et* $P(t)$ *quelconques*. Les principaux résultats sont la proposition 2.3 et la proposition 2.5. Sous des hypothèses supplémentaires convenables, on en tire des conséquences sur le groupe de Brauer de $X^c$ (corollaires 2.6 et 2.7, propositions 2.11 et 2.12).

Au paragraphe 3, dans le cas $P(t) = \alpha t^a(t-1)^b$, nous appliquons la méthode de la "descente ouverte" ([CTSk]) à la variété $X$ introduite au paragraphe 2. Dans [HBSk] et [CTPest] cette méthode avait été appliquée à l'ouvert de lissité de la variété définie par (1), ouvert qui est strictement contenu dans la variété $X$ ici considérée. C'est ce changement de modèle qui explique les progrès faits dans le présent article (théorème 3.1 et corollaire 3.2).

# 2 Calcul de groupes de Brauer

Soient $k$ un corps de caractéristique zéro et $\overline{k}$ une clôture algébrique de $k$. On note $\Gamma_k := \mathrm{Gal}(\overline{k}/k)$. On note $H^i(\Gamma_k, M)$ ou plus simplement $H^i(k, M)$ les groupes de cohomologie du groupe profini $\Gamma_k$ à valeurs dans un $\Gamma_k$-module continu discret $M$.

On supposera le lecteur familier avec la théorie des $k$-tores algébriques ([CTSa1], §2 ; [Vosk]).

Soit $X$ une $k$-variété, c'est-à-dire un $k$-schéma séparé de type fini. On note $\overline{X} = X \times_k \overline{k}$, puis $k[X]^* = H^0(X, \mathcal{O}_X^*)$ le groupe des unités de $X$ et $\overline{k}[X]^* = H^0(\overline{X}, \mathcal{O}_{\overline{X}}^*)$ celui de $\overline{X}$. On note $\mathrm{Pic}(X) = H^1_{\mathrm{Zar}}(X, \mathcal{O}_X^*)$ le groupe de Picard de $X$. Le théorème 90 de Hilbert revu par Grothendieck identifie ce groupe à $H^1_{\text{ét}}(X, \mathbb{G}_m)$.

Dans cet article, la notation $\mathrm{Br}(X)$ est utilisée pour le groupe de Brauer cohomologique $H^2_{\text{ét}}(X, \mathbb{G}_m)$ d'une $k$-variété $X$. Rappelons aussi les notations usuelles

$$\mathrm{Br}_0(X) = \mathrm{Im}[\mathrm{Br}(k) \to \mathrm{Br}(X)], \quad \mathrm{Br}_1(X) = \mathrm{Ker}[\mathrm{Br}(X) \to \mathrm{Br}(\overline{X})].$$

La suite spectrale de Leray $E_2^{pq} = H^p(k, H^q_{\text{ét}}(\overline{X}, \mathbb{G}_m)) \Longrightarrow H^n_{\text{ét}}(X, \mathbb{G}_m)$ donne naissance à la suite exacte (cf. [CTSa3], (1.5.0))

$$H^2(k, \overline{k}[X]^*) \to \mathrm{Br}_1(X) \to H^1(k, \mathrm{Pic}(\overline{X})) \to H^3(k, \overline{k}[X]^*).$$

Si l'on a $\overline{k}^* = \overline{k}[X]^*$ (c'est par exemple le cas pour $X$ propre, réduite, et géométriquement connexe), alors cette suite s'écrit

$$\mathrm{Br}(k) \to \mathrm{Br}_1(X) \to H^1(k, \mathrm{Pic}(\overline{X})) \to H^3(k, \overline{k}^*),$$

soit encore

$$0 \to \mathrm{Br}_0(X) \to \mathrm{Br}_1(X) \to H^1(k, \mathrm{Pic}(\overline{X})) \to H^3(k, \overline{k}^*).$$

Notons que $H^3(k, \overline{k}^*) = 0$ si $k$ est un corps de nombres ([CF], 7.11.4). Par ailleurs la flèche $H^1(k, \mathrm{Pic}(\overline{X})) \to H^3(k, \overline{k}^*)$ est nulle si $X(k) \neq \emptyset$; en effet tout $k$-point de $X$ définit une rétraction de la flèche

$$H^3(k, \overline{k}^*) = E_2^{30} \to E_\infty^{30} \subset H^3_{\text{ét}}(X, \mathbb{G}_m),$$

et donc toutes les différentielles de la suite spectrale aboutissant aux termes $E_n^{30}$ sont nulles.

Soit $P(t)$ un polynôme. Ecrivons-le $P(t) = \alpha \prod_{i=1}^m p_i(t)^{a_i}$ avec $\alpha \in k^*$ et avec les polynômes $p_i(t)$ irréductibles, unitaires et distincts deux à deux. On note $d_i$ le degré de $p_i(t)$ et $s = \sum_i a_i d_i$ le degré de $P$. On suppose $\prod_{i=1}^m p_i(t)$ de degré au moins égal à 2. Soit $k \subset K \subset \overline{k}$ une extension de degré $n$, et $N_{K/k} \colon K^* \to k^*$ l'application définie par la norme. Soit $\omega_1, \ldots, \omega_n$ une $k$-base de $K$. On note $N_{K/k}(\mathbf{z})$ la forme normique associée, où $\mathbf{z} = z_1\omega_1 + \cdots + z_n\omega_n$ est une $K$-variable.

Soit $V \subset \mathbf{A}_k^{n+1} \simeq \mathbf{A}_k^1 \times_k R_{K/k}(\mathbf{A}_K^1)$ l'ouvert de lissité de l'hypersurface affine définie par l'équation (1).

La projection $(t, \mathbf{z}) \mapsto t$ définit un morphisme surjectif $p \colon V \to \mathbf{A}_k^1$. Soit $U_0 \subset \mathbf{A}_k^1$ l'ouvert donné par $P(t) \neq 0$, et soit $U = p^{-1}(U_0) \subset V$. La $k$-variété $U$ est la variété affine définie par le système $P(t) = N_{K/k}(\mathbf{z}) \neq 0$. La restriction de $p$ à $U$ est un $U_0$-torseur sous le tore normique $T = R^1_{K/k}(\mathbb{G}_m)$, lequel est déployé par le passage de $k$ à $\overline{k}$. Comme le groupe de Picard de $\overline{U}_0$ est nul, il existe un isomorphisme de $\overline{k}$-variétés $\overline{U} \simeq \overline{U}_0 \times_{\overline{k}} \overline{T} \simeq \overline{U}_0 \times \mathbb{G}_m^{n-1}$. Il en résulte que $\mathrm{Pic}(\overline{U}) = 0$ et que le quotient $\overline{k}[U]^*/\overline{k}^*$ est engendré par les facteurs linéaires de $P(t)$ sur $\overline{k}$ et les caractères du tore $\overline{T}$.

Soit $T^c$ une compactification lisse équivariante de $T$ (voir [CTHaSk]). Le produit contracté $U \times^T T^c$ est une compactification partielle de $U$, propre et lisse sur $U_0$. Soit[1] $X$ la $k$-variété lisse obtenue par recollement de $V$ avec $U \times^T T^c$ le long de $U$. Soit $\pi \colon X \to \mathbf{A}_k^1$ le morphisme naturel, et $\overline{\pi} = \pi \times_k \overline{k}$. Comme toute fibre de $\pi$ est contenue soit dans $V$ soit dans $U \times^T T^c$, et que chacune des $k$-variétés $V$ et $U \times^T T^c$ est séparée, le $k$-schéma de type fini $X$ est séparé. C'est donc une $k$-variété.

La fibre générique $X_\eta$ de $\pi$ est projective et géométriquement intègre sur $k(t)$. Une fonction inversible sur $\overline{X}$ provient donc de $\overline{k}(t)^*$. Si un élément de

[1]Dans la construction qui suit, au lieu de recourir à une compactification lisse équivariante de $T$, on pourrait se contenter de l'existence d'une $k$-variété lisse et propre sur $U_0$ jouant le rôle de $U \times^T T^c$, ce qui est fourni par le théorème de Hironaka. Il faut alors modifier le diagramme de la proposition 2.2.

$\overline{k}(t)^*$ n'est pas constant, alors il possède un diviseur non trivial sur $\overline{V}$. Ainsi $\overline{k}[X]^* = \overline{k}^*$ (on a en fait l'énoncé plus précis $\overline{k}[V]^* = \overline{k}^*$).

Soit $X^c$ une $k$-compactification lisse de $X$ telle que $\pi$ s'étende à un morphisme (encore noté $\pi$) $X^c \to \mathbf{P}^1_k$. L'existence de $X^c$ découle du théorème de Hironaka. Le module galoisien $\mathrm{Div}_{\overline{X}\setminus\overline{U}}(\overline{X})$ des diviseurs de $\overline{X}$ à support hors de $\overline{U}$ est la somme directe du module des diviseurs "verticaux" $\mathrm{Div}_v := \mathrm{Div}_{\overline{V}\setminus\overline{U}}(\overline{V})$, librement engendré par les composantes irréductibles des fibres dégénérées, et du module $\mathrm{Div}_h$ des diviseurs de $\overline{U} \times^{\overline{T}} \overline{T^c}$ à support hors de $\overline{U}$, c'est-à-dire des diviseurs "horizontaux".

Le lemme suivant est sans doute bien connu (cf. [Sa], §6.b), mais nous ne l'avons pas trouvé explicitement dans la littérature.

**Lemme 2.1** *Soit $E$ un espace principal homogène sous un $k$-tore $T$. Soient $T^c$ une compactification équivariante de $T$ et $E^c$ le produit contracté $E \times^T T^c$. Soit $K/k$ une extension galoisienne de corps, de groupe de Galois $G$, déployant le $k$-tore $T$. Alors il existe un isomorphisme naturel de suites exactes de $G$-modules*

$$\begin{array}{ccccccccc}
0 & \longrightarrow & \widehat{T} & \longrightarrow & \mathrm{Div}_{T^c_K\setminus T_K}(T^c_K) & \longrightarrow & \mathrm{Pic}(T^c_K) & \longrightarrow & 0 \\
 & & \downarrow\wr & & \downarrow\wr & & \downarrow\wr & & \\
0 & \longrightarrow & K[E]^*/K^* & \longrightarrow & \mathrm{Div}_{E^c_K\setminus E_K}(E^c_K) & \longrightarrow & \mathrm{Pic}(E^c_K) & \longrightarrow & 0.
\end{array}$$

**Démonstration** On a la suite exacte de $G$-modules

$$0 \to K[E]^*/K^* \to \mathrm{Div}_{E^c_K\setminus E_K}(E^c_K) \to \mathrm{Pic}(E^c_K) \to 0. \qquad (\mathcal{E})$$

On peut écrire la suite exacte analogue pour $E_{k(E)} := E \times_k k(E)$, où $k(E)$ est le corps des fonctions de $E$, et l'extension galoisienne $K(E)/k(E)$ (de groupe $G$) au lieu de $K/k$ :

$$0 \to K(E)[E_{k(E)}]^*/K(E)^* \to \mathrm{Div}_{E^c_{K(E)}\setminus E_{K(E)}}(E^c_{K(E)}) \to \mathrm{Pic}(E^c_{K(E)}) \to 0$$

où $E_{K(E)} := E \times_k K(E)$, $E^c_{K(E)} := E^c \times_k K(E)$. La projection $E_{k(E)} \to E$ induit un isomorphisme galoisien de la première de ces suites dans la seconde, car $k$ est algébriquement clos dans $k(E)$.

Pour $E = T$, il est bien connu que $K[T]^*/K^* = \widehat{T}$. Pour les mêmes raisons que ci-dessus, la projection $T_{k(E)} \to T$ induit un isomorphisme galoisien de la première suite horizontale dans l'énoncé du lemme avec la suite

$$0 \to K(E)[T_{k(E)}]^*/K(E)^* \to \mathrm{Div}_{T^c_{K(E)}\setminus T_{K(E)}}(T^c_{K(E)}) \to \mathrm{Pic}(T^c_{K(E)}) \to 0.$$

Il suffit alors d'observer que le point générique de $E$ définit un isomorphisme de $T_{k(E)}$-espaces principaux homogènes $T_{k(E)} \simeq E_{k(E)}$, qui induit un

$T_{k(E)}$-isomorphisme équivariant $T^c_{k(E)} \simeq E^c_{k(E)}$, lequel induit un isomorphisme entre la suite pour $T_{k(E)}$ et celle pour $E_{k(E)}$. $\square$

Notons $\mathbb{Z}[K/k]$ le $\Gamma_k$-module induit $\mathbb{Z}[\Gamma_k/\Gamma_K]$. Au polynôme $P(t)$ nous attachons le $\Gamma_k$-module de permutation $\mathbb{Z}_P$, qui est la somme directe des $\mathbb{Z}[L_i/k]$, où $L_i = k[t]/p_i(t)$. Il est évident que le module $\overline{k}[U_0]^*/\overline{k}^*$, librement engendré en tant que groupe abélien par les classes des facteurs linéaires de $P(t)$ sur $\overline{k}$, est isomorphe à $\mathbb{Z}_P$. D'autre part le module galoisien $\mathrm{Div}_v$ est librement engendré par les composantes irréductibles du diviseur défini par $P(t) = N_{K/k}(\mathbf{z}) = 0$ sur $\overline{V}$, il est donc isomorphe à $\mathbb{Z}[K/k] \otimes \mathbb{Z}_P$. Quant au module galoisien $\mathrm{Div}_h$, on voit en utilisant le lemme 2.1 qu'il est isomorphe au module $\mathrm{Div}_{\overline{T^c}\setminus\overline{T}}(\overline{T^c})$ (voir la fin de la démonstration de la proposition suivante).

L'énoncé suivant permet de contrôler le $\Gamma_k$-module $\mathrm{Pic}(\overline{X})$ :

**Proposition 2.2** *Il existe un diagramme commutatif de $\Gamma_k$-modules, dont les lignes et les colonnes sont exactes*

$$\begin{array}{ccccccccc}
 & & 0 & & 0 & & 0 & & \\
 & & \downarrow & & \downarrow & & \downarrow & & \\
0 & \to & \mathbb{Z}_P & \to & \mathbb{Z}[K/k]\otimes\mathbb{Z}_P & \to & \widehat{T}\otimes\mathbb{Z}_P & \to & 0 \\
 & & \downarrow & & \downarrow & & \downarrow & & \\
0 & \to & \overline{k}[U]^*/\overline{k}^* & \to & (\mathbb{Z}[K/k]\otimes\mathbb{Z}_P)\oplus\mathrm{Div}_h & \to & \mathrm{Pic}(\overline{X}) & \to & 0 \\
 & & \downarrow & & \downarrow & & \downarrow & & \\
0 & \to & \widehat{T} & \to & \mathrm{Div}_h & \to & \mathrm{Pic}(\overline{T^c}) & \to & 0 \\
 & & \downarrow & & \downarrow & & \downarrow & & \\
 & & 0 & & 0 & & 0 & &
\end{array} \qquad (2)$$

**Démonstration** Expliquons d'abord comment ce diagramme est construit. La suite exacte horizontale supérieure est obtenue par tensorisation par $\mathbb{Z}_P$ de la suite exacte naturelle

$$0 \to \mathbb{Z} \to \mathbb{Z}[K/k] \to \widehat{T} \to 0,$$

où la flèche $\mathbb{Z} \to \mathbb{Z}[K/k]$ envoie 1 sur la norme $N_{K/k}$ (somme des éléments de $\Gamma_k/\Gamma_K$, i.e. des plongements de $K$ dans $\overline{k}$ ). Dans la suite exacte horizontale médiane, la flèche $\overline{k}[U]^*/\overline{k}^* \to (\mathbb{Z}[K/k]\otimes\mathbb{Z}_P)\oplus\mathrm{Div}_h$ associe à une fonction son diviseur sur $\overline{X}$.

La flèche $\mathbb{Z}[K/k]\otimes\mathbb{Z}_P \to (\mathbb{Z}[K/k]\otimes\mathbb{Z}_P)\oplus\mathrm{Div}_h$ est simplement la flèche d'inclusion via le premier facteur. La flèche $\mathbb{Z}_P \to \overline{k}[U]^*/\overline{k}^*$ est la flèche naturelle $\overline{k}[U_0]^*/\overline{k}^* \to \overline{k}[U]^*/\overline{k}^*$. Le carré en haut à gauche commute, comme on voit en calculant le diviseur, sur $\overline{X}$, des facteurs linéaires de $P(t)$ sur $\overline{k}$. Ceci définit la flèche $\widehat{T} \otimes \mathbb{Z}_P \to \mathrm{Pic}(\overline{X})$.

Les flèches de la suite horizontale médiane dans la suite horizontale inférieure sont définies par la restriction de $X$ à la fibre générique $X_\eta$ de $\pi$. Cette restriction tombe a priori dans la suite de $\Gamma_k$-modules $(\mathcal{E})$ relative à l'extension $\overline{k}(t)/k(t)$, à l'espace principal homogène $U_\eta$ du $k(t)$-tore $T_{k(t)}$, et à $X_\eta = U_\eta \times^{T_\eta} T^c_\eta$. D'après le lemme 2.1, cette suite est isomorphe à la suite de $\Gamma_k$-modules $(\mathcal{E})$ relative à l'extension $\overline{k}(t)/k(t)$, au $k(t)$-tore $T_{k(t)}$ et à $T^c_{k(t)}$. Mais les arguments du lemme 2.1 montrent aussi que cette suite de $\Gamma_k$-modules est isomorphe (par image réciproque de $k$ à $k(t)$) à la suite de $\Gamma_k$-modules $(\mathcal{E})$ relative à l'extension $\overline{k}/k$, au $k$-tore $T$ et à $T^c$. □

**Proposition 2.3** *Le groupe* $\mathrm{Pic}(\overline{X})$ *est sans torsion, et* $\mathrm{Br}(\overline{X})$ *est nul.*

**Démonstration** Comme $\mathrm{Pic}(\overline{T^c})$ est libre de type fini ([CTSa3], Cor. 2.A.2, p. 461), la colonne de droite du diagramme montre que $\mathrm{Pic}(\overline{X})$ est sans torsion. Pour le deuxième énoncé, notons que puisque $X$ est une $k$-variété lisse, $\mathrm{Br}(\overline{X})$ s'injecte dans le groupe de Brauer de la fibre générique $X_{\overline{k}(t)}$ de $\overline{\pi}\colon \overline{X} \to \mathbf{A}^1_{\overline{k}}$. Mais $X_{\overline{k}(t)}$ est une variété propre, lisse et rationnelle sur $\overline{k}(t)$. Comme le groupe de Brauer est, en caractéristique zéro, un invariant birationnel des variétés projectives et lisses ([Gr], Cor. 7.5) et que le groupe de Brauer de l'espace projectif sur un corps coïncide avec le groupe de Brauer du corps de base (énoncé facile à obtenir en caractéristique zéro, par réduction au cas de la droite affine), on a $\mathrm{Br}(X_{\overline{k}(t)}) = \mathrm{Br}(\overline{k}(t))$. Ce dernier groupe est trivial, comme il résulte du théorème de Tsen. □

Dans le diagramme (2), on dispose d'une section évidente de la projection

$$(\mathbb{Z}[K/k] \otimes \mathbb{Z}_P) \oplus \mathrm{Div}_h \to \mathrm{Div}_h\,.$$

Ceci définit (via la commutativité du carré de droite du diagramme (2), qui donne la flèche $\mathrm{Div}_h \to \mathrm{Pic}(\overline{X})$) un morphisme de suites exactes

$$\begin{array}{ccccccccc} 0 & \to & \widehat{T} & \to & \mathrm{Div}_h & \to & \mathrm{Pic}(\overline{T^c}) & \to & 0 \\ & & \downarrow & & \downarrow & & \| & & \\ 0 & \to & \widehat{T} \otimes \mathbb{Z}_P & \to & \mathrm{Pic}(\overline{X}) & \to & \mathrm{Pic}(\overline{T^c}) & \to & 0 \end{array} \tag{3}$$

Soit $N_i = N_{L_i/k} \in \mathbb{Z}[L_i/k]$ la somme $\sum_\sigma \sigma$ pour $\sigma$ parcourant les plongements de $L_i$ dans $\overline{k}$. Soit $\chi \in \widehat{T}$, image de $\tilde{\chi} \in \mathbb{Z}[K/k]$ par la surjection canonique. Si on ajoute à $\chi \in \mathrm{Div}_h$ l'élément de $\mathrm{Div}_v = \mathbb{Z}[K/k] \otimes \mathbb{Z}_P$ défini par $\tilde{\chi} \otimes (\sum a_i N_i)$, on obtient le diviseur de la fonction $\chi$ (inversible sur $\overline{U}$). Il en résulte que la flèche $\widehat{T} \to \widehat{T} \otimes \mathbb{Z}_P$ dans le diagramme (3) est induite, par tensorisation par $\widehat{T}$, par la flèche $j_P\colon \mathbb{Z} \to \mathbb{Z}_P$ envoyant 1 sur $-\sum_{i=1}^m a_i N_i$.

Observons ici que la flèche $j_P$ admet une rétraction Galois-équivariante si et seulement si les entiers $a_i d_i$ sont premiers entre eux dans leur ensemble.

Pour un $\Gamma_k$-module $M$ on définit le groupe

$$Ш^i_\omega(M) := \bigcap_{g \in \Gamma_k} \mathrm{Ker}[H^i(\Gamma_k, M) \to H^i(\langle g \rangle, M)],$$

où $\langle g \rangle \subset \Gamma_k$ est le sous-groupe procyclique engendré par $g$. Si $M$ est un $\Gamma_k$-module de permutation, ou bien si $M$ est déployé par une extension $K/k$ de groupe de Galois métacyclique (i.e. dont tous les sous-groupes de Sylow sont cycliques), alors $Ш^2_\omega(M) = 0$.

On sait ([Vosk], Chap. 2, §4.6 ; [CTSa1], preuve de la Prop. 6) que le module galoisien $\mathrm{Pic}(\overline{T^c})$ est $\hat{H}^{-1}$-trivial. En particulier, pour tout sous-groupe procyclique $\langle g \rangle \subset \Gamma_k$, on a l'égalité $H^1(\langle g \rangle, \mathrm{Pic}(\overline{T^c})) = 0$ via la 2-périodicité de la cohomologie d'un groupe (fini) cyclique ([CF], Chap. IV, §8, theorem 5). De la suite exacte inférieure du diagramme (2) on déduit alors classiquement ([CTSa2], Prop. 9.5) l'isomorphisme $H^1(k, \mathrm{Pic}(\overline{T^c})) \simeq Ш^2_\omega(\widehat{T})$.

**Définition 2.4** Pour un $\Gamma_k$-module $M$ on définit

$$Ш^2_\omega(M)_P = \mathrm{Ker}[j_{P*} \colon Ш^2_\omega(M) \to Ш^2_\omega(M \otimes \mathbb{Z}_P)].$$

**Proposition 2.5** *Soit $X/k$ comme ci-dessus.*

*a) Il y a une suite exacte naturelle*

$$0 \to H^1(k, \widehat{T} \otimes \mathbb{Z}_P)/j_{P*}H^1(k, \widehat{T}) \to H^1(k, \mathrm{Pic}(\overline{X})) \to Ш^2_\omega(\widehat{T})_P \to 0.$$

*b) Les éléments de* $\mathrm{Br}(X)$ *dont l'image dans* $H^1(k, \mathrm{Pic}(\overline{X}))$ *provient de* $H^1(k, \widehat{T} \otimes \mathbb{Z}_P)$ *sont précisément les éléments du groupe de Brauer vertical de $X$ par rapport à la projection $X \to \mathbf{A}^1_k$, c'est-à-dire les éléments de* $\mathrm{Br}(X)$ *dont la restriction à la fibre générique $X_\eta$ provient de* $\mathrm{Br}(k(t))$.

**Démonstration** De la suite inférieure du diagramme (3) on tire la suite exacte

$$\mathrm{Pic}(\overline{T^c})^{\Gamma_k} \to H^1(k, \widehat{T} \otimes \mathbb{Z}_P) \to H^1(k, \mathrm{Pic}(\overline{X})) \to H^1(k, \mathrm{Pic}(\overline{T^c})) \to H^2(k, \widehat{T} \otimes \mathbb{Z}_P).$$

D'autre part, de la suite supérieure on déduit, comme on a vu, un isomorphisme $H^1(k, \mathrm{Pic}(\overline{T^c})) \simeq Ш^2_\omega(\widehat{T})$, et du calcul de la flèche $j_P \colon \widehat{T} \to \widehat{T} \otimes \mathbb{Z}_P$ dans ce diagramme, et de la nullité de $H^1(k, \mathrm{Div}_h)$, on tire la suite exacte annoncée en a).

On a vu plus haut $\overline{k}^* = \overline{k}[X]^*$ et $\mathrm{Br}(\overline{X}) = 0$, donc $\mathrm{Br}_1(X) = \mathrm{Br}(X)$. On a donc la suite exacte

$$\mathrm{Br}(k) \to \mathrm{Br}(X) \to H^1(k, \mathrm{Pic}(\overline{X})) \to H^3(k, \overline{k}^*).$$

Pour la fibre générique $X_\eta = X_{k(t)}$, on a $\mathrm{Br}(X_{\overline{k}(t)}) = \mathrm{Br}(\overline{k}(t)) = 0$, et la suite spectrale de Hochschild-Serre pour la projection $X_{\overline{k}(t)} \to X_{k(t)}$ et le faisceau étale $\mathbb{G}_m$ donne naissance à la suite exacte

$$\mathrm{Br}(k(t)) \to \mathrm{Br}(X_{k(t)}) \to H^1(k, \mathrm{Pic}(X_{\overline{k}(t)})) \to H^3(k, \overline{k}(t)^*)$$

et la première suite s'envoie de façon naturelle dans la seconde.

La flèche $H^1(k, \mathrm{Pic}(\overline{X})) \to H^1(k, \mathrm{Pic}(X_{\overline{k}(t)}))$ s'identifie à la flèche

$$H^1(k, \mathrm{Pic}(\overline{X})) \to H^1(k, \mathrm{Pic}(\overline{T}^c)),$$

dont le noyau est précisément l'image de $H^1(k, \widehat{T} \otimes \mathbb{Z}_P)$. Ceci établit le point b). □

**Remarque** On peut en fait décrire précisément les éléments de $\mathrm{Br}(X)$ dont la classe dans $H^1(k, \mathrm{Pic}(\overline{X}))$ provient de $H^1(k, \widehat{T} \otimes \mathbb{Z}_P)$. Ce dernier groupe est naturellement isomorphe au groupe

$$\bigoplus_{i=1}^{m} \mathrm{Ker}[H^1(L_i, \mathbb{Q}/\mathbb{Z}) \to H^1(L_i \otimes_k K, \mathbb{Q}/\mathbb{Z})].$$

Soit $\theta_i$ la classe de $t$ dans $L_i = k[t]/p_i(t)$. Soit

$$\{\chi_i\}_{i=1,\dots,m} \in \bigoplus_{i=1}^{m} \mathrm{Ker}[H^1(L_i, \mathbb{Q}/\mathbb{Z}) \to H^1(L_i \otimes_k K, \mathbb{Q}/\mathbb{Z})].$$

On peut alors considérer

$$\mathcal{A} = \sum_{i=1}^{m} \mathrm{Cores}_{L_i(t)/k(t)}(t - \theta_i, \chi_i) \in \mathrm{Br}(k(t)),$$

et vérifier directement que la restriction de $A$ au corps des fonctions de $X$ est non ramifiée sur $X$.

**Corollaire 2.6** *Dans chacun des cas suivants, le groupe de Brauer* $\mathrm{Br}(X)$ *est vertical par rapport à* $X \to \mathbf{A}^1_k$, *et il en est de même, a fortiori, du groupe de Brauer* $\mathrm{Br}(X^c)$ *par rapport à* $X^c \to \mathbf{P}^1_k$ :

*a) le groupe de Brauer de* $T^c$ *est réduit au groupe de Brauer de* $k$ ;

*b) le tore* $T = R^1_{K/k}\mathbb{G}_m$ *est un facteur direct d'un* $k$*-tore* $k$*-rationnel;*

*c) le groupe de Galois de la clôture galoisienne de $K/k$ a tous ses sous-groupes de Sylow cycliques (ce qui est clairement le cas si l'extension $K/k$ est cyclique) ;*

*d) l'extension $K/k$ est de degré premier ;*

*e) le p.g.c.d. des $a_i.d_i$ est égal à 1 (ce qui est clairement le cas si l'un des $p_i$ est de degré $d_i = 1$ et de multiplicité $a_i = 1$).*

**Démonstration** On rappelle les faits suivants. Si $K/k$ satisfait c) ou d), alors $T$ satisfait b) ([CTSa1], Prop. 2 et Prop. 6, et [CTSa2], Prop. 9.1). Si $T$ satisfait b), il satisfait a) ([CTSa1], Prop. 6). Sous a), on a $H^1(k, \mathrm{Pic}(\overline{T^c})) = 0$ (voir le début du pagraphe 2, et observer que $T^c(k)$ contient $T(k)$ et donc est non vide). Ainsi ([CTSa2], Prop. 9.5) $Ш^2_\omega(\widehat{T}) = 0$. L'hypothèse e) implique que la flèche $j_P \colon \mathbb{Z} \to \mathbb{Z}_P$ admet une rétraction Galois-équivariante. On a alors $Ш^2_\omega(M)_P = 0$ pour tout module galoisien $M$, en particulier pour $\widehat{T}$. □

**Remarque** Supposons que $k$ est un corps de nombres, et admettons l'hypothèse de Schinzel (cf. [CTSD], §4). Lorsque $K/k$ est cyclique, on sait alors (extension due à Serre d'un ancien résultat de Sansuc et de l'un des auteurs, voir [CTSD], Th. 4.2) que l'obstruction de Brauer-Manin au principe de Hasse et à l'approximation faible est la seule obstruction pour $X^c$. Mais pour $K/k$ plus général, on ne sait le faire sous aucune des hypothèses a), b), c), d) ci-dessus, bien que chacune de ces hypothèses implique la validité du principe de Hasse et de l'approximation faible pour les fibres lisses de $X^c \to \mathbf{P}^1_k$.

**Corollaire 2.7** *Supposons que $P(t)$ est un produit de facteurs linéaires sur $k$ dont les multiplicités sont premières entre elles dans leur ensemble, et que l'extension $K/k$ ne contient pas de sous-extension cyclique non triviale. Alors le groupe de Brauer de $X$ est réduit à l'image* $\mathrm{Br}_0(X)$ *de* $\mathrm{Br}(k)$. *A fortiori le groupe de Brauer de $X^c$ est-il réduit à l'image* $\mathrm{Br}_0(X^c)$ *de* $\mathrm{Br}(k)$.

**Démonstration** L'hypothèse sur les multiplicités implique $Ш^2_\omega(\widehat{T})_P = 0$ (corollaire précédent, cas e)). Sous l'hypothèse faite sur $K/k$, le noyau de la restriction $H^1(k, \mathbb{Q}/\mathbb{Z}) \to H^1(K, \mathbb{Q}/\mathbb{Z})$ est trivial, soit encore $H^1(k, \widehat{T}) = 0$. Comme $P$ est déployé, on a alors

$$H^1(k, \widehat{T} \otimes \mathbb{Z}_P) = 0.$$

On a donc $H^1(k, \mathrm{Pic}(\overline{X})) = 0$, et donc $\mathrm{Br}(X) = \mathrm{Br}_0(X)$. □

**Exemple** Supposons l'extension $K/k$ de degré premier, non cyclique, et le polynôme $P(t)$ déployé. Alors, sans hypothèse sur les $a_i$, on a l'égalité $\mathrm{Br}_0(X) = \mathrm{Br}(X)$ et donc $\mathrm{Br}_0(X^c) = \mathrm{Br}(X^c)$. On a en effet dans ce cas

$H^1(k,\widehat{T}) = 0$, donc $H^1(k,\widehat{T} \otimes \mathbb{Z}_P) = 0$ et par ailleurs $T$ est un facteur direct d'un $k$-tore $k$-rationnel ([CTSa2], Prop. 9.1), donc $H^1(k, \mathrm{Pic}(\overline{T^c})) = 0$ ([CTSa1], Prop. 6). Ainsi $H^1(k, \mathrm{Pic}(\overline{X})) = 0$.

Il serait souhaitable de décrire le sous-groupe $\mathrm{Br}(X^c) \subset \mathrm{Br}(X)$. Nous donnons quelques résultats concernant la partie verticale, par rapport au morphisme $\pi\colon X^c \to \mathbf{P}^1_k$, de $\mathrm{Br}(X^c)$.

Notons $F = k(U) = k(X^c)$ le corps des fonctions de $X^c$. Etant donné $A \in \mathrm{Br}(k(t))$, on note $A_F$ son image dans $\mathrm{Br}(F)$ par l'inclusion $k(t) \subset F$ induite par $X^c \to \mathbf{P}^1_k$.

**Proposition 2.8** *Supposons que le polynôme $P(t)$, de degré $s$, est de la forme $t^aQ(t)$ avec $a > 0$ et $Q(0) \neq 0$. Soit $\chi \in \mathrm{Ker}[H^1(k,\mathbb{Q}/\mathbb{Z}) \to H^1(K,\mathbb{Q}/\mathbb{Z})]$.*

*a) La classe d'algèbre cyclique sur $k(t)$ définie par le cup-produit $(t^a,\chi)$ définit un élément de $\mathrm{Br}(k(X^c))$ non ramifié sur l'image réciproque de $\mathbf{A}^1_k$ dans $X^c$.*

*b) Pour tout polynôme $R(t) \in k[t]$, la classe de l'algèbre $s(R(t),\chi)$ définit un élément de $\mathrm{Br}(k(X^c))$ qui est non ramifié aux points de codimension 1 de $X^c$ situés au-dessus du point à l'infini de $\mathbf{P}^1_k$.*

**Démonstration** Rappelons tout d'abord la notation employée. Soient $K$ un corps, $K_s$ une clôture séparable de $K$, $\rho \in K^*$ et $\xi \in H^1(\mathrm{Gal}(K_s/K),\mathbb{Q}/\mathbb{Z})$. On note $(\rho,\xi) \in \mathrm{Br}(K)$ le cup-produit de $\rho \in K^* = H^0(\mathrm{Gal}(K_s/K),K_s^*)$ avec le bord $\delta(\xi) \in H^2(\mathrm{Gal}(K_s/K),\mathbb{Z})$, le bord $\delta$ étant pris pour la suite exacte évidente $0 \to \mathbb{Z} \to \mathbb{Q} \to \mathbb{Q}/\mathbb{Z} \to 0$.

Pour établir la proposition, il suffit de vérifier que la restriction de $(t^a,\chi)$ à $F = k(X_c)$ est non ramifiée au-dessus du point $O$ défini par $t = 0$ dans $\mathbf{A}^1_k$. L'égalité $t^aQ(t) = N_{K/k}(\mathbf{z}) \in F$ implique l'égalité

$$(t^aQ(t),\chi)_F = (N_{K/k}(\mathbf{z}),\chi) \in \mathrm{Br}(F).$$

On en déduit $(N_{K/k}(\mathbf{z}),\chi) = N_{K/k}(\mathbf{z},\chi_K) \in \mathrm{Br}(F)$ par la formule de projection. Par hypothèse, la restriction $\chi_K$ de $\chi$ à $K$ est nulle. Ainsi

$$(t^aQ(t),\chi)_F = 0 \in \mathrm{Br}(F)$$

et $(t^a,\chi) = a(t,\chi)_F = -(Q(t),\chi)_F$ est clairement non ramifié aux points de $X^c$ au-dessus de $O$. Soit $u = 1/t$. L'égalité $P(t) = N_{K/k}(\mathbf{z})$ dans $F$ donne une égalité $H(u) = u^sN_{K/k}(\mathbf{z})$, avec $H \in k[u]$ satisfaisant $H(0) \neq 0$. Procédant comme ci-dessus, on voit que pour tout $\chi \in \mathrm{Ker}[H^1(k,\mathbb{Q}/\mathbb{Z}) \to H^1(K,\mathbb{Q}/\mathbb{Z})]$, et tout $R(t) \in k[t]$, la classe de l'algèbre $s(R(t),\chi)$ est non ramifiée aux points de codimension 1 de $X^c$ situés au-dessus du point à l'infini ($u = 0$) de $\mathbf{P}^1_k$. □

**Proposition 2.9** *Supposons que le polynôme $P(t)$, de degré $s$, est de la forme $t^a Q(t)$ avec $a > 0$ et $Q(0) \neq 0$. Soit $\chi \in H^1(k, \mathbb{Q}/\mathbb{Z})$ quelconque. Supposons que $(t, \chi)_F \in \mathrm{Br}(F)$ est non ramifié aux points de $X^c$ au-dessus de $t = 0$. Alors :*

*a)* $\chi \in \mathrm{Ker}[H^1(k, \mathbb{Q}/\mathbb{Z}) \to H^1(K, \mathbb{Q}/\mathbb{Z})]$.

*b) Pour $r = n/(n, a)$, on a $r\chi = 0$ dans $H^1(k, \mathbb{Q}/\mathbb{Z})$.*

**Démonstration** Pour l'énoncé a), il suffit de comparer le résidu de $(t, \chi)$ en le point $t = 0$ de $\mathbf{A}^1_k$ et le résidu de $(t, \chi)_F$ au point générique du diviseur de $X$ défini par $t = 0$ ; en effet la clôture algébrique de $k$ dans le corps des fonctions de la $k$-variété d'équation $N_{K/k}(\mathbf{z}) = 0$ est $K$ (pour les propriétés bien connues des résidus, on renvoie au §1 de [CTSD]). Considérons l'énoncé b). Soit $m = a/(n, a)$. Effectuons le changement de base $t = v^r$. On obtient l'équation $v^{nm} Q(v^r) = N_{K/k}(\mathbf{z})$. Le changement de variables $\mathbf{z}' = \mathbf{z}/v^m$ donne une équivalence birationnelle entre la variété d'équation $v^{nm} Q(v^r) = N_{K/k}(\mathbf{z})$ et celle d'équation $Q(v^r) = N_{K/k}(\mathbf{z}')$, changement qui respecte la projection sur la droite affine $\mathrm{Spec}(k[v])$. L'hypothèse implique que $(v^r, \chi)_{F'}$, où $F'$ désigne le corps des fonctions de la nouvelle variété, est non ramifié au-dessus de $v = 0$. Mais la fibre de cette nouvelle fibration au-dessus du point $v = 0$ est géométriquement intègre. Ceci implique ([CTSD], Prop. 1.1.1) que $(v^r, \chi) \in \mathrm{Br}(k(v))$ est non ramifié en $v = 0$, et ceci équivaut à la condition $r\chi = 0 \in H^1(k, \mathbb{Q}/\mathbb{Z})$. □

Supposons que $P(t)$ s'écrit $P(t) = \alpha \prod_{i=1}^m (t - e_i)^{a_i}$, avec $a_i > 0$, chaque $e_i$ dans $k$ et $e_i \neq e_j$ pour $i \neq j$. On a alors $s = \sum_{i=1}^m a_i$. La structure de $\mathrm{Br}(k(t))$, le fait que les fibres de $X \to \mathbf{A}^1_k$ au-dessus des points de $U_0$ soient géométriquement intègres, et la proposition 2.9 ci-dessus impliquent ([CTSD], Prop. 1.1.1) que tout élément vertical de $\mathrm{Br}(X) \subset \mathrm{Br}(F)$ est, à addition près d'un élément de $\mathrm{Br}(k)$, l'image réciproque d'un élément de la forme $A = \sum_{i=1}^m (t - e_i, \chi_i)$, avec chaque $\chi_i \in \mathrm{Ker}[H^1(k, \mathbb{Q}/\mathbb{Z}) \to H^1(K, \mathbb{Q}/\mathbb{Z})]$, satisfaisant de plus $(n/(n, a_i))\chi_i = 0$.

Considérons le cas où tous les $a_i$ sont égaux à 1, i.e. $P(t) = \alpha \prod_{i=1}^s (t - e_i)$ avec $e_i \neq e_j$ pour $i \neq j$. Alors $m = s$. Dans ce cas, d'après le corollaire 2.6, tout le groupe de Brauer de $X$, et donc de $X^c$, est vertical. D'après la proposition 2.8, tout élément de la forme $\mathcal{A} = \sum_{i=1}^m (t - e_i, \chi_i) \in \mathrm{Br}(k(t))$, avec chaque caractère $\chi_i \in \mathrm{Ker}[H^1(k, \mathbb{Q}/\mathbb{Z}) \to H^1(K, \mathbb{Q}/\mathbb{Z})]$, définit un élément $\mathcal{A}_F \in \mathrm{Br}(F) = \mathrm{Br}(k(X^c))$ non ramifié sur l'ouvert de $X^c$ image réciproque de $\mathbf{A}^1_k$. Au-dessus du point à l'infini $u = 0$, la même proposition assure que $s\mathcal{A}_F$ est non ramifié. Comme par ailleurs $n\chi_i = 0$ pour tout $i$, on voit que $(n, s)\mathcal{A}_F$ est non ramifié sur $X^c$. Au voisinage du point $u = 0$, l'algèbre $\mathcal{A}$

diffère de l'opposé de $(u, \sum_i \chi_i)$ par une algèbre non ramifiée. Si $\mathcal{A}_F$ est non ramifiée, la proposition 2.9 implique $(n/(n,s)).(\sum_i \chi_i) = 0$.

Notons le cas particulier : si $s$ et $n$ sont premiers entre eux, alors les éléments de $\mathrm{Br}(X^c)$ sont les sommes d'un élément de $\mathrm{Br}(k)$ et d'éléments $(t - e_i, \chi_i)_F$, avec $\chi_i \in \mathrm{Ker}[H^1(k, \mathbb{Q}/\mathbb{Z}) \to H^1(K, \mathbb{Q}/\mathbb{Z})]$.

Nous avons donc établi la :

**Proposition 2.10** *Soit* $P(t) = \alpha \prod_{i=1}^{s} (t - e_i)$ *séparable et déployé.*

*a) Tout élément de* $\mathrm{Br}(X^c) \subset \mathrm{Br}(k(X^c))$ *s'écrit* $\rho_F + \sum_{i=1}^{s} (t - e_i, \chi_i)_F$ *avec* $\rho \in \mathrm{Br}(k)$ *et chaque* $\chi_i \in \mathrm{Ker}[H^1(k, \mathbb{Q}/\mathbb{Z}) \to H^1(K, \mathbb{Q}/\mathbb{Z})]$, *avec* $(n/(n,s)).(\sum_i \chi_i) = 0$.

*b) Tout élément de* $\mathrm{Br}(k(X^c))$ *de la forme* $\rho_F + \sum_{i=1}^{s} (t - e_i, \chi_i)_F$ *avec* $\rho \in \mathrm{Br}(k)$ *et chaque* $\chi_i \in \mathrm{Ker}[H^1(k, \mathbb{Q}/\mathbb{Z}) \to H^1(K, \mathbb{Q}/\mathbb{Z})]$ *est non ramifié au-dessus de* $\mathbf{A}^1_k$ *et il est non ramifié sur tout* $X^c$ *si* $s$ *est premier à* $n = [K : k]$.

Considérons maintenant le cas où $P(t)$ est de la forme $P(t) = \alpha t^a (t-1)^b$, avec $(a, b) = 1$. Cette dernière hypothèse assure (corollaire 2.6) que $\mathrm{Br}(X)$ est vertical par rapport à $X \to \mathbf{A}^1_k$, et donc que $\mathrm{Br}(X^c)$ est vertical par rapport à $X^c \to \mathbf{P}^1_k$. Soient $c$ et $d$ tels que que $ad - bc = 1$. D'après la proposition 2.9, tout élément (vertical) de $\mathrm{Br}(X^c)$ provient, à addition près d'un élément de $\mathrm{Br}(k)$, d'un élément de la forme $(t, \chi) + (t-1, \psi) \in \mathrm{Br}(k(t))$, avec $\chi, \psi$ dans $\mathrm{Ker}[H^1(k, \mathbb{Q}/\mathbb{Z}) \to H^1(K, \mathbb{Q}/\mathbb{Z})]$. Le sous-groupe de $k(t)^*$ engendré par $t$ et $(t-1)$ coïncide avec le sous-groupe engendré par $t^a(t-1)^b$ et $t^c(t-1)^d$. Ainsi tout élément de $\mathrm{Br}(X^c)$ provient, à addition près d'un élément de $\mathrm{Br}(k)$, d'un élément de la forme $(t^a(t-1)^b, \chi) + (t^c(t-1)^d, \psi) \in \mathrm{Br}(k(t))$. De l'égalité $\alpha t^a (t-1)^b = N_{K/k}(\mathbf{z})$ dans $F$ et de la formule de projection on déduit que $(t^a(t-1)^b, \gamma)_F$ est dans l'image de $\mathrm{Br}(k)$ pour tout $\gamma$ dans le groupe $\mathrm{Ker}[H^1(k, \mathbb{Q}/\mathbb{Z}) \to H^1(K, \mathbb{Q}/\mathbb{Z})]$. Ainsi tout élément de $\mathrm{Br}(X^c)$ provient, à addition près d'un élément de $\mathrm{Br}(k)$, d'un élément de la forme $\mathcal{A} = (t^c(t-1)^d, \chi) \in \mathrm{Br}(k(t))$, avec $\chi \in \mathrm{Ker}[H^1(k, \mathbb{Q}/\mathbb{Z}) \to H^1(K, \mathbb{Q}/\mathbb{Z})]$.

Soit $\mathcal{A} \in \mathrm{Br}(k(t))$ un élément arbitraire de cette forme. Soient $A = (a, n)$, $B = (b, n)$, $C = (a + b, n)$. On a $(A, B) = 1$, $(B, C) = 1$, $(A, C) = 1$, ce qui implique que le triplet $(AB, AC, BC)$ n'a pas de diviseur commun. Posons $N = n/ABC \in \mathbb{N}$. Pour tout $\chi \in \mathrm{Ker}[H^1(k, \mathbb{Q}/\mathbb{Z}) \to H^1(K, \mathbb{Q}/\mathbb{Z})]$, la proposition 2.8 assure que $a\mathcal{A}_F$ est non ramifiée au-dessus de $t = 0$, que $b\mathcal{A}_F$ est non ramifiée au-dessus de $t = 1$, et que $(a+b)\mathcal{A}_F$ est non ramifiée au-dessus de $t = \infty$. La proposition 2.9 montre que si $\mathcal{A}_F$ est non ramifiée au-dessus de $t = 0$, alors $(n/A).\chi = 0$ ; que si $\mathcal{A}_F$ est non ramifiée au-dessus de $t = 1$, alors $(n/B).\chi = 0$ ; enfin que si $\mathcal{A}_F$ est non ramifiée au-dessus de $t = \infty$, alors $(n/C).\chi = 0$. Si donc $\mathcal{A}_F$ est non ramifiée sur $X^c$, on a nécessairement $N\chi = 0$.

Supposons inversement $N\chi = 0$. On voit alors que $(N,a)\mathcal{A}_F$ est non ramifiée au-dessus de $t=0$, $(N,b)\mathcal{A}_F$ est non ramifiée au-dessus de $t=1$ et $(N,a+b)\mathcal{A}_F$ est non ramifiée au-dessus de $t=\infty$.

Nous pouvons énoncer la :

**Proposition 2.11** *Supposons $P(t) = \alpha t^a(t-1)^b$, avec $(a,b)=1$. Soient $c,d$ avec $ad-bc=1$. Supposons que le quotient $N$ de $n$ par le produit $(a,n).(b,n).(a+b,n)$ est premier à $a.b.(a+b)$. Alors tout élément du groupe de Brauer de $X^c$ est de la forme $\rho_F+\sigma_F$, avec $\rho$ dans $\mathrm{Br}(k)$ et $\sigma = (t^c(t-1)^d,\chi)$ dans $\mathrm{Br}(k(t))$, où $\chi$ est un caractère dans $\mathrm{Ker}[H^1(k,\mathbb{Z}/N) \to H^1(K,\mathbb{Z}/N)]$.*

**Exemples** 1) Soit $a=2$, $b=3$, $n=30$. Alors $N=1$. Dans ce cas, $\mathrm{Br}(X^c)/\mathrm{Br}(k)=0$.

2) Soit $a=1$, $b=1$, $n$ impair. Alors $N=n$, et $\mathrm{Br}(X^c)/\mathrm{Br}(k)$ est formé des éléments $(t,\chi)_F$ avec $\chi \in \mathrm{Ker}[H^1(k,\mathbb{Z}/n) \to H^1(K,\mathbb{Z}/n)]$. (Pour un exemple concret, voir [CTSal], p. 541.)

3) Soit $a=1, b=1, n=2m$ avec $m$ impair. Alors $N=m$ et $\mathrm{Br}(X^c)/\mathrm{Br}(k)$ est formé des éléments $(t,\chi)_F$ avec

$$\chi \in \mathrm{Ker}[H^1(k,\mathbb{Z}/m) \to H^1(K,\mathbb{Z}/m)].$$

On comparera les deux derniers exemples avec la proposition suivante.

**Proposition 2.12** *Supposons $P(t) = \alpha t(t-1)$.*

*a) Soit $\chi \in H^1(k,\mathbb{Q}/\mathbb{Z})$. Supposons $(t,\chi)_F$ non ramifié au-dessus du point $t=\infty$. Alors pour toute sous-extension $L/k$ de $K/k$ avec $[K:L]=2$, on a $\chi_L = 0 \in H^1(L,\mathbb{Q}/\mathbb{Z})$.*

*b) Si l'extension $K/k$ est galoisienne de groupe $G$, et tout caractère de $G$ trivial sur les éléments d'ordre 2 de $G$ est trivial, alors*

$$\mathrm{Br}(X^c)/\mathrm{Br}_0(X^c) = 0.$$

*c) Si l'extension $K/k$ est galoisienne de groupe $G$, et que le groupe $G$ est engendré par ses éléments d'ordre 2, alors $\mathrm{Br}(X^c)/\mathrm{Br}_0(X^c)=0$.*

**Démonstration** La $L$-algèbre $K\otimes_k L$ se décompose en un produit $K\times M$, où $M$ est une $L$-algèbre séparable. Sur le corps $L$, la variété qui nous intéresse est définie par l'équation

$$\alpha t(t-1) = N_{K/L}(\mathbf{z}_1).N_{M/L}(\mathbf{w}).$$

Si l'on pose $t=1/u$, puis $\mathbf{z}_2 = u.\mathbf{z}_1$, on obtient l'équation

$$\alpha(1-u) = N_{K/L}(\mathbf{z}_2).N_{M/L}(\mathbf{w}).$$

Ainsi la fibration $X^c_L \to \mathbf{P}^1_L$ est birationnelle, au-dessus de $\mathbf{P}^1_L$, à une fibration dont la fibre en $u = 0$ est géométriquement intègre. L'hypothèse que $(t,\chi)_F$ est non ramifié au-dessus du point $t = \infty$ implique alors que le résidu de $(t,\chi)_L = (1/u,\chi)_L$ en $u = 0$ est nul, i.e. $\chi_L = 0$, et établit le point a). Ainsi $\chi$ est dans l'intersection des noyaux des restrictions $H^1(k,\mathbb{Q}/\mathbb{Z}) \to H^1(L,\mathbb{Q}/\mathbb{Z})$ pour toutes les sous-extensions $K/L/k$ avec $[K : L] = 2$. Si $K/k$ satisfait l'hypothèse de b), cette intersection est nulle.

D'après la proposition 2.10, tout élément de $\mathrm{Br}(X^c) \subset \mathrm{Br}(F)$ s'écrit comme une somme $\rho_F + (t,\chi)_F + (t-1,\psi)_F$, avec $\chi$ et $\psi$ dans le noyau de la restriction $H^1(k,\mathbb{Q}/\mathbb{Z}) \to H^1(K,\mathbb{Q}/\mathbb{Z})$. L'équation $\alpha t(t-1) = N_{K/k}(\mathbf{z})$ et la formule de projection montrent qu'un tel élément peut s'écrire sous la forme plus simple $\rho_F + (t,\chi)_F$ avec $\chi \in \ker[H^1(k,\mathbb{Q}/\mathbb{Z}) \to H^1(K,\mathbb{Q}/\mathbb{Z})]$. L'énoncé b) résulte alors de a).

L'énoncé c) est une conséquence immédiate. □

Cette proposition s'applique par exemple lorsque $K/k$ est une extension multiquadratique, ou lorsque le groupe $G$ est un groupe symétrique. Elle s'applique aussi lorsque $G$ est un groupe simple différent de $\mathbb{Z}/p$ avec $p$ premier impair, mais on n'obtient alors qu'un cas particulier du corollaire 2.7.

Il semble difficile de décrire des éléments non verticaux de $\mathrm{Br}(X^c)$. Nous devons pour l'instant nous contenter de :

**Questions** Soit $K/k$ une extension biquadratique, i.e. galoisienne de groupe de Galois $G = \mathbb{Z}/2\times\mathbb{Z}/2$. Soit $T = R^1_{K/k}\mathbb{G}_m$. De la suite exacte de $G$-modules

$$0 \to \mathbb{Z} \to \mathbb{Z}[G] \to \hat{T} \to 0,$$

on déduit $H^2(H,\hat{T}) \simeq H^3(H,\mathbb{Z})$ pour tout sous-groupe $H \subset G$, donc

$$Ш^2_\omega(\widehat{T}) = \mathbb{Z}/2.$$

a) Soit $P(t) = \alpha\prod_{i=1}^m (t-e_i)^2$, avec $e_i \in k$. On a alors $Ш^2_\omega(\widehat{T})_P = \mathbb{Z}/2$. Sous l'hypothèse $H^3(k,\overline{k}^*) = 0$, satisfaite si $k$ est un corps de nombres, il existe des éléments de $\mathrm{Br}(X)$ qui sont non verticaux par rapport à $X \to \mathbf{A}^1_k$. Ces éléments semblent difficiles à expliciter. Existe-t-il dans $\mathrm{Br}(X^c)$ des éléments non verticaux par rapport à la flèche $X^c \to \mathbf{P}^1_k$ ?

b) Supposons maintenant que tout facteur irréductible $R(t)$ du polynôme $P(t)$ définisse une extension $k[t]/R(t)$ contenant l'une des trois sous-extensions quadratiques de $K/k$. On a $Ш^2_\omega(\widehat{T}) = \mathbb{Z}/2$. Par contre pour tout corps $L$ et tout $L$-tore $M$ déployé par une extension quadratique de $L$, on a $Ш^2_\omega(\widehat{M}) = 0$. En particulier, avec les hypothèses ci-dessus sur $P$, on a $Ш^2_\omega(\widehat{T}\otimes\mathbb{Z}_P) = 0$ car $\mathbb{Z}_P = \bigoplus_i \mathbb{Z}[L_i/k]$. On a donc $Ш^2_\omega(\widehat{T})_P = \mathbb{Z}/2$. Il existe donc (Prop. 2.5) un élément de $H^1(k,\mathrm{Pic}(\overline{X}))$ d'image non nulle dans $Ш^2_\omega(\widehat{T})_P$. Sous l'hypothèse $H^3(k, \overline{k}^*) = 0$, un tel élément se relève en un élément (difficile à expliciter) de $\mathrm{Br}(X)$ qui est non vertical par rapport à $X \to \mathbf{A}^1_k$. Existe-t-il un tel élément qui provienne de $\mathrm{Br}(X^c)$ ? Il est facile de donner un exemple lorsque la $k$-variété $X$ est $k$-birationnelle au produit d'un espace principal homogène sous $T$ et d'une droite, mais on aimerait avoir un exemple moins trivial. Il conviendrait par exemple d'étudier l'équation $\alpha.(t^2 - a) = N_{K/k}(\mathbf{z})$, avec $K = k(\sqrt{a}, \sqrt{b})$. Peut-on donner un exemple d'obstruction de Brauer-Manin non verticale (pour $X^c \to \mathbf{P}^1_k$) ?

# 3 Descente

Soient $k$ un corps des nombres et $\mathbb{A}_k$ l'anneau des adèles de $k$. Soit $X$ une $k$-variété et $X(\mathbb{A}_k)$ l'espace topologique des adèles de $X$. Par somme des invariants locaux, on définit un accouplement continu à gauche

$$X(\mathbb{A}_k) \times \mathrm{Br}(X) \to \mathbb{Q}/\mathbb{Z}.$$

C'est l'accouplement de Manin. Pour tout sous-ensemble $B \subset \mathrm{Br}(X)$ on note $X(\mathbb{A}_k)^B \subset X(\mathbb{A}_k)$ le fermé de $X(\mathbb{A}_k)$ formé des adèles orthogonales à $B$ par rapport à l'accouplement ci-dessus. Pour $B = \mathrm{Br}(X)$, on écrit simplement $X(\mathbb{A}_k)^{\mathrm{Br}(X)} = X(\mathbb{A}_k)^{\mathrm{Br}}$. Comme l'observa Manin, la loi de réciprocité de la théorie du corps de classes assure que l'inclusion diagonale $X(k) \subset X(\mathbb{A}_k)$ se factorise par $X(k) \subset X(\mathbb{A}_k)^{\mathrm{Br}} \subset X(\mathbb{A}_k)^B$. L'adhérence de $X(k)$ dans $X(\mathbb{A}_k)$ est donc contenue dans le fermé $X(\mathbb{A}_k)^{\mathrm{Br}}$ de $X(\mathbb{A}_k)$. On dit que l'obstruction de Brauer–Manin (resp. l'obstruction de Brauer–Manin attachée à $B \subset \mathrm{Br}(X)$) est la seule obstruction au principe de Hasse et à l'approximation faible sur $X$ si $X(k)$ est dense dans $X(\mathbb{A}_k)^{\mathrm{Br}}$ (resp. si $X(k)$ est dense dans $X(\mathbb{A}_k)^B$).

Le principal résultat arithmétique de cet article est le suivant.

**Théorème 3.1** *Soient $k$ le corps $\mathbb{Q}$ des rationnels et $K/k$ une extension finie de corps. Soient $\alpha \in k^*$ et $a, b \in \mathbb{Z}$. L'obstruction de Brauer–Manin est la seule obstruction au principe de Hasse et à l'approximation faible pour tout modèle propre et lisse de la variété donnée par l'équation*

$$\alpha t^a (t-1)^b = N_{K/k}(\mathbf{z}), \tag{4}$$

*où $t$ est une variable dans $k$ et $\mathbf{z}$ une variable dans $K$.*

**Démonstration** Si $a$ ou $b$ est nul, l'équation (4) définit une $k$-variété $k$-birationnelle à un espace principal homogène sous un tore. Dans ce cas le résultat est bien connu ([Sa], Cor. 8.13). Soit $n$ le degré de $K$ sur $k$. Un changement de variable birationnel permet de remplacer le couple $(a, b)$ par tout couple d'entiers $(a', b')$ avec $a'$ congru à $a$ modulo $n$ et $b'$ congru à $b$ modulo $n$. On suppose désormais $a > 0$ et $b > 0$, et l'on reprend les notations et définitions du début du §2 : ouverts $U \subset V$ de la variété définie par (4), tore $T = R^1_{K/k}\mathbb{G}_m$, compactification lisse équivariante $T^c$, compactification lisse partielle $X$ de $V$, compactification lisse $X^c$ de $X$. Il suffit de démontrer notre énoncé pour un modèle donné, par exemple, pour $X^c$. Soit $\{M_v\} \in X^c(\mathbb{A}_k)^{\mathrm{Br}}$. Soit $\Sigma$ un ensemble fini de places de $k$. On cherche à montrer l'existence d'un point rationnel $M \in X^c(k)$, qui soit de plus arbitrairement proche de chaque $M_v$ pour $v \in \Sigma$.

Comme on a $\overline{k}[X]^* = \overline{k}^*$, que le groupe abélien $\mathrm{Pic}(\overline{X})$ est libre de type fini et que $\mathrm{Br}(\overline{X}) = 0$, et donc en particulier est fini (proposition 2.3), le corollaire 1.2 de [CTSk] montre que l'ensemble $X(\mathbb{A}_k)^{\mathrm{Br}}$ est dense dans $X^c(\mathbb{A}_k)^{\mathrm{Br}}$. On peut donc supposer $\{M_v\} \in X(\mathbb{A}_k)^{\mathrm{Br}}$.

Comme $\mathrm{Br}(\overline{X})$ est fini et $\mathrm{Pic}(\overline{X})$ est libre de type fini (donc $H^1(k, \mathrm{Pic}(\overline{X}))$ est fini), le quotient $\mathrm{Br}(X)/\mathrm{Br}_0(X)$ est fini, donc quotient d'un sous-groupe fini $B$ de $\mathrm{Br}(X)$. Soit $U$ l'ouvert de $X$ d'équation $P(t) = N_{K/k}(\mathbf{z}) \neq 0$.

On peut trouver un ensemble fini de places $\Sigma' \supset \Sigma$ (contenant les places archimédiennes de $k$) tel que $X$ (resp. $U$) s'étende en un $\mathcal{O}_{\Sigma'}$-schéma lisse $\mathcal{X}$ (resp. $\mathcal{U}$), et les éléments de $B$ appartiennent à $\mathrm{Br}(\mathcal{X})$, où $\mathcal{O}_{\Sigma'} \subset k$ est l'anneau des entiers en dehors de $\Sigma'$. Quitte à agrandir $\Sigma'$, on peut supposer que, pour $v \notin \Sigma'$, $M_v \in \mathcal{X}(\mathcal{O}_v)$ et $\mathcal{U}(\mathcal{O}_v) \neq \emptyset$ (la dernière propriété résulte du théorème de Lang-Weil et du lemme de Hensel).

Par continuité de l'accouplement $X(k_v) \times \mathrm{Br}(X) \to \mathbb{Q}/\mathbb{Z}$, on peut, pour $v \in \Sigma'$, trouver un point $M'_v$ de $U(k_v)$ qui est proche de $M_v$ et tel que l'on ait $\alpha(M_v) = \alpha(M'_v)$ pour tout $\alpha \in B$. Choisissons $M'_v$ arbitraire dans $\mathcal{U}(\mathcal{O}_v)$ pour $v \notin \Sigma'$, alors $\alpha(M'_v) = \alpha(M_v) = 0$ pour toute place $v \notin \Sigma'$ et tout $\alpha \in B$, donc $\{M'_v\} \in X(\mathbb{A}_k)^{\mathrm{Br}}$. Finalement on voit que l'on peut supposer $M_v$ dans $U(k_v)$ pour toute place $v$ de $k$, quitte à remplacer $M_v$ par $M'_v$.

Le théorème principal de la théorie de la descente ([CTSa3] ; [CTSk] Prop. 1.3 ; [Sk] Thm. 6.1.2) montre qu'il existe un torseur universel $\mathcal{T}_0$ sur $X$, i.e. un torseur de type id: $\mathrm{Pic}(\overline{X}) \to \mathrm{Pic}(\overline{X})$ et une adèle $\{N_v\} \in \mathcal{T}_0(\mathbb{A}_k)$ qui se projette sur $\{M_v\}$ par la flèche structurale $\mathcal{T}_0 \to X$.

Pour le torseur universel $\mathcal{T}_0$, on a $\overline{k}^* = \overline{k}[\mathcal{T}_0]^*$ et $\mathrm{Pic}(\overline{\mathcal{T}}_0) = 0$ ([CTSa3], Prop. 2.1.1), a fortiori $H^1(k, \mathrm{Pic}(\overline{\mathcal{T}}_0)) = 0$. On a donc $\mathrm{Br}_0(\mathcal{T}_0) = \mathrm{Br}_1(\mathcal{T}_0)$. (En utilisant [HaSk], on peut montrer le résultat plus précis $\mathrm{Br}_0(\mathcal{T}_0) = \mathrm{Br}(\mathcal{T}_0)$, mais on n'a pas besoin de cet énoncé pour la démonstration qui suit.)

Avec les notations du diagramme principal du §2, on a ici $\mathbb{Z}_P = \mathbb{Z}^2$ et donc $\widehat{T} \otimes \mathbb{Z}_P = \widehat{T}^2$. Le $\Gamma_k$-homomorphisme $\widehat{T}^2 \to \mathrm{Pic}(\overline{X}) = \widehat{T}_0$ se dualise en un homomorphisme de $k$-tores $T_0 \to T^2$. Soit $\mathcal{T}_1 = \mathcal{T}_0 \times^{T_0} T^2$ le torseur sur $X$ sous $T^2$ obtenu par changement de groupe.

Soit $\{P_v\} \in \mathcal{T}_1(\mathbb{A}_k)$ l'image de $\{N_v\} \in \mathcal{T}_0(\mathbb{A}_k)$ via la projection naturelle

$\mathcal{T}_0 \to \mathcal{T}_1$. Par fonctorialité de l'accouplement de Manin, cette projection envoie $\mathcal{T}_0(\mathbb{A}_k) = \mathcal{T}_0(\mathbb{A}_k)^{\mathrm{Br}_1}$ (cette dernière égalité provenant de $\mathrm{Br}_0(\mathcal{T}_0) = \mathrm{Br}_1(\mathcal{T}_0)$) dans $\mathcal{T}_1(\mathbb{A}_k)^{\mathrm{Br}_1}$.

En résumé : l'adèle $\{P_v\} \in \mathcal{T}_1(\mathbb{A}_k)$ est orthogonale à $\mathrm{Br}_1(\mathcal{T}_1)$, pour chaque $v$, on a $P_v \in \mathcal{T}_{1,U}(k_v)$, et $P_v$ a pour image $M_v$ par la projection $\mathcal{T}_{1,U} \to U$.

La description locale des torseurs ([CTSa3], Thm. 2.3.1 p. 421 ; [Sk], Thm. 4.3.1), les deux lignes supérieures du diagramme (2) (avec $\hat{T} \otimes \mathbb{Z}_P = \hat{T}^2$) jouant le rôle du diagramme (2.3.2) de [CTSa3] (resp. du diagramme (4.21) de [Sk]) (noter que $\mathrm{Pic}(\overline{U}) = 0$), montre que la restriction $\mathcal{T}_{1,U}$ de $\mathcal{T}_1$ au-dessus de l'ouvert $U \subset X$ est donnée par un système d'équations :

$$0 \neq \alpha t^a (t-1)^b = N_{K/k}(\mathbf{z}), \quad 0 \neq t = \beta_1 N_{K/k}(\mathbf{x}), \quad 0 \neq t - 1 = \beta_2 N_{K/k}(\mathbf{y}),$$

avec $\beta_1, \beta_2 \in k^*$ convenables et $\mathbf{x}, \mathbf{y}$ variables dans $K$. Un changement de variables évident montre que $\mathcal{T}_{1,U}$ est $k$-isomorphe au produit de la sous-variété lisse $Y \subset \mathbf{A}_k^{2n}$ donnée par l'équation

$$\beta_1 N_{K/k}(\mathbf{x}) - \beta_2 N_{K/k}(\mathbf{y}) = 1, \quad N_{K/k}(\mathbf{x}) \neq 0, \quad N_{K/k}(\mathbf{y}) \neq 0$$

avec $\mathbf{x}, \mathbf{y}$ variables dans $K$, et de l'espace principal homogène $E$ du tore $T$ donné par l'équation

$$\alpha \beta_1^a \beta_2^b = N_{K/k}(\mathbf{z}'),$$

avec $\mathbf{z}'$ variable dans $K$.

Soit $E^c$, resp. $Y^c$, une $k$-compactification lisse de $E$, resp. $Y$. On peut prendre $E^c = E \times^T T^c$.

Le $k$-morphisme $q \colon \mathcal{T}_{1,U} \to E$ induit une application $k$-rationnelle de la $k$-variété lisse $\mathcal{T}_1$ vers la $k$-variété projective $E^c$. Une telle application est automatiquement définie sur un ouvert $W \subset \mathcal{T}_1$ contenant tous les points de codimension 1 de $\mathcal{T}_1$. Soit $\mathbf{a} \in \mathrm{Br}(E^c)$. L'élément $q^*(\mathbf{a}) \in \mathrm{Br}(\mathcal{T}_{1,U})$ appartient donc à $\mathrm{Br}(W) \subset \mathrm{Br}(\mathcal{T}_{1,U})$. Comme $\mathcal{T}_1$ est lisse sur un corps de caractéristique zéro, et que $W$ contient tous les points de codimension 1 de $\mathcal{T}_1$, le théorème de pureté pour le groupe de Brauer ([Gr], Thm. 6.1 (c) et Cor. 6.2) assure que l'inclusion $\mathrm{Br}(\mathcal{T}_1) \subset \mathrm{Br}(W)$ est une égalité. Il existe donc $\mathbf{b} \in \mathrm{Br}(\mathcal{T}_1)$ tel que $\mathbf{b}_U = q^*(\mathbf{a}) \in \mathrm{Br}(\mathcal{T}_{1,U})$. Comme $\mathrm{Br}(E^c) = \mathrm{Br}_1(E^c)$ (cette égalité provenant du fait que $T$, et donc $E$ est une variété géométriquement rationnelle), on a $q^*(\mathbf{a}) \in \mathrm{Br}_1(\mathcal{T}_{1,U})$, et donc $\mathbf{b} \in \mathrm{Br}_1(\mathcal{T}_1)$.

Comme l'adèle $\{P_v\}$ est orthogonale à $\mathrm{Br}_1(\mathcal{T}_1)$, on voit alors que l'adèle $\{q(P_v)\} \in E^c(\mathbf{A}_k)$ est orthogonale à $\mathrm{Br}_1(E^c) = \mathrm{Br}(E^c)$. On sait ([Sa], Cor. 8.13) que l'obstruction de Brauer-Manin au principe de Hasse et à l'approximation faible est la seule pour $E^c$, compactification lisse d'un espace principal homogène sous un $k$-tore. On a donc $E^c(k) \neq \emptyset$, et il existe $M_1 \in E(k)$ arbitrairement proche de $q(P_v)$ pour chaque $v \in \Sigma$.

Par ailleurs, sous l'hypothèse que le corps de nombres $k$ est le corps $\mathbb{Q}$ des rationnels, on sait ([HBSk]) que le principe de Hasse et l'approximation faible valent pour $Y^c$. On peut donc trouver un point $M_2 \in Y(k)$ arbitrairement proche des images de $P_v \in \mathcal{T}_{1,U}(k_v) \simeq Y(k_v) \times E(k_v)$ par la première projection, ceci pour chaque $v \in \Sigma$. L'application composée $Y(k) \times E(k) \simeq \mathcal{T}_{1,U}(k) \to U(k)$ envoie alors le point $(M_2, M_1)$ sur un $k$-point $M$ de $U$ proche de chaque $M_v$ pour $v \in \Sigma$. □

**Corollaire 3.2** *Soit $k$ le corps $\mathbb{Q}$ des rationnels. Soient $\alpha \in k^*$ et $a, b \in \mathbb{N}$. Soit $V$ la $k$-variété définie par*

$$\alpha t^a (t-1)^b = N_{K/k}(\mathbf{z}) \neq 0,$$

*avec $t$ variable dans $k$ et $\mathbf{z}$ variable dans $K$, et soit $p\colon V \to \mathbf{A}^1_k$ le morphisme défini par $t$. Soit $X^c$ une $k$-compactification lisse de $V$ équipée d'un $k$-morphisme $p\colon X^c \to \mathbf{P}^1_k$ étendant $p$. Soit*

$$\mathrm{Br}_{\mathrm{vert}}(X^c) = \mathrm{Br}(X^c) \cap p^*(\mathrm{Br}(k(\mathbf{P}^1))) \subset \mathrm{Br}(k(X)).$$

*a) Si $K/k$ est cyclique, ou si $a$ et $b$ sont premiers entre eux, l'obstruction de Brauer-Manin verticale est la seule pour $X^c$, autrement dit : $X^c(k)$ est dense dans $X^c(\mathbb{A}_k)^{\mathrm{Br}_{\mathrm{vert}}(X^c)}$.*

*b) Si $a$ et $b$ sont premiers entre eux, et si $K/k$ ne contient pas de sous-extension cyclique, alors le principe de Hasse et l'approximation faible valent pour $X^c$.*

Il suffit de combiner le théorème 3.1 avec les corollaires 2.6 et 2.7. Le corollaire 2.6 donne d'ailleurs d'autres exemples où l'énoncé a) vaut.

**Questions d'effectivité** Comme on l'a mentionné au paragraphe 2, pour $a, b$ et $K/k$ arbitraires, on ne sait pas calculer explicitement le sous-groupe $\mathrm{Br}(X^c) \subset \mathrm{Br}(X)$. Le théorème 3.1 n'est donc pas effectif.

Sous l'hypothèse que $a$ et $b$ sont premiers entre eux (hypothèse de [HBSk]), le théorème est effectif dans certains cas. Il en est ainsi lorsqu'on peut assurer $\mathrm{Br}(X^c)/\mathrm{Br}(k) = 0$, comme c'est le cas sous l'hypothèse du corollaire 3.2 b). La proposition 2.11 fournit d'autres cas où le calcul de $\mathrm{Br}(X^c)$ est possible. Les exemples suivant cette proposition donnent, par application du théorème 3.1 :

**Corollaire 3.3** *Soit $K$ un corps extension finie de $k = \mathbb{Q}$ de degré 30, et soit $\alpha \in k^*$. Le principe de Hasse et l'approximation faible valent pour tout modèle projectif et lisse de la variété donnée par*

$$\alpha t^2 (t-1)^3 = N_{K/k}(\mathbf{z}).$$

Cet exemple est intéressant, car c'est en quelque sorte le premier cas de couple $(a, b)$ pour lequel l'existence d'un point $k$-rationnel sur un modèle projectif et lisse de la variété définie par $\alpha t^a(t-1)^b = N_{K/k}(\mathbf{z})$ n'est pas automatique.

Pour $P(t)$ de la forme $\alpha t(t-1)$, il existe un point $k$-rationnel sur tout modèle projectif et lisse de $U$ ([HBSk], §2, Rem. 1). La seule question à considérer est donc celle de l'approximation faible.

**Corollaire 3.4** *Soit $K$ un corps extension finie de $k = \mathbb{Q}$, de degré $m$ ou $2m$, avec $m$ impair. Soit $\alpha \in k^*$. Soit $U$ la $k$-variété définie par les équations*

$$\alpha t(t-1) = N_{K/k}(\mathbf{z}) \neq 0.$$

*Soit $X$ le groupe fini $\mathrm{Ker}[H^1(k, \mathbb{Z}/m) \to H^1(K, \mathbb{Z}/m)]$. Pour $\chi \in X$, soit $\mathcal{A}_\chi = (t, \chi) \in \mathrm{Br}(k(t))$.*

*a) Il existe un ensemble fini $S_0$ de places de $k$ tel que pour $v \notin S_0$, tout $\mathcal{A}_\chi$ s'annule sur $U(k_v)$.*

*b) Pour tout ensemble fini $S$ de places de $k$, un point $\{P_v\} \in \prod_{v \in S} U(k_v)$ est dans l'adhérence de $U(k)$ si et seulement si on peut trouver des points $P_v \in U(k_v)$ pour $v \in S_0 \setminus S$ tels que pour tout $\chi \in X$, on ait*

$$\sum_{v \in S_0} \mathcal{A}_\chi(P_v) = \sum_{v \in S_0} (t_v, \chi)_v = 0,$$

*où $t_v = t(P_v)$ et $(t_v, \chi) \in \mathrm{Br}(k_v) \subset \mathbb{Q}/\mathbb{Z}$.*

Le cas particulier $m = 3$ du corollaire ci-dessus avait été établi dans [CTSal], Thm. 6.2. Même dans ce cas, l'approximation faible ne vaut pas toujours, comme le montre un exemple de D. Coray ([CTSal] p. 541).

Le théorème 3.1, la proposition 2.12 et l'existence automatique d'un point rationnel sur un modèle projectif lisse ([HBSk], §2, Rem. 1) impliquent enfin le :

**Corollaire 3.5** *Soit $K$ un corps extension galoisienne finie de $k = \mathbb{Q}$, de groupe $G$, et soit $\alpha \in k^*$. Si tout caractère de $G$ trivial sur les éléments d'ordre 2 de $G$ est trivial, alors l'approximation faible vaut pour tout modèle projectif et lisse de la variété donnée par*

$$\alpha t(t-1) = N_{K/k}(\mathbf{z}).$$

L'hypothèse sur $G$ est satisfaite dans les cas suivants : le groupe $G$ est simple non cyclique ; le groupe $G$ est engendré par ses éléments d'ordre 2, ce qui est le cas si $G$ est un groupe symétrique, et aussi si $G$ est un produit de groupes d'ordre 2.

**Remerciements** Cet article est issu de discussions lors de la conférence *Higher dimensional varieties and rational points* (Institut Alfréd Rényi, Budapest, septembre 2001). Nous en remercions les organisateurs. Nous remercions aussi le rapporteur pour sa lecture attentive du tapuscrit.

# Bibliographie

[CF] J. W. S. Cassels and A. Fröhlich (ed.), *Algebraic number theory*, Academic Press, London and New York, 1967

[CTPest] J.-L. Colliot-Thélène, Points rationnels sur les fibrations, in *Higher dimensional varieties and rational points* (K. J. Böröczky, J. Kollár, T. Szamuely, ed.), Bolyai Society Math. Studies **12**, Springer, 2003

[CTHaSk] J.-L. Colliot-Thélène, D. Harari et A. N. Skorobogatov, *Compactification équivariante d'un tore (d'après Brylinski et Künnemann)*, prépublication

[CTSal] J.-L. Colliot-Thélène et P. Salberger, *Arithmetic on some singular cubic surfaces*, Proc. London Math. Soc. (3) **58** (1989) 519–549

[CTSa1] J.-L. Colliot-Thélène et J.-J. Sansuc, *La R-équivalence sur les tores*, Ann. Sci. École Norm. Sup. **10** (1977) 175–230

[CTSa2] J.-L. Colliot-Thélène et J.-J. Sansuc, *Principal homogeneous spaces under flasque tori : applications*, J. Algebra **106** (1987) 148–205

[CTSa3] J.-L. Colliot-Thélène et J.-J. Sansuc, *La descente sur les variétés rationnelles*. II, Duke Math. J. **54** (1987) 375–492

[CTSk] J.-L. Colliot-Thélène and A. N. Skorobogatov, *Descent on fibrations over* $\mathbf{P}^1_k$ *revisited*, Math. Proc. Camb. Phil. Soc. **128** (2000) 383–393

[CTSD] J.-L. Colliot-Thélène et Sir Peter Swinnerton-Dyer, *Hasse principle and weak approximation for pencils of Severi-Brauer and similar varieties*, J. reine angew. Math. **453** (1994) 49–112

[Gr] A. Grothendieck, Le groupe de Brauer III : exemples et compléments, in *Dix exposés sur la cohomologie des schémas*, Adv. Stud. Pure Math., **3**, Masson et North-Holland, 1968, pp. 88–188

[HaSk] D. Harari and A. N. Skorobogatov, *The Brauer group of torsors and its arithmetic applications*, Max Planck Institut Preprint **33** (2002)

[HBSk] R. Heath-Brown and A. N. Skorobogatov, *Rational solutions of certain equations involving norms*, Acta Math. **189** (2002), 161–177

[Sa] J.-J. Sansuc, *Groupe de Brauer et arithmétique des groupes algébriques linéaires sur un corps de nombres*, J. reine angew. Math. **327** (1981) 12–80

[Sk] A.N. Skorobogatov, *Torsors and rational points*, Cambridge Tracts in Mathematics **144**, Cambridge University Press, 2001

[Vosk] V. E. Voskresenskiĭ, *Algebraic groups and their birational invariants*, translated from the Russian manuscript by Boris Kunyavskiĭ, Translations of Mathematical Monographs, **179**, AMS, Providence, RI, 1998

Jean-Louis Colliot-Thélène,
C.N.R.S., U.M.R. 8628,
Mathématiques, Bâtiment 425,
Université de Paris-Sud,
F-91405 Orsay, France
e-mail : colliot@math.u-psud.fr
web : www.math.u-psud.fr/~colliot

David Harari,
D.M.A., E.N.S.,
45 rue d'Ulm,
F-75005 Paris, France
e-mail : harari@dma.ens.fr
web : www.dma.ens.fr/ harari

Alexei N. Skorobogatov,
Department of Mathematics,
Imperial College,
180 Queen's Gate,
London SW7 2BZ, United Kingdom
e-mail : a.skorobogatov@ic.ac.uk
web : www.ma.ic.ac.uk/~anskor

# Constructing elements in Shafarevich–Tate groups of modular motives

Neil Dummigan William Stein Mark Watkins

**Abstract**

We study Shafarevich–Tate groups of motives attached to modular forms on $\Gamma_0(N)$ of weight $> 2$. We deduce a criterion for the existence of nontrivial elements of these Shafarevich–Tate groups, and give 16 examples in which a strong form of the Beilinson–Bloch conjecture would imply the existence of such elements. We also use modular symbols and observations about Tamagawa numbers to compute nontrivial conjectural lower bounds on the orders of the Shafarevich–Tate groups of modular motives of low level and weight $\leq 12$. Our methods build upon the idea of visibility due to Cremona and Mazur, but in the context of motives rather than abelian varieties.

# 1 Introduction

Let $E$ be an elliptic curve defined over $\mathbb{Q}$ and $L(E,s)$ the associated $L$-function. The conjecture of Birch and Swinnerton-Dyer [BS-D] predicts that the order of vanishing of $L(E,s)$ at $s=1$ is the rank of the group $E(\mathbb{Q})$ of rational points, and also gives an interpretation of the leading term in the Taylor expansion in terms of various quantities, including the order of the Shafarevich–Tate group of $E$.

Cremona and Mazur [CM1] look, among all strong Weil elliptic curves over $\mathbb{Q}$ of conductor $N \leq 5500$, at those with nontrivial Shafarevich–Tate group (according to the Birch and Swinnerton-Dyer conjecture). Suppose that the Shafarevich–Tate group has predicted elements of prime order $p$. In most cases they find another elliptic curve, often of the same conductor, whose $p$-torsion is Galois-isomorphic to that of the first one, and which has positive rank. The rational points on the second elliptic curve produce classes in the common $H^1(\mathbb{Q}, E[p])$. They show [CM2] that these lie in the Shafarevich–Tate group of the first curve, so rational points on one curve explain elements of the Shafarevich–Tate group of the other curve.

The Bloch–Kato conjecture [BK] is the generalisation to arbitrary motives of the leading term part of the Birch and Swinnerton-Dyer conjecture. The Beilinson–Bloch conjecture [B, Be] generalises the part about the order of vanishing at the central point, identifying it with the rank of a certain Chow group.

This paper is a partial generalisation of [CM1] and [AS] from abelian varieties over $\mathbb{Q}$ associated to modular forms of weight 2 to the motives attached to modular forms of higher weight. It also does for congruences between modular forms of equal weight what [Du2] did for congruences between modular forms of different weights.

We consider the situation where two newforms $f$ and $g$, both of even weight $k > 2$ and level $N$, are congruent modulo a maximal ideal $\mathfrak{q}$ of odd residue characteristic, and $L(g, k/2) = 0$ but $L(f, k/2) \neq 0$. It turns out that this forces $L(g, s)$ to vanish to order $\geq 2$ at $s = k/2$. In Section 7, we give sixteen such examples (all with $k = 4$ and $k = 6$), and in each example, we find that $\mathfrak{q}$ divides the numerator of the algebraic number $L(f, k/2)/\mathrm{vol}_\infty$, where $\mathrm{vol}_\infty$ is a certain canonical period.

In fact, we show how this divisibility may be deduced from the vanishing of $L(g, k/2)$ using recent work of Vatsal [V]. The point is, the congruence between $f$ and $g$ leads to a congruence between suitable "algebraic parts" of the special values $L(f, k/2)$ and $L(g, k/2)$. In slightly more detail, a multiplicity one result of Faltings and Jordan shows that the congruence of Fourier expansions leads to a congruence of certain associated cohomology classes. These are then identified with the modular symbols which give rise to the algebraic parts of special values. If $L(g, k/2)$ vanishes then the congruence implies that $L(f, k/2)/\mathrm{vol}_\infty$ must be divisible by $\mathfrak{q}$.

The Bloch–Kato conjecture sometimes then implies that the Shafarevich–Tate group Ш attached to $f$ has nonzero $\mathfrak{q}$-torsion. Under certain hypotheses and assumptions, the most substantial of which is the Beilinson–Bloch conjecture relating the vanishing of $L(g, k/2)$ to the existence of algebraic cycles, we are able to construct some of the predicted elements of Ш using the Galois-theoretic interpretation of the congruence to transfer elements from a Selmer group for $g$ to a Selmer group for $f$. One might say that algebraic cycles for one motive explain elements of Ш for the other, or that we use the congruence to link the Beilinson–Bloch conjecture for one motive with the Bloch–Kato conjecture for the other.

We also compute data which, assuming the Bloch–Kato conjecture, provides lower bounds for the orders of numerous Shafarevich–Tate groups (see Section 7.3). We thank the referee for many constructive comments.

# 2 Motives and Galois representations

This section and the next provide definitions of some of the quantities appearing later in the Bloch–Kato conjecture. Let $f = \sum a_n q^n$ be a newform of weight $k \geq 2$ for $\Gamma_0(N)$, with coefficients in an algebraic number field $E$, which is necessarily totally real. Let $\lambda$ be any finite prime of $E$, and let $\ell$ denote its residue characteristic. A theorem of Deligne [De1] implies the existence of a two-dimensional vector space $V_\lambda$ over $E_\lambda$, and a continuous representation

$$\rho_\lambda \colon \mathrm{Gal}(\overline{\mathbb{Q}}/\mathbb{Q}) \to \mathrm{Aut}(V_\lambda),$$

such that

1. $\rho_\lambda$ is unramified at $p$ for all primes $p$ not dividing $\ell N$, and

2. if $\mathrm{Frob}_p$ is an arithmetic Frobenius element at such a $p$ then the characteristic polynomial of $\mathrm{Frob}_p^{-1}$ acting on $V_\lambda$ is $x^2 - a_p x + p^{k-1}$.

Following Scholl [Sc], we can construct $V_\lambda$ as the $\lambda$-adic realisation of a Grothendieck motive $M_f$. There are also Betti and de Rham realisations $V_B$ and $V_{\mathrm{dR}}$, both 2-dimensional $E$-vector spaces. For details of the construction see [Sc]. The de Rham realisation has a Hodge filtration $V_{\mathrm{dR}} = F^0 \supset F^1 = \cdots = F^{k-1} \supset F^k = \{0\}$. The Betti realisation $V_B$ comes from singular cohomology, while $V_\lambda$ comes from étale $\ell$-adic cohomology. For each prime $\lambda$, there is a natural isomorphism $V_B \otimes E_\lambda \simeq V_\lambda$. We may choose a $\mathrm{Gal}(\overline{\mathbb{Q}}/\mathbb{Q})$-stable $O_\lambda$-module $T_\lambda$ inside each $V_\lambda$. Define $A_\lambda = V_\lambda / T_\lambda$. Let $A[\lambda]$ denote the $\lambda$-torsion in $A_\lambda$. There is the Tate twist $V_\lambda(j)$ (for any integer $j$), which amounts to multiplying the action of $\mathrm{Frob}_p$ by $p^j$.

Following [BK], Section 3, for $p \neq \ell$ (including $p = \infty$), we let

$$H^1_f(\mathbb{Q}_p, V_\lambda(j)) = \ker\Big(H^1(D_p, V_\lambda(j)) \to H^1(I_p, V_\lambda(j))\Big).$$

The subscript $f$ stands for “finite part”; $D_p$ is a decomposition subgroup at a prime above $p$, $I_p$ is the inertia subgroup, and the cohomology is for continuous cocycles and coboundaries. For $p = \ell$, let

$$H^1_f(\mathbb{Q}_\ell, V_\lambda(j)) = \ker\Big(H^1(D_\ell, V_\lambda(j)) \to H^1(D_\ell, V_\lambda(j) \otimes_{\mathbb{Q}_\ell} B_{\mathrm{cris}})\Big)$$

(see [BK], Section 1 for definitions of Fontaine's rings $B_{\mathrm{cris}}$ and $B_{\mathrm{dR}}$). Let $H^1_f(\mathbb{Q}, V_\lambda(j))$ be the subspace of elements of $H^1(\mathbb{Q}, V_\lambda(j))$ whose local restrictions lie in $H^1_f(\mathbb{Q}_p, V_\lambda(j))$ for all primes $p$.

There is a natural exact sequence

$$0 \to T_\lambda(j) \to V_\lambda(j) \xrightarrow{\pi} A_\lambda(j) \to 0.$$

Let $H^1_f(\mathbb{Q}_p, A_\lambda(j)) = \pi_* H^1_f(\mathbb{Q}_p, V_\lambda(j))$. We then define the $\lambda$-Selmer group $H^1_f(\mathbb{Q}, A_\lambda(j))$ as the subgroup of elements of $H^1(\mathbb{Q}, A_\lambda(j))$ whose local restrictions lie in $H^1_f(\mathbb{Q}_p, A_\lambda(j))$ for all primes $p$. Note that the condition at $p = \infty$ is superfluous unless $\ell = 2$. Define the Shafarevich–Tate group

$$Ш(j) = \bigoplus_\lambda H^1_f(\mathbb{Q}, A_\lambda(j))/\pi_* H^1_f(\mathbb{Q}, V_\lambda(j)).$$

Define an ideal $\#Ш(j)$ of $O_E$, in which the exponent of any prime ideal $\lambda$ is the length of the $\lambda$-component of $Ш(j)$. We shall only concern ourselves with the case $j = k/2$, and write $Ш$ for $Ш(k/2)$. It depends on the choice of $\mathrm{Gal}(\overline{\mathbb{Q}}/\mathbb{Q})$-stable $O_\lambda$-module $T_\lambda$ inside each $V_\lambda$. But if $A[\lambda]$ is irreducible then $T_\lambda$ is unique up to scaling and the $\lambda$-part of $Ш$ is independent of choices.

In the case $k = 2$ the motive comes from a (self-dual) isogeny class of abelian varieties over $\mathbb{Q}$, with endomorphism algebra containing $E$. We can choose an abelian variety $B$ in the isogeny class whose endomorphism ring contains the full ring of integers $O_E$. If one takes all the $T_\lambda(1)$ to be $\lambda$-adic Tate modules, then what we have defined above coincides with the usual Shafarevich–Tate group of $B$ (here we assume finiteness of the latter, or just take the quotient by its maximal divisible subgroup). To see this one uses [BK], 3.11 for $\ell = p$. For $\ell \neq p$, $H^1_f(\mathbb{Q}_p, V_\ell) = 0$. Considering the formal group, we can represent every class in $B(\mathbb{Q}_p)/\ell B(\mathbb{Q}_p)$ by an $\ell$-power torsion point in $B(\mathbb{Q}_p)$, so that it maps to zero in $H^1(\mathbb{Q}_p, A_\ell)$.

Define the group of global torsion points

$$\Gamma_{\mathbb{Q}} = \bigoplus_\lambda H^0(\mathbb{Q}, A_\lambda(k/2)).$$

This is analogous to the group of rational torsion points on an elliptic curve. Define an ideal $\#\Gamma_{\mathbb{Q}}$ of $O_E$, in which the exponent of any prime ideal $\lambda$ is the length of the $\lambda$-component of $\Gamma_{\mathbb{Q}}$.

# 3 Canonical periods

From now on, we assume for convenience that $N \geq 3$. We need to choose convenient $O_E$-lattices $T_B$ and $T_{\mathrm{dR}}$ in the Betti and de Rham realisations $V_B$ and $V_{\mathrm{dR}}$ of $M_f$. We do this in such a way that $T_B$ and $T_{\mathrm{dR}} \otimes_{O_E} O_E[1/Nk!]$ agree respectively with the $O_E$-lattice $\mathfrak{M}_{f,B}$ and the $O_E[1/Nk!]$-lattice $\mathfrak{M}_{f,\mathrm{dR}}$ defined in [DFG1] using cohomology, with nonconstant coefficients, of modular curves. (See especially [DFG1], Sections 2.2 and 5.4, and the paragraph preceding Lemma 2.3.)

For any finite prime $\lambda$ of $O_E$, define the $O_\lambda$ module $T_\lambda$ inside $V_\lambda$ to be the image of $T_B \otimes O_\lambda$ under the natural isomorphism $V_B \otimes E_\lambda \simeq V_\lambda$. Then the $O_\lambda$-module $T_\lambda$ is $\mathrm{Gal}(\overline{\mathbb{Q}}/\mathbb{Q})$-stable.

Let $M(N)$ be the modular curve over $\mathbb{Z}[1/N]$ parametrising generalised elliptic curves with full level-$N$ structure. Let $\mathfrak{E}$ be the universal generalised elliptic curve over $M(N)$. Let $\mathfrak{E}^{k-2}$ be the $(k-2)$-fold fibre product of $\mathfrak{E}$ over $M(N)$. (The motive $M_f$ is constructed using a projector on the cohomology of a desingularisation of $\mathfrak{E}^{k-2}$). We realise $M(N)(\mathbb{C})$ as the disjoint union of $\varphi(N)$ copies of the quotient $\Gamma(N)\backslash\mathfrak{H}^*$ (where $\mathfrak{H}^*$ is the completed upper half plane), and let $\tau$ be a variable on $\mathfrak{H}$, so that the fibre $\mathfrak{E}_\tau$ is isomorphic to the elliptic curve with period lattice generated by 1 and $\tau$. Let $z_i \in \mathbb{C}/\langle 1, \tau \rangle$ be a variable on the $i$th copy of $\mathfrak{E}_\tau$ in the fibre product. Then $2\pi i f(\tau)\, d\tau \wedge dz_1 \wedge \cdots \wedge dz_{k-2}$ is a well-defined differential form on (a desingularisation of) $\mathfrak{E}^{k-2}$ and naturally represents a generating element of $F^{k-1}T_{\mathrm{dR}}$. (At least, we can make our choices locally at primes dividing $Nk!$ so that this is the case.) We shall call this element $e(f)$.

Under the de Rham isomorphism between $V_{\mathrm{dR}} \otimes \mathbb{C}$ and $V_B \otimes \mathbb{C}$, $e(f)$ maps to some element $\omega_f$. There is a natural action of complex conjugation on $V_B$, breaking it up into one-dimensional $E$-vector spaces $V_B^+$ and $V_B^-$. Let $\omega_f^+$ and $\omega_f^-$ be the projections of $\omega_f$ to $V_B^+ \otimes \mathbb{C}$ and $V_B^- \otimes \mathbb{C}$ respectively. Let $T_B^\pm$ be the intersections of $V_B^\pm$ with $T_B$. These are rank one $O_E$-modules, but not necessarily free, since the class number of $O_E$ may be greater than one. Choose nonzero elements $\delta_f^\pm$ of $T_B^\pm$ and let $\mathfrak{a}^\pm$ be the ideals $[T_B^\pm : O_E\delta_f^\pm]$. Define complex numbers $\Omega_f^\pm$ by $\omega_f^\pm = \Omega_f^\pm \delta_f^\pm$.

# 4 The Bloch–Kato conjecture

In this section we extract from the Bloch–Kato conjecture for $L(f, k/2)$ a prediction about the order of the Shafarevich–Tate group, by analysing the other terms in the formula.

Let $L(f, s)$ be the $L$-function attached to $f$. For $\Re(s) > \frac{k+1}{2}$ it is defined by the Dirichlet series with Euler product $\sum_{n=1}^{\infty} a_n n^{-s} = \prod_p (P_p(p^{-s}))^{-1}$, but there is an analytic continuation given by an integral, as described in the next section. Suppose that $L(f, k/2) \neq 0$. The Bloch–Kato conjecture for the motive $M_f(k/2)$ predicts the following equality of fractional ideals of $E$:

$$\frac{L(f, k/2)}{\mathrm{vol}_\infty} = \left( \prod_p c_p(k/2) \right) \frac{\#\text{Ш}}{\mathfrak{a}^\pm (\#\Gamma_{\mathbb{Q}})^2}.$$

Here, **and from this point onwards,** $\pm$ represents the parity of $(k/2) - 1$. The quantity $\mathrm{vol}_\infty$ is equal to $(2\pi i)^{k/2}$ multiplied by the determinant of the isomorphism $V_B^\pm \otimes \mathbb{C} \simeq (V_{\mathrm{dR}}/F^{k/2}) \otimes \mathbb{C}$, calculated with respect to the lattices

$O_E\delta_f^{\pm}$ and the image of $T_{\mathrm{dR}}$. For $l \neq p$, $\mathrm{ord}_\lambda(c_p(j))$ is defined to be

$$\begin{aligned}\mathrm{length}\, H^1_f(\mathbb{Q}_p, T_\lambda(j))_{\mathrm{tors}} - \mathrm{ord}_\lambda(P_p(p^{-j}))\\ = \mathrm{length}\left(H^0(\mathbb{Q}_p, A_\lambda(j))/H^0\left(\mathbb{Q}_p, V_\lambda(j)^{I_p}/T_\lambda(j)^{I_p}\right)\right).\end{aligned}$$

We omit the definition of $\mathrm{ord}_\lambda(c_p(j))$ for $\lambda \mid p$, which requires one to assume Fontaine's de Rham conjecture ([Fo1], Appendix A6), and depends on the choices of $T_{\mathrm{dR}}$ and $T_B$, locally at $\lambda$. (We shall mainly be concerned with the $q$-part of the Bloch–Kato conjecture, where $q$ is a prime of good reduction. For such primes, the de Rham conjecture follows from Faltings [Fa], Theorem 5.6.)

Strictly speaking, the conjecture in [BK] is only given for $E = \mathbb{Q}$. We have taken here the obvious generalisation of a slight rearrangement of [BK], (5.15.1). The Bloch–Kato conjecture has been reformulated and generalised by Fontaine and Perrin-Riou, who work with general $E$, though that is not really the point of their work. [Fo2], Section 11 sketches how to deduce the original conjecture from theirs, in the case $E = \mathbb{Q}$.

**Lemma 4.1** $\mathrm{vol}_\infty/\mathfrak{a}^{\pm} = c(2\pi i)^{k/2}\mathfrak{a}^{\pm}\Omega_{\pm}$, *with* $c \in E$ *and* $\mathrm{ord}_\lambda(c) = 0$ *for* $\lambda \nmid Nk!$.

**Proof** We note that $\mathrm{vol}_\infty$ is equal to $(2\pi i)^{k/2}$ times the determinant of the period map from $F^{k/2}V_{\mathrm{dR}} \otimes \mathbb{C}$ to $V_B^{\pm} \otimes \mathbb{C}$, with respect to lattices dual to those we used above in the definition of $\mathrm{vol}_\infty$ (cf. [De2], last paragraph of 1.7). Here we are using natural pairings. Meanwhile, $\Omega_{\pm}$ is the determinant of the same map with respect to the lattices $F^{k/2}T_{\mathrm{dR}}$ and $O_E\delta_f^{\pm}$. Recall that the index of $O_E\delta_f^{\pm}$ in $T_B^{\pm}$ is the ideal $\mathfrak{a}^{\pm}$. Then the proof is completed by noting that, locally away from primes dividing $Nk!$, the index of $T_{\mathrm{dR}}$ in its dual is equal to the index of $T_B$ in its dual, both being equal to the ideal denoted $\eta$ in [DFG2]. □

**Remark 4.2** Note that the "quantities" $\mathfrak{a}^{\pm}\Omega_{\pm}$ and $\mathrm{vol}_\infty/\mathfrak{a}^{\pm}$ are independent of the choice of $\delta_f^{\pm}$.

**Lemma 4.3** *Let* $p \nmid N$ *be a prime and* $j$ *an integer. Then the fractional ideal* $c_p(j)$ *is supported at most on divisors of* $p$.

**Proof** As on [Fl2], p. 30, for odd $l \neq p$, $\mathrm{ord}_\lambda(c_p(j))$ is the length of the finite $O_\lambda$-module $H^0(\mathbb{Q}_p, H^1(I_p, T_\lambda(j))_{\mathrm{tors}})$, where $I_p$ is an inertia group at $p$. But $T_\lambda(j)$ is a trivial $I_p$-module, so $H^1(I_p, T_\lambda(j))$ is torsion free. □

**Lemma 4.4** *Let $q \nmid N$ be a prime satisfying $q > k$. Suppose that $A[\mathfrak{q}]$ is an irreducible representation of* $\mathrm{Gal}(\overline{\mathbb{Q}}/\mathbb{Q})$, *where $\mathfrak{q} \mid q$. Let $p \mid N$ be a prime, and if $p^2 \mid N$ suppose that $p \not\equiv \pm 1 \pmod{q}$. Suppose also that $f$ is not congruent modulo $\mathfrak{q}$ (for Fourier coefficients of index coprime to $Nq$) to any newform of weight $k$, trivial character, and level dividing $N/p$. Then* $\mathrm{ord}_{\mathfrak{q}}(c_p(j)) = 0$ *for all integers $j$.*

**Proof** There is a natural injective map from $V_{\mathfrak{q}}(j)^{I_p}/T_{\mathfrak{q}}(j)^{I_p}$ to $H^0(I_p, A_{\mathfrak{q}}(j))$ (i.e., $A_{\mathfrak{q}}(j)^{I_p}$). Consideration of $\mathfrak{q}$-torsion shows that

$$\dim_{O_E/\mathfrak{q}} H^0(I_p, A[\mathfrak{q}](j)) \geq \dim_{E_{\mathfrak{q}}} H^0(I_p, V_{\mathfrak{q}}(j)).$$

To prove the lemma it suffices to show that

$$\dim_{O_E/\mathfrak{q}} H^0(I_p, A[\mathfrak{q}](j)) = \dim_{E_{\mathfrak{q}}} H^0(I_p, V_{\mathfrak{q}}(j)),$$

since this ensures that $H^0(I_p, A_{\mathfrak{q}}(j)) = V_{\mathfrak{q}}(j)^{I_p}/T_{\mathfrak{q}}(j)^{I_p}$, and therefore that $H^0(\mathbb{Q}_p, A_{\mathfrak{q}}(j)) = H^0(\mathbb{Q}_p, V_{\mathfrak{q}}(j)^{I_p}/T_{\mathfrak{q}}(j)^{I_p})$.

Suppose that Condition (b) of [L], Proposition 2.3 is not satisfied. Then there exists a character $\chi\colon \mathrm{Gal}(\overline{\mathbb{Q}}/\mathbb{Q}) \to O_{\mathfrak{q}}^{\times}$ of $q$-power order such that the $p$-part of the conductor of $V_{\mathfrak{q}} \otimes \chi$ is strictly smaller than that of $V_{\mathfrak{q}}$. Let $f_\chi$ denote the *newform*, of level dividing $N/p$, associated with $V_{\mathfrak{q}} \otimes \chi$. The character of $f_\chi$ has conductor at worst $p$. Since $\chi$ has conductor $p$ and $q$-power order, $p \equiv 1 \pmod{q}$, so by hypothesis $p^2 \nmid N$. Hence $f_\chi$ has level coprime to $p$ and must have trivial character. Then the existence of $f_\chi$ contradicts our hypotheses.

Suppose now that

$$\dim_{O_E/\mathfrak{q}} H^0(I_p, A[\mathfrak{q}](j)) > \dim_{E_{\mathfrak{q}}} H^0(I_p, V_{\mathfrak{q}}(j)),$$

(if not, there is nothing to prove). If Condition (a) of [L], Proposition 2.3 were not satisfied then [L], Proposition 2.2 would imply the existence of an impossible twist, as in the previous paragraph. (Here we are also using [L], Proposition 1.1.)

Since Condition (c) is clearly also satisfied, we are in a situation covered by one of the three cases in [L], Proposition 2.3. Since $p \not\equiv -1 \pmod{q}$ if $p^2 \mid N$, Case 3 is excluded, so $A[\mathfrak{q}](j)$ is unramified at $p$ and $\mathrm{ord}_p(N) = 1$. (Here we are using Carayol's result that $N$ is the prime-to-$q$ part of the conductor of $V_{\mathfrak{q}}$ [Ca1].) But then [JL], Theorem 1 (which uses the condition $q > k$) implies the existence of a newform of weight $k$, trivial character and level dividing $N/p$, congruent to $g$ modulo $\mathfrak{q}$, for Fourier coefficients of index coprime to $Nq$. This contradicts our hypotheses. □

**Remark 4.5** For an example of what can be done when $f$ is congruent to a form of lower level, see the first example in Section 7.4 below.

**Lemma 4.6** *If $\mathfrak{q} \mid q$ is a prime of $E$ such that $q \nmid Nk!$, then* $\mathrm{ord}_{\mathfrak{q}}(c_q) = 0$.

**Proof** It follows from [DFG1], Lemma 5.7 (whose proof relies on an application, at the end of Section 2.2, of the results of [Fa]) that $T_{\mathfrak{q}}$ is the $O_{\mathfrak{q}}[\mathrm{Gal}(\overline{\mathbb{Q}}_q/\mathbb{Q}_q)]$-module associated to the filtered module $T_{\mathrm{dR}} \otimes O_{\mathfrak{q}}$ by the functor they call $\mathbb{V}$. (This property is part of the definition of an $S$-integral premotivic structure given in [DFG1], Section 1.2.) Given this, the lemma follows from [BK], Theorem 4.1(iii). (That $\mathbb{V}$ is the same as the functor used in [BK], Theorem 4.1 follows from [Fa], first paragraph of 2(h).) □

**Lemma 4.7** *If $A[\lambda]$ is an irreducible representation of* $\mathrm{Gal}(\overline{\mathbb{Q}}/\mathbb{Q})$, *then*

$$\mathrm{ord}_\lambda(\#\Gamma_{\mathbb{Q}}) = 0.$$

**Proof** This follows trivially from the definition. □

Putting together the above lemmas we arrive at the following:

**Proposition 4.8** *Let $q \nmid N$ be a prime satisfying $q > k$ and suppose that $A[\mathfrak{q}]$ is an irreducible representation of* $\mathrm{Gal}(\overline{\mathbb{Q}}/\mathbb{Q})$, *where $\mathfrak{q} \mid q$. Assume the same hypotheses as in Lemma 4.4 for all $p \mid N$. Choose $T_{\mathrm{dR}}$ and $T_B$ which locally at $\mathfrak{q}$ are as in the previous section. If $L(f, k/2)\mathfrak{a}^{\pm}/\mathrm{vol}_\infty \neq 0$ then the Bloch–Kato conjecture predicts that*

$$\mathrm{ord}_{\mathfrak{q}}(\#\text{Ш}) = \mathrm{ord}_{\mathfrak{q}}(L(f, k/2)\mathfrak{a}^{\pm}/\mathrm{vol}_\infty).$$

# 5 Congruences of special values

Let $f = \sum a_n q^n$ and $g = \sum b_n q^n$ be newforms of equal weight $k \geq 2$ for $\Gamma_0(N)$. Let $E$ be a number field large enough to contain all the coefficients $a_n$ and $b_n$. Suppose that $\mathfrak{q} \mid q$ is a prime of $E$ such that $f \equiv g \pmod{\mathfrak{q}}$, i.e. $a_n \equiv b_n \pmod{\mathfrak{q}}$ for all $n$. Assume that $A[\mathfrak{q}]$ is an irreducible representation of $\mathrm{Gal}(\overline{\mathbb{Q}}/\mathbb{Q})$ and that $q \nmid N\varphi(N)k!$. Choose $\delta_f^{\pm} \in T_B^{\pm}$ in such a way that $\mathrm{ord}_{\mathfrak{q}}(\mathfrak{a}^{\pm}) = 0$, i.e., $\delta_f^{\pm}$ generates $T_B^{\pm}$ locally at $\mathfrak{q}$. Make two further assumptions:

$$L(f, k/2) \neq 0 \qquad \text{and} \qquad L(g, k/2) = 0.$$

**Proposition 5.1** *With assumptions as above,* $\mathrm{ord}_{\mathfrak{q}}(L(f, k/2)/\mathrm{vol}_\infty) > 0$.

**Proof** This is based on some of the ideas used in [V], Section 1. Note the apparent typographical error in [V], Theorem 1.13 which should presumably refer to "Condition 2". Since $\mathrm{ord}_{\mathfrak{q}}(\mathfrak{a}^{\pm}) = 0$, we just need to show that $\mathrm{ord}_{\mathfrak{q}}\big(L(f, k/2)/((2\pi i)^{k/2}\Omega_{\pm})\big) > 0$, where $\pm 1 = (-1)^{(k/2)-1}$. It is well known, and easy to prove, that

$$\int_0^\infty f(iy)y^{s-1}dy = (2\pi)^{-s}\Gamma(s)L(f, s).$$

Hence, if for $0 \leq j \leq k-2$ we define the $j$th period

$$r_j(f) = \int_0^{i\infty} f(z) z^j dz,$$

where the integral is taken along the positive imaginary axis, then

$$r_j(f) = j!(-2\pi i)^{-(j+1)} L_f(j+1).$$

Thus we are reduced to showing that $\mathrm{ord}_{\mathfrak{q}}(r_{(k/2)-1}(f)/\Omega_{\pm}) > 0$.

Let $\mathcal{D}_0$ be the group of divisors of degree zero supported on $\mathbb{P}^1(\mathbb{Q})$. For a $\mathbb{Z}$-algebra $R$ and integer $r \geq 0$, let $P_r(R)$ be the additive group of homogeneous polynomials of degree $r$ in $R[X,Y]$. Both these groups have a natural action of $\Gamma_1(N)$. Let $S_{\Gamma_1(N)}(k,R) := \mathrm{Hom}_{\Gamma_1(N)}(\mathcal{D}_0, P_{k-2}(R))$ be the $R$-module of weight $k$ modular symbols for $\Gamma_1(N)$.

Via the isomorphism (8) of [V], Section 1.5 combined with the argument of [V], 1.7, the cohomology class $\omega_f^{\pm}$ corresponds to a modular symbol $\Phi_f^{\pm} \in S_{\Gamma_1(N)}(k,\mathbb{C})$, and $\delta_f^{\pm}$ corresponds to an element $\Delta_f^{\pm} \in S_{\Gamma_1(N)}(k, O_{E,\mathfrak{q}})$. We are now dealing with cohomology over $X_1(N)$ rather than $M(N)$, which is why we insist that $q \nmid \varphi(N)$. It follows from the last line of [St], Section 4.2 that, up to some small factorials which do not matter locally at $\mathfrak{q}$,

$$\Phi_f^{\pm}([\infty]-[0]) = \sum_{\substack{j=0,\\ j\equiv(k/2)-1 \pmod 2}}^{k-2} r_f(j) X^j Y^{k-2-j}.$$

Since $\omega_f^{\pm} = \Omega_f^{\pm}\delta_f^{\pm}$, we see that

$$\Delta_f^{\pm}([\infty]-[0]) = \sum_{\substack{j=0,\\ j\equiv(k/2)-1 \pmod 2}}^{k-2} (r_f(j)/\Omega_f^{\pm}) X^j Y^{k-2-j}.$$

The coefficient of $X^{(k/2)-1}Y^{(k/2)-1}$ is what we would like to show is divisible by $\mathfrak{q}$. Similarly

$$\Phi_g^{\pm}([\infty]-[0]) = \sum_{\substack{j=0,\\ j\equiv(k/2)-1 \pmod 2}}^{k-2} r_g(j) X^j Y^{k-2-j}.$$

The coefficient of $X^{(k/2)-1}Y^{(k/2)-1}$ in this is 0, since $L(g,k/2)=0$. Therefore it would suffice to show that, for some $\mu \in O_E$, the element $\Delta_f^{\pm} - \mu\Delta_g^{\pm}$ is divisible by $\mathfrak{q}$ in $S_{\Gamma_1(N)}(k, O_{E,\mathfrak{q}})$. It suffices to show that, for some $\mu \in O_E$, the element $\delta_f^{\pm} - \mu\delta_g^{\pm}$ is divisible by $\mathfrak{q}$, considered as an element of $\mathfrak{q}$-adic cohomology of $X_1(N)$ with nonconstant coefficients. This would be the case if $\delta_f^\pm$ and $\delta_g^\pm$ generate the same one-dimensional subspace upon reduction modulo $\mathfrak{q}$. But this is a consequence of [FJ], Theorem 2.1(1) (for which we need the irreducibility of $A[\mathfrak{q}]$). □

**Remark 5.2** The signs in the functional equations of $L(f,s)$ and $L(g,s)$ are equal. They are determined by the eigenvalue of the Atkin–Lehner involution $W_N$, which is determined by $a_N$ and $b_N$ modulo $\mathfrak{q}$, because $a_N$ and $b_N$ are each $N^{k/2-1}$ times this sign and $\mathfrak{q}$ has residue characteristic coprime to $2N$. The common sign in the functional equation is $(-1)^{k/2}w_N$, where $w_N$ is the common eigenvalue of $W_N$ acting on $f$ and $g$.

This is analogous to [CM1], remark at the end of Section 3, which shows that if $\mathfrak{q}$ has odd residue characteristic and $L(f,k/2) \neq 0$ but $L(g,k/2) = 0$ then $L(g,s)$ must vanish to order at least two at $s = k/2$. Note that Maeda's conjecture implies that there are no examples of $g$ of level one with positive sign in their functional equation such that $L(g,k/2) = 0$ (see [CF]).

# 6 Constructing elements of the Shafarevich–Tate group

Let $f$, $g$ and $\mathfrak{q}$ be as in the first paragraph of the previous section. In the previous section we showed how the congruence between $f$ and $g$ relates the vanishing of $L(g,k/2)$ to the divisibility by $\mathfrak{q}$ of an "algebraic part" of $L(f,k/2)$. Conjecturally the former is associated with the existence of certain algebraic cycles (for $M_g$) while the latter is associated with the existence of certain elements of the Shafarevich–Tate group (for $M_f$, as we saw in §4). In this section we show how the congruence, interpreted in terms of Galois representations, provides a direct link between algebraic cycles and the Shafarevich–Tate group.

For $f$ we have defined $V_\lambda$, $T_\lambda$ and $A_\lambda$. Let $V'_\lambda$, $T'_\lambda$ and $A'_\lambda$ be the corresponding objects for $g$. Since $a_p$ is the trace of $\mathrm{Frob}_p^{-1}$ on $V_\lambda$, it follows from the Chebotarev Density Theorem that $A[\mathfrak{q}]$ and $A'[\mathfrak{q}]$, if irreducible, are isomorphic as $\mathrm{Gal}(\overline{\mathbb{Q}}/\mathbb{Q})$-modules.

Recall that $L(g,k/2) = 0$ and $L(f,k/2) \neq 0$. Since the sign in the functional equation for $L(g,s)$ is positive (this follows from $L(f,k/2) \neq 0$, see Remark 5.2), the order of vanishing of $L(g,s)$ at $s = k/2$ is at least 2. According to the Beilinson–Bloch conjecture [B, Be], the order of vanishing of $L(g,s)$ at $s = k/2$ is the rank of the group $\mathrm{CH}_0^{k/2}(M_g)(\mathbb{Q})$ of $\mathbb{Q}$-rational rational equivalence classes of null-homologous, algebraic cycles of codimension $k/2$ on the motive $M_g$. (This generalises the part of the Birch–Swinnerton-Dyer conjecture which says that for an elliptic curve $E/\mathbb{Q}$, the order of vanishing of $L(E,s)$ at $s=1$ is equal to the rank of the Mordell-Weil group $E(\mathbb{Q})$.)

Via the $\mathfrak{q}$-adic Abel–Jacobi map, $\mathrm{CH}_0^{k/2}(M_g)(\mathbb{Q})$ maps to $H^1(\mathbb{Q}, V'_{\mathfrak{q}}(k/2))$, and its image is contained in the subspace $H^1_f(\mathbb{Q}, V'_{\mathfrak{q}}(k/2))$, by [Ne], 3.1 and 3.2. If, as expected, the $\mathfrak{q}$-adic Abel–Jacobi map is injective, we get (assuming also the Beilinson–Bloch conjecture) a subspace of $H^1_f(\mathbb{Q}, V'_{\mathfrak{q}}(k/2))$ of dimension equal to the order of vanishing of $L(g,s)$ at $s=k/2$. In fact, one could simply conjecture that the dimension of $H^1_f(\mathbb{Q}, V'_{\mathfrak{q}}(k/2))$ is equal to the order of vanishing of $L(g,s)$ at $s=k/2$. This would follow from the "conjectures" $C_r(M)$ and $C^i_\lambda(M)$ of [Fo2], Sections 1 and 6.5. We shall call it the "strong" Beilinson–Bloch conjecture.

Similarly, if $L(f,k/2)\neq 0$ then we expect that $H^1_f(\mathbb{Q}, V_{\mathfrak{q}}(k/2)) = 0$, so that $H^1_f(\mathbb{Q}, A_{\mathfrak{q}}(k/2))$ coincides with the $\mathfrak{q}$-part of Ш.

**Theorem 6.1** *Let $q \nmid N$ be a prime satisfying $q > k$. Let $r$ be the dimension of $H^1_f(\mathbb{Q}, V'_{\mathfrak{q}}(k/2))$. Suppose that $A[\mathfrak{q}]$ is an irreducible representation of* $\mathrm{Gal}(\overline{\mathbb{Q}}/\mathbb{Q})$ *and that for no prime $p \mid N$ is $f$ congruent modulo $\mathfrak{q}$ (for Fourier coefficients of index coprime to $Nq$) to a newform of weight $k$, trivial character and level dividing $N/p$. Suppose that, for all primes $p \mid N$, $p \not\equiv -w_p$* (mod $q$), *with $p \not\equiv \pm 1$* (mod $q$) *if $p^2 \mid N$. (Here $w_p$ is the common eigenvalue of the Atkin–Lehner involution $W_p$ acting on $f$ and $g$.) Then the $\mathfrak{q}$-torsion subgroup of $H^1_f(\mathbb{Q}, A_{\mathfrak{q}}(k/2))$ has $\mathbb{F}_{\mathfrak{q}}$-rank at least $r$.*

**Proof** The theorem is trivially true if $r=0$, so we assume that $r>0$. It follows easily from our hypothesis that the rank of the free part of $H^1_f(\mathbb{Q}, T'_{\mathfrak{q}}(k/2))$ is $r$. The natural map from $H^1_f(\mathbb{Q}, T'_{\mathfrak{q}}(k/2))/\mathfrak{q}H^1_f(\mathbb{Q}, T'_{\mathfrak{q}}(k/2))$ to $H^1(\mathbb{Q}, A'[\mathfrak{q}](k/2))$ is injective. Take a nonzero class $c$ in the image, which has $\mathbb{F}_{\mathfrak{q}}$-rank $r$. Choose $d \in H^1_f(\mathbb{Q}, T'_{\mathfrak{q}}(k/2))$ mapping to $c$. Consider the $\mathrm{Gal}(\overline{\mathbb{Q}}/\mathbb{Q})$-cohomology of the short exact sequence

$$0 \to A[\mathfrak{q}](k/2) \to A_{\mathfrak{q}}(k/2) \xrightarrow{\pi} A_{\mathfrak{q}}(k/2) \to 0,$$

where $\pi$ is multiplication by a uniformising element of $O_{\mathfrak{q}}$. By irreducibility, $H^0(\mathbb{Q}, A[\mathfrak{q}](k/2))$ is trivial. Hence $H^0(\mathbb{Q}, A_{\mathfrak{q}}(k/2))$ is trivial, so that $H^1(\mathbb{Q}, A[\mathfrak{q}](k/2))$ injects into $H^1(\mathbb{Q}, A_{\mathfrak{q}}(k/2))$, and we get a nonzero $\mathfrak{q}$-torsion class $\gamma \in H^1(\mathbb{Q}, A_{\mathfrak{q}}(k/2))$.

Our aim is to show that $\mathrm{res}_p(\gamma) \in H^1_f(\mathbb{Q}_p, A_{\mathfrak{q}}(k/2))$, for all (finite) primes $p$. We consider separately the cases $p \nmid qN$, $p \mid N$ and $p = q$.

**Case 1,** $p \nmid qN$:

Consider the $I_p$-cohomology of the short exact sequence

$$0 \to A'[\mathfrak{q}](k/2) \to A'_{\mathfrak{q}}(k/2) \xrightarrow{\pi} A'_{\mathfrak{q}}(k/2) \to 0$$

(the analogue for $g$ of the above).

Since in this case $A'_{\mathfrak{q}}(k/2)$ is unramified at $p$, $H^0(I_p, A'_{\mathfrak{q}}(k/2)) = A'_{\mathfrak{q}}(k/2)$, which is $\mathfrak{q}$-divisible. Therefore $H^1(I_p, A'[\mathfrak{q}](k/2))$ (which, remember, is the same as $H^1(I_p, A[\mathfrak{q}](k/2)))$ injects into $H^1(I_p, A'_{\mathfrak{q}}(k/2))$. It follows from the fact that $d \in H^1_f(\mathbb{Q}, T'_{\mathfrak{q}}(k/2))$ that the image in $H^1(I_p, A'_{\mathfrak{q}}(k/2))$ of the restriction of $c$ is zero, hence that the restriction of $c$ to $H^1(I_p, A'[\mathfrak{q}](k/2)) \simeq H^1(I_p, A[\mathfrak{q}](k/2))$ is zero. Hence the restriction of $\gamma$ to $H^1(I_p, A_{\mathfrak{q}}(k/2))$ is also zero. By [Fl1], line 3 of p. 125, $H^1_f(\mathbb{Q}_p, A_{\mathfrak{q}}(k/2))$ is equal to (not just contained in) the kernel of the map from $H^1(\mathbb{Q}_p, A_{\mathfrak{q}}(k/2))$ to $H^1(I_p, A_{\mathfrak{q}}(k/2))$, so we have shown that $\mathrm{res}_p(\gamma) \in H^1_f(\mathbb{Q}_p, A_{\mathfrak{q}}(k/2))$.

**Case 2, $p \mid N$:**

We first show that $H^0(I_p, A'_{\mathfrak{q}}(k/2))$ is $\mathfrak{q}$-divisible. It suffices to show that

$$\dim H^0(I_p, A'[\mathfrak{q}](k/2)) = \dim H^0(I_p, V'_{\mathfrak{q}}(k/2)),$$

since then the natural map from $H^0(I_p, V'_{\mathfrak{q}}(k/2))$ to $H^0(I_p, A'_{\mathfrak{q}}(k/2))$ is surjective; this may be done as in the proof of Lemma 4.4. It follows as above that the image of $c \in H^1(\mathbb{Q}, A[\mathfrak{q}](k/2))$ in $H^1(I_p, A[\mathfrak{q}](k/2))$ is zero. Then $\mathrm{res}_p(c)$ comes from $H^1(D_p/I_p, H^0(I_p, A[\mathfrak{q}](k/2)))$, by inflation-restriction. The order of this group is the same as the order of the group $H^0(\mathbb{Q}_p, A[\mathfrak{q}](k/2))$ (this is [W], Lemma 1), which we claim is trivial. By the work of Carayol [Ca1], the level $N$ is the conductor of $V_{\mathfrak{q}}(k/2)$, so $p \mid N$ implies that $V_{\mathfrak{q}}(k/2)$ is ramified at $p$, hence $\dim H^0(I_p, V_{\mathfrak{q}}(k/2)) = 0$ or $1$. As above, we see that $\dim H^0(I_p, V_{\mathfrak{q}}(k/2)) = \dim H^0(I_p, A[\mathfrak{q}](k/2))$, so we need only consider the case where this common dimension is 1. The (motivic) Euler factor at $p$ for $M_f$ is $(1 - \alpha p^{-s})^{-1}$, where $\mathrm{Frob}_p^{-1}$ acts as multiplication by $\alpha$ on the one-dimensional space $H^0(I_p, V_{\mathfrak{q}})$. It follows from [Ca1], Theoréme A that this is the same as the Euler factor at $p$ of $L(f, s)$. By [AL], Theorems 3(ii) and 5, it then follows that $p^2 \nmid N$ and $\alpha = -w_p p^{(k/2)-1}$, where $w_p = \pm 1$ is such that $W_p f = w_p f$. We twist by $k/2$, so that $\mathrm{Frob}_p^{-1}$ acts on $H^0(I_p, V_{\mathfrak{q}}(k/2))$ (hence also on $H^0(I_p, A[\mathfrak{q}](k/2)))$ as $-w_p p^{-1}$. Since $p \not\equiv -w_p \pmod{q}$, we see that $H^0(\mathbb{Q}_p, A[\mathfrak{q}](k/2))$ is trivial. Hence $\mathrm{res}_p(c) = 0$ so $\mathrm{res}_p(\gamma) = 0$ and certainly lies in $H^1_f(\mathbb{Q}_p, A_{\mathfrak{q}}(k/2))$.

**Case 3, $p = q$:**

Since $q \nmid N$ is a prime of good reduction for the motive $M_g$, $V'_{\mathfrak{q}}$ is a crystalline representation of $\mathrm{Gal}(\overline{\mathbb{Q}}_q/\mathbb{Q}_q)$, which means that $D_{\mathrm{cris}}(V'_{\mathfrak{q}})$ and $V'_{\mathfrak{q}}$ have the same dimension, where $D_{\mathrm{cris}}(V'_{\mathfrak{q}}) := H^0(\mathbb{Q}_q, V'_{\mathfrak{q}} \otimes_{\mathbb{Q}_q} B_{\mathrm{cris}})$. (This is a consequence of [Fa], Theorem 5.6.) As already noted in the proof of Lemma 4.6, $T_{\mathfrak{q}}$ is the $O_{\mathfrak{q}}[\mathrm{Gal}(\overline{\mathbb{Q}}_q/\mathbb{Q}_q)]$-module associated to the filtered module $T_{\mathrm{dR}} \otimes O_{\mathfrak{q}}$.

Since also $q > k$, we may now prove, in the same manner as [Du1], Proposition 9.2, that $\mathrm{res}_q(\gamma) \in H^1_f(\mathbb{Q}_q, A_{\mathfrak{q}}(k/2))$. For the convenience of the reader, we give some details.

In [BK], Lemma 4.4, a cohomological functor $\{h^i\}_{i\geq 0}$ is constructed on the Fontaine–Lafaille category of filtered Dieudonné modules over $\mathbb{Z}_q$. $h^i(D) = 0$ for all $i \geq 2$ and all $D$, and $h^i(D) = \mathrm{Ext}^i(1_{FD}, D)$ for all $i$ and $D$, where $1_{FD}$ is the "unit" filtered Dieudonné module.

Now let $D = T_{\mathrm{dR}} \otimes O_{\mathfrak{q}}$ and $D' = T'_{\mathrm{dR}} \otimes O_{\mathfrak{q}}$. By [BK], Lemma 4.5(c),

$$h^1(D) \simeq H^1_e(\mathbb{Q}_q, T_{\mathfrak{q}}),$$

where

$$H^1_e(\mathbb{Q}_q, T_{\mathfrak{q}}) = \ker(H^1(\mathbb{Q}_q, T_{\mathfrak{q}}) \to H^1(\mathbb{Q}_q, V_{\mathfrak{q}})/H^1_e(\mathbb{Q}_q, V_{\mathfrak{q}}))$$

and

$$H^1_e(\mathbb{Q}_q, V_{\mathfrak{q}}) = \ker(H^1(\mathbb{Q}_q, V_{\mathfrak{q}}) \to H^1(\mathbb{Q}_q, B^{f=1}_{\mathrm{cris}} \otimes_{\mathbb{Q}_q} V_{\mathfrak{q}})).$$

Likewise $h^1(D') \simeq H^1_e(\mathbb{Q}_q, T'_{\mathfrak{q}})$. When applying results of [BK] we view $D$, $T_{\mathfrak{q}}$ etc. simply as $\mathbb{Z}_q$-modules, forgetting the $O_{\mathfrak{q}}$-structure.

For an integer $j$, let $D(j)$ be $D$ with the Hodge filtration shifted by $j$. Then

$$h^1(D(j)) \simeq H^1_e(\mathbb{Q}_q, T_{\mathfrak{q}}(j))$$

(provided that $k - p + 1 < j < p - 1$, so that $D(j)$ satisfies the hypotheses of [BK], Lemma 4.5). By [BK], Corollary 3.8.4,

$$H^1_f(\mathbb{Q}_q, V_{\mathfrak{q}}(j))/H^1_e(\mathbb{Q}_q, V_{\mathfrak{q}}(j)) \simeq (D(j) \otimes_{\mathbb{Z}_q} \mathbb{Q}_q)/(1 - f)(D(j) \otimes_{\mathbb{Z}_q} \mathbb{Q}_q),$$

where $f$ is the Frobenius operator on crystalline cohomology. By Scholl [Sc], 1.2.4(ii), and the Weil conjectures, $H^1_e(\mathbb{Q}_q, V_{\mathfrak{q}}(j)) = H^1_f(\mathbb{Q}_q, V_{\mathfrak{q}}(j))$, since $j \neq (k-1)/2$. Similarly $H^1_e(\mathbb{Q}_q, V'_{\mathfrak{q}}(j)) = H^1_f(\mathbb{Q}_q, V'_{\mathfrak{q}}(j))$.

We have

$$h^1(D(k/2)) \simeq H^1_f(\mathbb{Q}_q, T_{\mathfrak{q}}(k/2)) \quad \text{and} \quad h^1(D'(k/2)) \simeq H^1_f(\mathbb{Q}_q, T'_{\mathfrak{q}}(k/2)).$$

The exact sequence in [BK], middle of p. 366, gives a commutative diagram

$$\begin{array}{ccccc} h^1(D'(k/2)) & \xrightarrow{\pi} & h^1(D'(k/2)) & \longrightarrow & h^1(D'(k/2)/\mathfrak{q}D'(k/2)) \\ \downarrow & & \downarrow & & \downarrow \\ H^1(\mathbb{Q}_q, T'_{\mathfrak{q}}(k/2)) & \xrightarrow{\pi} & H^1(\mathbb{Q}_q, T'_{\mathfrak{q}}(k/2)) & \longrightarrow & H^1(\mathbb{Q}_q, A'[\mathfrak{q}](k/2)). \end{array}$$

The vertical arrows are all inclusions, and we know the image of $h^1(D'(k/2))$ in $H^1(\mathbb{Q}_q, T'_{\mathfrak{q}}(k/2))$ is exactly $H^1_f(\mathbb{Q}_q, T'_{\mathfrak{q}}(k/2))$. The top right horizontal map is surjective since $h^2(D'(k/2)) = 0$.

The class $\mathrm{res}_q(c) \in H^1(\mathbb{Q}_q, A'[\mathfrak{q}](k/2))$ is in the image of $H^1_f(\mathbb{Q}_q, T'_{\mathfrak{q}}(k/2))$, by construction, and therefore is in the image of $h^1(D'(k/2)/\mathfrak{q}D'(k/2))$. By the fullness and exactness of the Fontaine–Lafaille functor [FL] (see [BK], Theorem 4.3), $D'(k/2)/\mathfrak{q}D'(k/2)$ is isomorphic to $D(k/2)/\mathfrak{q}D(k/2)$.

It follows that the class $\mathrm{res}_q(c) \in H^1(\mathbb{Q}_q, A[\mathfrak{q}](k/2))$ is in the image of $h^1(D(k/2)/\mathfrak{q}D(k/2))$ by the vertical map in the exact sequence analogous to the above. Since the map from $h^1(D(k/2))$ to $h^1(D(k/2)/\mathfrak{q}D(k/2))$ is surjective, $\mathrm{res}_q(c)$ lies in the image of $H^1_f(\mathbb{Q}_q, T_{\mathfrak{q}}(k/2))$. From this it follows that $\mathrm{res}_q(\gamma) \in H^1_f(\mathbb{Q}_q, A_{\mathfrak{q}}(k/2))$, as desired. □

[AS], Theorem 2.7 is concerned with verifying local conditions in the case $k = 2$, where $f$ and $g$ are associated with abelian varieties $A$ and $B$. (Their theorem also applies to abelian varieties over number fields.) Our restriction outlawing congruences modulo $\mathfrak{q}$ with cusp forms of lower level is analogous to theirs forbidding $q$ from dividing Tamagawa factors $c_{A,l}$ and $c_{B,l}$. (In the case where $A$ is an elliptic curve with $\mathrm{ord}_l(j(A)) < 0$, consideration of a Tate parametrisation shows that if $q \mid c_{A,l}$, i.e., if $q \mid \mathrm{ord}_l(j(A))$, then it is possible that $A[q]$ is unramified at $l$.)

In this paper we have encountered two technical problems which we dealt with in quite similar ways:

1. dealing with the $\mathfrak{q}$-part of $c_p$ for $p \mid N$;

2. proving local conditions at primes $p \mid N$, for an element of $\mathfrak{q}$-torsion.

If our only interest was in testing the Bloch–Kato conjecture at $\mathfrak{q}$, we could have made these problems cancel out, as in [DFG1], Lemma 8.11, by weakening the local conditions. However, we have chosen not to do so, since we are also interested in the Shafarevich–Tate group, and since the hypotheses we had to assume are not particularly strong. Note that, since $A[\mathfrak{q}]$ is irreducible, the $\mathfrak{q}$-part of Ш does not depend on the choice of $T_{\mathfrak{q}}$.

# 7 Examples and Experiments

This section contains tables and numerical examples illustrating the main themes of this paper. In Section 7.1, we explain Table 1, which contains 16 examples of pairs $f, g$ such that the strong Beilinson–Bloch conjecture and Theorem 6.1 together imply the existence of nontrivial elements of the Shafarevich–Tate group of the motive attached to $f$. Section 7.2 outlines the higher-weight modular symbol computations used in making Table 1. Section 7.3 discusses Table 2, which summarizes the results of an extensive computation of conjectural orders of Shafarevich–Tate groups for modular motives of low level and weight. Section 7.4 gives specific examples in which various hypotheses fail.

Note that in §7 "modular symbol" has a different meaning from in §5, being related to homology rather than cohomology. For precise definitions see [SV].

## 7.1 Table 1: visible Ш

Table 1 lists sixteen pairs of newforms $f$ and $g$ (of equal weights and levels)

| $g$ | deg $g$ | $f$ | deg $f$ | possible $q$ |
|---|---|---|---|---|
| **127k4A** | 1 | **127k4C** | 17 | 43 |
| **159k4B** | 1 | **159k4E** | 16 | 5, 23 |
| **365k4A** | 1 | **365k4E** | 18 | 29 |
| **369k4B** | 1 | **369k4I** | 9 | 13 |
| **453k4A** | 1 | **453k4E** | 23 | 17 |
| **465k4B** | 1 | **465k4I** | 7 | 11 |
| **477k4B** | 1 | **477k4L** | 12 | 73 |
| **567k4B** | 1 | **567k4H** | 8 | 23 |
| **581k4A** | 1 | **581k4E** | 34 | $19^2$ |
| **657k4A** | 1 | **657k4C** | 7 | 5 |
| **657k4A** | 1 | **657k4G** | 12 | 5 |
| **681k4A** | 1 | **681k4D** | 30 | 59 |
| **684k4C** | 1 | **684k4K** | 4 | $7^2$ |
| **95k6A** | 1 | **95k6D** | 9 | 31, 59 |
| **122k6A** | 1 | **122k6D** | 6 | 73 |
| **260k6A** | 1 | **260k6E** | 4 | 17 |

Table 1: Visible Ш

along with at least one prime $q$ such that there is a prime $\mathfrak{q} \mid q$ with $f \equiv g \pmod{\mathfrak{q}}$. In each case, $\mathrm{ord}_{s=k/2} L(g, k/2) \geq 2$ while $L(f, k/2) \neq 0$. It uses the following notation: the first column contains a label whose structure is

**[Level]k[Weight][GaloisOrbit]**

This label determines a newform $g = \sum a_n q^n$ up to Galois conjugacy. For example, **127k4C** denotes a newform in the third Galois orbit of newforms in

$S_4(\Gamma_0(127))$. Galois orbits are ordered first by the degree of $\mathbb{Q}(\dots, a_n, \dots)$, then by the sequence of absolute values $|\mathrm{Tr}(a_p(g))|$ for $p$ not dividing the level, with positive trace being first in the event that the two absolute values are equal, and the first Galois orbit is denoted **A**, the second **B**, and so on. The second column contains the degree of the field $\mathbb{Q}(\dots, a_n, \dots)$. The third and fourth columns contain $f$ and its degree, respectively. The fifth column contains at least one prime $q$ such that there is a prime $\mathfrak{q} \mid q$ with $f \equiv g \pmod{\mathfrak{q}}$, and such that the hypotheses of Theorem 6.1 are satisfied for $f$, $g$, and $\mathfrak{q}$.

For the two examples **581k4E** and **684k4K**, the square of a prime $q$ appears in the $q$-column, which means that $q^2$ divides the order of the group $S_k(\Gamma_0(N), \mathbb{Z})/(W + W^{\perp})$ defined at the end of 7.3 below.

We describe the first line of Table 1 in more detail. The next section gives further details on how the computations were performed.

Using modular symbols, we find that there is a newform

$$g = q - q^2 - 8q^3 - 7q^4 - 15q^5 + 8q^6 - 25q^7 + \cdots \in S_4(\Gamma_0(127))$$

with $L(g, 2) = 0$. Because $W_{127}(g) = g$, the functional equation has sign $+1$, so $L'(g, 2) = 0$ as well. We also find a newform $f \in S_4(\Gamma_0(127))$ whose Fourier coefficients generate a number field $K$ of degree 17, and by computing the image of the modular symbol $XY\{0, \infty\}$ under the period mapping, we find that $L(f, 2) \neq 0$. The newforms $f$ and $g$ are congruent modulo a prime $\mathfrak{q}$ of $K$ of residue characteristic 43. The mod $\mathfrak{q}$ reductions of $f$ and $g$ are both equal to

$$\overline{f} = q + 42q^2 + 35q^3 + 36q^4 + 28q^5 + 8q^6 + 18q^7 + \cdots \in \mathbb{F}_{43}[[q]].$$

There is no form in the Eisenstein subspaces of $M_4(\Gamma_0(127))$ whose Fourier coefficients of index $n$, with $(n, 127) = 1$, are congruent modulo 43 to those of $\overline{f}$, so $\rho_{f,\mathfrak{q}} \approx \rho_{g,\mathfrak{q}}$ is irreducible. Since 127 is prime and $S_4(\mathrm{SL}_2(\mathbb{Z})) = 0$, $\overline{f}$ does not arise from a level 1 form of weight 4. Thus we have checked the hypotheses of Theorem 6.1, so if $r$ is the dimension of $H^1_f(\mathbb{Q}, V'_{\mathfrak{q}}(k/2))$ then the $\mathfrak{q}$-torsion subgroup of $H^1_f(\mathbb{Q}, A_{\mathfrak{q}}(k/2))$ has $\mathbb{F}_{\mathfrak{q}}$-rank at least $r$.

Recall that since $\mathrm{ord}_{s=k/2} L(g, s) \geq 2$, we expect that $r \geq 2$. Then, since $L(f, k/2) \neq 0$, we expect that the $\mathfrak{q}$-torsion subgroup of $H^1_f(\mathbb{Q}, A_{\mathfrak{q}}(k/2))$ is equal to the $\mathfrak{q}$-torsion subgroup of Ш. Admitting these assumptions, we have constructed the $\mathfrak{q}$-torsion in Ш predicted by the Bloch–Kato conjecture.

For particular examples of elliptic curves one can often find and write down rational points predicted by the Birch and Swinnerton-Dyer conjecture. It would be nice if likewise one could explicitly produce algebraic cycles predicted by the Beilinson–Bloch conjecture in the above examples. Since $L'(g, k/2) = 0$, Heegner cycles have height zero (see [Z], Corollary 0.3.2), so ought to be trivial in $\mathrm{CH}_0^{k/2}(M_g) \otimes \mathbb{Q}$.

## 7.2 How the computation was performed

We give a brief summary of how the computation was performed. The algorithms we used were implemented by the second author, and most are a standard part of MAGMA (see [BCP]).

Let $g$, $f$, and $q$ be some data from a line of Table 1 and let $N$ denote the level of $g$. We verified the existence of a congruence modulo $q$, that $L(g,k/2) = L'(g,k/2) = 0$ and $L(f,k/2) \neq 0$, and that $\rho_{f,\mathfrak{q}} = \rho_{g,\mathfrak{q}}$ is irreducible and does not arise from any $S_k(\Gamma_0(N/p))$, as follows:

To prove there is a congruence, we showed that the corresponding *integral* spaces of modular symbols satisfy an appropriate congruence, which forces the existence of a congruence on the level of Fourier expansions. We showed that $\rho_{g,\mathfrak{q}}$ is irreducible by computing a set that contains all possible residue characteristics of congruences between $g$ and any Eisenstein series of level dividing $N$, where by congruence, we mean a congruence for all Fourier coefficients of index $n$ with $(n, N) = 1$. Similarly, we checked that $g$ is not congruent to any form $h$ of level $N/p$ for any $p$ that exactly divides $N$ by listing a basis of such $h$ and finding the possible congruences, where again we disregard the Fourier coefficients of index not coprime to $N$.

To verify that $L(g,k/2) = 0$, we computed the image of the modular symbol $\mathbf{e} = X^{\frac{k}{2}-1}Y^{\frac{k}{2}-1}\{0,\infty\}$ under a map with the same kernel as the period mapping, and found that the image was 0. The period mapping sends the modular symbol $\mathbf{e}$ to a nonzero multiple of $L(g, \frac{k}{2})$, so that $\mathbf{e}$ maps to 0 implies that $L(g,k/2) = 0$. In a similar way, we verified that $L(f,k/2) \neq 0$. Next, we checked that $W_N(g) = (-1)^{k/2}g$ which, because of the functional equation, implies that $L'(g,k/2) = 0$. Table 1 is of independent interest because it includes examples of modular forms of even weight $> 2$ with a zero at $k/2$ that is not forced by the functional equation. We found no such examples of weights $\geq 8$.

## 7.3 Conjecturally nontrivial Ш

In this section we apply some of the results of Section 4 to compute lower bounds on conjectural orders of Shafarevich–Tate groups of many modular motives. The results of this section suggest that Ш of a modular motive is usually "not visible at level $N$", that is, not explained by congruences at level $N$ (compare with the observations of [CM1] and [AS]). For example, when $k > 6$ we find many examples of conjecturally nontrivial Ш but no examples of nontrivial visible Ш.

For any newform $f$, let $L(M_f/\mathbb{Q}, s) = \prod_{i=1}^{d} L(f^{(i)}, s)$ where $f^{(i)}$ runs over the $\mathrm{Gal}(\overline{\mathbb{Q}}/\mathbb{Q})$-conjugates of $f$. Let $T$ be the complex torus $\mathbb{C}^d/(2\pi i)^{k/2}\mathcal{L}$, where $\mathcal{L}$ is the lattice defined by integrating integral cuspidal modular symbols for $\Gamma_0(N)$ against the conjugates of $f$. Let $\Omega_{M_f/\mathbb{Q}}$ denote the volume of the $(-1)^{(k/2)-1}$ eigenspace $T^{\pm} = \{z \in T : \overline{z} = (-1)^{(k/2)-1}z\}$ for complex conjugation on $T$.

**Lemma 7.1** *Suppose that $p \nmid Nk!$ is such that $f$ is not congruent to any of its Galois conjugates modulo a prime dividing $p$. Then the $p$-parts of*

$$\frac{L(M_f/\mathbb{Q}, k/2)}{\Omega_{M_f/\mathbb{Q}}} \quad \textit{and} \quad \mathrm{Norm}\left(\frac{L(f,k/2)}{\mathrm{vol}_\infty}\mathfrak{a}^{\pm}\right)$$

*are equal, where* $\mathrm{vol}_\infty$ *is as in Section 4.*

**Proof** Let $H$ be the $\mathbb{Z}$-module of all integral cuspidal modular symbols for $\Gamma_0(N)$. Let $I$ be the image of $H$ under projection into the submodule of $H \otimes \mathbb{Q}$ corresponding to $f$ and its Galois conjugates. Note that $I$ is not necessarily contained in $H$, but it is contained in $H \otimes \mathbb{Z}[\frac{1}{m}]$ where $m$ is divisible by the residue characteristics of any primes of congruence between $f$ and cuspforms of weight $k$ for $\Gamma_0(N)$ which are not Galois conjugate to $f$.

The lattice $\mathcal{L}$ defined before the lemma is obtained (up to divisors of $Nk!$) by pairing the cohomology modular symbols $\Phi^{\pm}_{f^{(i)}}$ (as in §5) with the homology modular symbols in $H$; equivalently, since the pairing factors through the map $H \to I$, the lattice $\mathcal{L}$ is obtained by pairing with the elements of $I$. For $1 \leq i \leq d$ let $I_i$ be the $O_E$-module generated by the image of the projection of $I$ into $I \otimes E$ corresponding to $f^{(i)}$. The finite index of $I \otimes O_E$ in $\bigoplus_{i=1}^{d} I_i$ is divisible only by primes of congruence between $f$ and its Galois conjugates. Up to these primes, $\Omega_{M_f/\mathbb{Q}}/(2\pi i)^{((k/2)-1)d}$ is then a product of the $d$ quantities obtained by pairing $\Phi^{\pm}_{f^{(i)}}$ with $I_i$, for $1 \leq i \leq d$. (These quantities inhabit a kind of tensor product of $\mathbb{C}^*$ over $E^*$ with the group of fractional ideals of $E$.) Bearing in mind the last line of §3, we see that these quantities are the $\mathfrak{a}^{\pm}\Omega^{\pm}_{f^{(i)}}$, up to divisors of $Nk!$. Now we may apply Lemma 4.1. We then have a factorisation of the left hand side which shows it to be equal to the right hand side, to the extent claimed by the lemma. Note that $\frac{L(f,k/2)}{\mathrm{vol}_\infty}\mathfrak{a}^{\pm}$ has an interpretation in terms of integral modular symbols, as in §5, and just gets Galois conjugated when one replaces $f$ by some $f^{(i)}$. □

**Remark 7.2** The newform $f = \mathbf{319k4C}$ is congruent to one of its Galois conjugates modulo 17 and 17 divides $L(M_f/\mathbb{Q}, k/2)/\Omega_{M_f/\mathbb{Q}}$, so the lemma and our computations say nothing about whether 17 divides $\mathrm{Norm}\left(\frac{L(f,k/2)}{\mathrm{vol}_\infty}\mathfrak{a}^{\pm}\right)$ or otherwise.

Let $\mathcal{S}$ be the set of newforms with level $N$ and weight $k$ satisfying either $k = 4$ and $N \leq 321$, or $k = 6$ and $N \leq 199$, or $k = 8$ and $N \leq 149$, or $k = 10$ and $N \leq 72$, or $k = 12$ and $N \leq 49$. Given $f \in \mathcal{S}$, let $B$ be defined as follows:

1. Let $L_1$ be the numerator of the rational number $L(M_f/\mathbb{Q}, k/2)/\Omega_{M_f/\mathbb{Q}}$. If $L_1 = 0$ let $B = 1$ and terminate.

2. Let $L_2$ be the part of $L_1$ that is coprime to $Nk!$.

3. Let $L_3$ be the part of $L_2$ that is coprime to $p \pm 1$ for every prime $p$ such that $p^2 \mid N$.

4. Let $L_4$ be the part of $L_3$ coprime to the residue characteristic of any prime of congruence between $f$ and a form of weight $k$, trivial character and lower level. (By congruence here, we mean a congruence for coefficients $a_n$ with $n$ coprime to the level of $f$.)

5. Let $L_5$ be the part of $L_4$ coprime to the residue characteristic of any prime of congruence between $f$ and an Eisenstein series. (This eliminates residue characteristics of reducible representations.)

6. Let $B$ be the part of $L_5$ coprime to the residue characteristic of any prime of congruence between $f$ and any one of its Galois conjugates.

Proposition 4.8 and Lemma 7.1 imply that if $\mathrm{ord}_p(B) > 0$ then, according to the Bloch–Kato conjecture, $\mathrm{ord}_p(\#\text{Ш}) = \mathrm{ord}_p(B) > 0$.

We computed $B$ for every newform in $\mathcal{S}$. There are many examples in which $L_3$ is large, but $B$ is not, and this is because of Tamagawa factors. For example, **39k4C** has $L_3 = 19$, but $B = 1$ because of a 19-congruence with a form of level 13; in this case we must have $19 \mid c_3(2)$, where $c_3(2)$ is as in Section 4. See Section 7.4 for more details. Also note that in every example $B$ is a perfect square, which, away from congruence primes, is as predicted by the existence of Flach's generalised Cassels–Tate pairing [Fl1]. (Note that if $A[\lambda]$ is irreducible then the lattice $T_\lambda$ is at worst a scalar multiple of its dual, so the pairing shows that the order of the $\lambda$-part of Ш, if finite, is a square.) That our computed value of $B$ should be a square is not *a priori* obvious.

For simplicity, we discard residue characteristics instead of primes of rings of integers, so our definition of $B$ is overly conservative. For example, 5 occurs in row 2 of Table 1 but not in Table 2, because **159k4E** is Eisenstein at some prime above 5, but the prime of congruences of characteristic 5 between **159k4B** and **159k4E** is not Eisenstein.

The newforms for which $B > 1$ are given in Table 2 on pp. 112–115. The second column of the table records the degree of the field generated by the Fourier coefficients of $f$. The third contains $B$. Let $W$ be the intersection of the span of all conjugates of $f$ with $S_k(\Gamma_0(N), \mathbb{Z})$ and $W^\perp$ the Petersson orthogonal complement of $W$ in $S_k(\Gamma_0(N), \mathbb{Z})$. The fourth column contains the odd prime divisors of $\#(S_k(\Gamma_0(N), \mathbb{Z})/(W + W^\perp))$, which are exactly the possible primes of congruence between $f$ and nonconjugate cusp forms of the same weight and level. We place a $*$ next to the four entries of Table 2 that also occur in Table 1.

## 7.4 Examples in which hypotheses fail

We have some other examples where forms of different levels are congruent (for Fourier coefficients of index coprime to the levels). However, Remark 5.2 does not apply, so that one of the forms could have an odd functional equation, and the other could have an even functional equation. For instance, we have a 19-congruence between the newforms $g =$ **13k4A** and $f =$ **39k4C** of Fourier coefficients of index coprime to 39. Here $L(f,2) \neq 0$, while $L(g,2) = 0$ since $L(g,s)$ has *odd* functional equation. Here $f$ fails the condition about not being congruent to a form of lower level, so in Lemma 4.4 it is possible that $\mathrm{ord}_{\mathfrak{q}}(c_3(2)) > 0$. In fact this does happen. Because $V'_{\mathfrak{q}}$ (attached to $g$ of level 13) is unramified at $p = 3$, $H^0(I_p, A[\mathfrak{q}])$ (the same as $H^0(I_p, A'[\mathfrak{q}])$) is two-dimensional. As in (2) of the proof of Theorem 6.1, one of the eigenvalues of $\mathrm{Frob}_p^{-1}$ acting on this two-dimensional space is $\alpha = -w_p p^{(k/2)-1}$, where $W_p f = w_p f$. The other must be $\beta = -w_p p^{k/2}$, so that $\alpha\beta = p^{k-1}$. Twisting by $k/2$, we see that $\mathrm{Frob}_p^{-1}$ acts as $-w_p$ on the quotient of $H^0(I_p, A[\mathfrak{q}](k/2))$ by the image of $H^0(I_p, V_{\mathfrak{q}}(k/2))$. Hence $\mathrm{ord}_{\mathfrak{q}}(c_p(k/2)) > 0$ when $w_p = -1$, which is the case in our example here with $p = 3$. Likewise $H^0(\mathbb{Q}_p, A[\mathfrak{q}](k/2))$ is nontrivial when $w_p = -1$, so (2) of the proof of Theorem 6.1 does not work. This is just as well, since had it worked we would have expected $\mathrm{ord}_{\mathfrak{q}}(L(f,k/2)/\mathrm{vol}_\infty) \geq 3$, which computation shows not to be the case.

In the following example, the divisibility between the levels is the other way round. There is a 7-congruence between $g =$ **122k6A** and $f =$ **61k6B**, both $L$-functions have even functional equation, and $L(g,3) = 0$. In the proof of Theorem 6.1, there is a problem with the local condition at $p = 2$. The map from $H^1(I_2, A'[\mathfrak{q}](3))$ to $H^1(I_2, A'_{\mathfrak{q}}(3))$ is not necessarily injective, but its kernel is at most one dimensional, so we still get the $\mathfrak{q}$-torsion subgroup of $H^1_f(\mathbb{Q}, A_{\mathfrak{q}}(2))$ having $\mathbb{F}_{\mathfrak{q}}$-rank at least 1 (assuming $r \geq 2$), and thus get elements of Ш for **61k6B** (assuming all along the strong Beilinson–Bloch conjecture). In particular, these elements of Ш are *invisible* at level 61. When the levels are different we are no longer able to apply [FJ], Theorem 2.1. However, we still have the congruences of integral modular symbols required to make the proof of Proposition 5.1 go through. Indeed, as noted above, the congruences of modular forms were found by producing congruences of modular symbols. Despite these congruences of modular symbols, Remark 5.2 does not apply, since there is no reason to suppose that $w_N = w_{N'}$, where $N$ and $N'$ are the distinct levels.

Finally, there are two examples where we have a form $g$ with even functional equation such that $L(g,k/2) = 0$, and a congruent form $f$ which has odd functional equation; these are a 23-congruence between $g =$ **453k4A** and $f =$ **151k4A**, and a 43-congruence between $g =$ **681k4A** and $f =$ **227k4A**. If $\mathrm{ord}_{s=2} L(f,s) = 1$, it ought to be the case that $\dim(H^1_f(\mathbb{Q}, V_{\mathfrak{q}}(2))) = 1$. If we assume this is so, and similarly that $r = \mathrm{ord}_{s=2}(L(g,s)) \geq 2$, then unfortunately the appropriate modification of Theorem 6.1 (with strong Beilinson–Bloch conjecture) does not necessarily provide us with nontrivial $\mathfrak{q}$-torsion in Ш. It only tells us that the $\mathfrak{q}$-torsion subgroup of $H^1_f(\mathbb{Q}, A_{\mathfrak{q}}(2))$ has $\mathbb{F}_{\mathfrak{q}}$-rank at least 1. It could all be in the image of $H^1_f(\mathbb{Q}, V_{\mathfrak{q}}(2))$. Ш appears in the conjectural formula for the first derivative of the complex $L$ function, evaluated at $s = k/2$, but in combination with a regulator that we have no way of calculating.

Let $L_q(f,s)$ and $L_q(g,s)$ be the $q$-adic $L$ functions associated with $f$ and $g$ by the construction of Mazur, Tate and Teitelbaum [MTT], each divided by a suitable canonical period. We may show that $\mathfrak{q} \mid L'_q(f, k/2)$, though it is not quite clear what to make of this. This divisibility may be proved as follows. The measures $d\mu_{f,\alpha}$ and (a $q$-adic unit times) $d\mu_{g,\alpha'}$ in [MTT] (again, suitably normalised) are congruent $\bmod\, \mathfrak{q}$, as a result of the congruence between the modular symbols out of which they are constructed. Integrating an appropriate function against these measures, we find that $L'_q(f, k/2)$ is congruent $\bmod\, \mathfrak{q}$ to $L'_q(g, k/2)$. It remains to observe that $L'_q(g, k/2) = 0$, since $L(g, k/2) = 0$ forces $L_q(g, k/2) = 0$, but we are in a case where the signs in the functional equations of $L(g,s)$ and $L_q(g,s)$ are the same, positive in this instance. (According to the proposition in [MTT], Section 18, the signs differ precisely when $L_q(g,s)$ has a "trivial zero" at $s = k/2$.)

We also found some examples for which the conditions of Theorem 6.1 were not met. For example, we have a 7-congruence between **639k4B** and **639k4H**, but $w_{71} = -1$, so that $71 \equiv -w_{71} \pmod 7$. There is a similar problem with a 7-congruence between **260k6A** and **260k6E** — here $w_{13} = 1$ so that $13 \equiv -w_{13} \pmod 7$. According to Propositions 5.1 and 4.8, Bloch–Kato still predicts that the $\mathfrak{q}$-part of Ш is nontrivial in these examples. Finally, there is a 5-congruence between **116k6A** and **116k6D**, but here the prime 5 is less than the weight 6 so Propositions 5.1 and 4.8 (and even Lemma 7.1) do not apply.

| $f$ | $\deg f$ | $B$ (bound for Ш) | all odd congruence primes |
|---|---|---|---|
| **127k4C**$*$ | 17 | $43^2$ | 43, 127 |
| **159k4E**$*$ | 8 | $23^2$ | 3, 5, 11, 23, 53, 13605689 |
| **263k4B** | 39 | $41^2$ | 263 |
| **269k4C** | 39 | $23^2$ | 269 |
| **271k4B** | 39 | $29^2$ | 271 |
| **281k4B** | 40 | $29^2$ | 281 |
| **295k4C** | 16 | $7^2$ | 3, 5, 11, 59, 101, 659, 70791023 |
| **299k4C** | 20 | $29^2$ | 13, 23, 103, 20063, 21961 |
| **321k4C** | 16 | $13^2$ | 3, 5, 107, 157, 12782373452377 |
| **95k6D**$*$ | 9 | $31^2{\cdot}59^2$ | 3, 5, 17, 19, 31, 59, 113, 26701 |
| **101k6B** | 24 | $17^2$ | 101 |
| **103k6B** | 24 | $23^2$ | 103 |
| **111k6C** | 9 | $11^2$ | 3, 37, 2796169609 |
| **122k6D**$*$ | 6 | $73^2$ | 3, 5, 61, 73, 1303196179 |
| **153k6G** | 5 | $7^2$ | 3, 17, 61, 227 |
| **157k6B** | 34 | $251^2$ | 157 |
| **167k6B** | 40 | $41^2$ | 167 |
| **172k6B** | 9 | $7^2$ | 3, 11, 43, 787 |
| **173k6B** | 39 | $71^2$ | 173 |
| **181k6B** | 40 | $107^2$ | 181 |
| **191k6B** | 46 | $85091^2$ | 191 |
| **193k6B** | 41 | $31^2$ | 193 |
| **199k6B** | 46 | $200329^2$ | 199 |

| $f$ | deg $f$ | $B$ (bound for Ш) | all odd congruence primes |
|---|---|---|---|
| **47k8B** | 16 | $19^2$ | 47 |
| **59k8B** | 20 | $29^2$ | 59 |
| **67k8B** | 20 | $29^2$ | 67 |
| **71k8B** | 24 | $379^2$ | 71 |
| **73k8B** | 22 | $197^2$ | 73 |
| **74k8C** | 6 | $23^2$ | 37, 127, 821, 8327168869 |
| **79k8B** | 25 | $307^2$ | 79 |
| **83k8B** | 27 | $1019^2$ | 83 |
| **87k8C** | 9 | $11^2$ | 3, 5, 7, 29, 31, 59, 947, 22877, 3549902897 |
| **89k8B** | 29 | $44491^2$ | 89 |
| **97k8B** | 29 | $11^2 \cdot 277^2$ | 97 |
| **101k8B** | 33 | $19^2 \cdot 11503^2$ | 101 |
| **103k8B** | 32 | $75367^2$ | 103 |
| **107k8B** | 34 | $17^2 \cdot 491^2$ | 107 |
| **109k8B** | 33 | $23^2 \cdot 229^2$ | 109 |
| **111k8C** | 12 | $127^2$ | 3, 7, 11, 13, 17, 23, 37, 6451, 18583, 51162187 |
| **113k8B** | 35 | $67^2 \cdot 641^2$ | 113 |
| **115k8B** | 12 | $37^2$ | 3, 5, 19, 23, 572437, 5168196102449 |
| **117k8I** | 8 | $19^2$ | 3, 13, 181 |
| **118k8C** | 8 | $37^2$ | 5, 13, 17, 59, 163, 3923085859759909 |
| **119k8C** | 16 | $1283^2$ | 3, 7, 13, 17, 109, 883, 5324191, 91528147213 |
| **121k8F** | 6 | $71^2$ | 3, 11, 17, 41 |
| **121k8G** | 12 | $13^2$ | 3, 11 |
| **121k8H** | 12 | $19^2$ | 5, 11 |
| **125k8D** | 16 | $179^2$ | 5 |
| **127k8B** | 39 | $59^2$ | 127 |

| $f$ | deg $f$ | $B$ (bound for Ш) | all odd congruence primes |
|---|---|---|---|
| **128k8F** | 4 | $11^2$ | 1 |
| **131k8B** | 43 | $241^2 \cdot 817838201^2$ | 131 |
| **134k8C** | 11 | $61^2$ | 11, 17, 41, 67, 71, 421, 2356138931854759 |
| **137k8B** | 42 | $71^2 \cdot 749093^2$ | 137 |
| **139k8B** | 43 | $47^2 \cdot 89^2 \cdot 1021^2$ | 139 |
| **141k8C** | 14 | $13^2$ | 3, 5, 7, 47, 4639, 43831013, 4047347102598757 |
| **142k8B** | 10 | $11^2$ | 3, 53, 71, 56377, 1965431024315921873 |
| **143k8C** | 19 | $307^2$ | 3, 11, 13, 89, 199, 409, 178397, 639259, 17440535 97287 |
| **143k8D** | 21 | $109^2$ | 3, 7, 11, 13, 61, 79, 103, 173, 241, 769, 36583 |
| **145k8C** | 17 | $29587^2$ | 5, 11, 29, 107, 251623, 393577, 518737, 9837145 699 |
| **146k8C** | 12 | $3691^2$ | 11, 73, 269, 503, 1673540153, 11374452082219 |
| **148k8B** | 11 | $19^2$ | 3, 37 |
| **149k8B** | 47 | $11^4 \cdot 40996789^2$ | 149 |
| **43k10B** | 17 | $449^2$ | 43 |
| **47k10B** | 20 | $2213^2$ | 47 |
| **53k10B** | 21 | $673^2$ | 53 |
| **55k10D** | 9 | $71^2$ | 3, 5, 11, 251, 317, 61339, 19869191 |
| **59k10B** | 25 | $37^2$ | 59 |
| **62k10E** | 7 | $23^2$ | 3, 31, 101, 523, 617, 41192083 |
| **64k10K** | 2 | $19^2$ | 3 |
| **67k10B** | 26 | $191^2 \cdot 617^2$ | 67 |
| **68k10B** | 7 | $83^2$ | 3, 7, 17, 8311 |
| **71k10B** | 30 | $1103^2$ | 71 |

| $f$ | deg $f$ | $B$ (bound for Ш) | all odd congruence primes |
|---|---|---|---|
| **19k12B** | 9 | $67^2$ | 5, 17, 19, 31, 571 |
| **31k12B** | 15 | $67^2 \cdot 71^2$ | 31, 13488901 |
| **35k12C** | 6 | $17^2$ | 5, 7, 23, 29, 107, 8609, 1307051 |
| **39k12C** | 6 | $73^2$ | 3, 13, 1491079, 3719832979693 |
| **41k12B** | 20 | $54347^2$ | 7, 41, 3271, 6277 |
| **43k12B** | 20 | $212969^2$ | 43, 1669, 483167 |
| **47k12B** | 23 | $24469^2$ | 17, 47, 59, 2789 |
| **49k12H** | 12 | $271^2$ | 7 |

Table 2: Conjecturally nontrivial Ш (mostly invisible)

# References


[AL] A. O. L. Atkin, J. Lehner, Hecke operators on $\Gamma_0(m)$, *Math. Ann.* **185** (1970) 135–160

[AS] A. Agashe, W. Stein, Visibility of Shafarevich–Tate groups of abelian varieties, preprint

[BS-D] B. J. Birch, H. P. F. Swinnerton-Dyer, Notes on elliptic curves. I and II, *J. reine angew. Math.* **212** (1963) 7–25 and **218** (1965) 79–108

[B] S. Bloch, Algebraic cycles and values of $L$-functions, *J. reine angew. Math.* **350** (1984) 94–108

[BCP] W. Bosma, J. Cannon, and C. Playoust, *The Magma algebra system. I. The user language*, J. Symbolic Comput. **24** (1997) no. 3-4, 235–265, Computational algebra and number theory (London, 1993)

[Be] A. Beilinson, Height pairing between algebraic cycles, *in* Current trends in arithmetical algebraic geometry (K. Ribet, ed.) *Contemp. Math.* **67** (1987) 1–24

[BK] S. Bloch, K. Kato, L-functions and Tamagawa numbers of motives, The Grothendieck Festschrift Volume I, 333–400, Progress in Mathematics, 86, Birkhäuser, Boston, 1990

[Ca1] H. Carayol, Sur les représentations $\ell$-adiques associées aux formes modulaires de Hilbert, *Ann. Sci. École Norm. Sup. (4)***19** (1986) 409–468

[Ca2] H. Carayol, Sur les représentations Galoisiennes modulo $\ell$ attachées aux formes modulaires, *Duke Math. J.* **59** (1989) 785–801

[CM1] J. E. Cremona, B. Mazur, Visualizing elements in the Shafarevich–Tate group, *Experiment. Math.* **9** (2000) 13–28

[CM2] J. E. Cremona, B. Mazur, Appendix to A. Agashe, W. Stein, Visible evidence for the Birch and Swinnerton-Dyer conjecture for modular abelian varieties of rank zero, preprint

[CF] B. Conrey, D. Farmer, On the non-vanishing of $L_f(s)$ at the center of the critical strip, preprint

[De1] P. Deligne, Formes modulaires et représentations $\ell$-adiques. Sém. Bourbaki, éxp. 355, Lect. Notes Math. **179**, 139–172, Springer, 1969

[De2] P. Deligne, Valeurs de Fonctions $L$ et Périodes d'Intégrales, *AMS Proc. Symp. Pure Math.*, Vol. 33 (1979) part 2, 313–346

[DFG1] F. Diamond, M. Flach, L. Guo, Adjoint motives of modular forms and the Tamagawa number conjecture, preprint available from: http://www.andromeda.rutgers.edu/~liguo/lgpapers.html

[DFG2] F. Diamond, M. Flach, L. Guo, The Bloch–Kato conjecture for adjoint motives of modular forms, *Math. Res. Lett.* **8** (2001) 437–442

[Du1] N. Dummigan, Symmetric square $L$-functions and Shafarevich–Tate groups, *Experiment. Math.* **10** (2001) 383–400

[Du2] N. Dummigan, Congruences of modular forms and Selmer groups, *Math. Res. Lett.* **8** (2001) 479–494

[Fa] G. Faltings, Crystalline cohomology and $p$-adic Galois representations, *in* Algebraic analysis, geometry and number theory (J. Igusa, ed.) 25–80, Johns Hopkins University Press, Baltimore, 1989

[FJ] G. Faltings, B. Jordan, Crystalline cohomology and $\mathrm{GL}(2,\mathbb{Q})$, *Israel J. Math.* **90** (1995) 1–66

[Fl1] M. Flach, A generalisation of the Cassels-Tate pairing, *J. reine angew. Math.* **412** (1990) 113–127

[Fl2] M. Flach, On the degree of modular parametrisations, Séminaire de Théorie des Nombres, Paris 1991-92 (S. David, ed.) 23–36, Progress in mathematics, 116, Birkhäuser, Basel Boston Berlin, 1993

[Fo1] J.-M. Fontaine, Sur certains types de représentations $p$-adiques du groupe de Galois d'un corps local, construction d'un anneau de Barsotti–Tate, *Ann. Math.* **115** (1982) 529–577

[Fo2] J.-M. Fontaine, Valeurs spéciales des fonctions $L$ des motifs, Sém. Bourbaki, 1991/92, Exp. No. 751 *Astérisque* **206** (1992) 205–249

[FL] J.-M. Fontaine, G. Lafaille, Construction de représentations $p$-adiques, *Ann. Sci. E.N.S.* **15** (1982) 547–608

[JL] B. W. Jordan, R. Livné, Conjecture "epsilon" for weight $k > 2$, *Bull. Amer. Math. Soc.* **21** (1989) 51–56

[L] R. Livné, On the conductors of mod $\ell$ Galois representations coming from modular forms, *J. Number Theory* **31** (1989) 133–141

[MTT] B. Mazur, J. Tate, J. Teitelbaum, On $p$-adic analogues of the conjectures of Birch and Swinnerton-Dyer, *Invent. Math.* **84** (1986) 1–48

[Ne] J. Nekovár, $p$-adic Abel–Jacobi maps and $p$-adic heights. The arithmetic and geometry of algebraic cycles (Banff, AB, 1998) 367–379, CRM Proc. Lecture Notes, 24, Amer. Math. Soc., Providence, RI, 2000

[Sc] A. J. Scholl, Motives for modular forms, *Invent. Math.* **100** (1990) 419–430

[SV] W. A. Stein, H. A. Verrill, Cuspidal modular symbols are transportable, *L.M.S. Journal of Computational Mathematics* **4** (2001) 170–181

[St] G. Stevens, $\Lambda$-adic modular forms of half-integral weight and a $\Lambda$-adic Shintani lifting. Arithmetic geometry (Tempe, AZ, 1993) 129–151, Contemp. Math., **174**, Amer. Math. Soc., Providence, RI, 1994

[V] V. Vatsal, Canonical periods and congruence formulae, *Duke Math. J.* **98** (1999) 397–419

[W] L. C. Washington, Galois cohomology, *in* Modular Forms and Fermat's Last Theorem, (G. Cornell, J. H. Silverman, G. Stevens, eds.) 101–120, Springer-Verlag, New York, 1997

[Z] S. Zhang, Heights of Heegner cycles and derivatives of $L$-series, *Invent. Math.* **130** (1997) 99–152

Neil Dummigan,
University of Sheffield,
Department of Pure Mathematics,
Hicks Building,
Hounsfield Road,
Sheffield, S3 7RH, U.K.
e-mail: n.p.dummigan@shef.ac.uk

William Stein,
Harvard University,
Department of Mathematics,
One Oxford Street,
Cambridge, MA 02138, U.S.A.
e-mail: was@math.harvard.edu

Mark Watkins,
Penn State Mathematics Department,
University Park,
State College, PA 16802, U.S.A.
e-mail: watkins@math.psu.edu

# A counterexample to a conjecture of Selmer

Tom Fisher

**Abstract**

We present a counterexample to a conjecture cited by Cassels [CaI] and attributed to Selmer. The issues raised have been given new significance by the recent work of Heath-Brown [HB] and Swinnerton-Dyer [SwD] on the arithmetic of diagonal cubic surfaces.

## 1 Introduction

Let $E$ be an elliptic curve over a number field $k$, with complex multiplication by $\mathbf{Z}[\omega]$ where $\omega$ is a primitive cube root of unity. Let $K = k(\omega)$, so that $[K : k] = 1$ or 2 according as $\omega \in k$ or $\omega \notin k$. In his work on cubic surfaces, Heath-Brown [HB] makes implicit use of the following statement.

**Theorem 1.1** *If $[K : k] = 2$ and the Tate–Shafarevich group $Ш(E/k)$ is finite, then the order of $Ш(E/K)[\sqrt{-3}]$ is a perfect square.*

We explain how this result follows from the work of Cassels [CaIV], and give an example to show that the condition $[K : k] = 2$ is necessary.

For the application to cubic surfaces, we only need a special case of the theorem, namely that $Ш(E/K)[\sqrt{-3}]$ cannot have order 3. This result, still conditional on the finiteness of the Tate–Shafarevich group, has already appeared in [BF] and [SwD]. In fact Swinnerton-Dyer [SwD] vastly generalises Heath-Brown's results. In the case $[K : k] = 2$ he proves the Hasse principle for diagonal cubic 3-folds over $k$, conditional only on the finiteness of the Tate–Shafarevich group for elliptic curves over $k$. The condition $[K : k] = 2$ is unnatural, and conjecturally should not appear. However, the counterexample presented in this article suggests that, if we are to follow the methods of Heath-Brown and Swinnerton-Dyer, then this condition on $k$ is unavoidable.

In §2 we recall how it is possible to pass between the fields $k$ and $K$. Then in §3 we give a modern treatment of the descent by 3-isogeny studied by Selmer [S1] and Cassels [CaI]. In §§4–5 we recall how the conjectures of

Selmer may be deduced from properties of the Cassels–Tate pairing. This culminates in a proof of Theorem 1.1. Finally in §6 we present our new example.

## 2 Decomposition into Galois eigenspaces

Let $E$ be an elliptic curve over $k$ with complex multiplication by $\mathbf{Z}[\omega]$. The isogeny $[\sqrt{-3}]\colon E \to E$ is defined over $K = k(\omega)$. But the kernel $E[\sqrt{-3}]$ is defined over $k$. It follows that there is a 3-isogeny $\phi\colon E \to \widetilde{E}$ defined over $k$ with $E[\sqrt{-3}] = E[\phi]$. Here $\widetilde{E}$ is a second elliptic curve defined over $k$, which we immediately recognise as the $-3$-twist of $E$. The dual isogeny $\widehat{\phi}\colon \widetilde{E} \to E$ satisfies $\phi \circ \widehat{\phi} = [3]$ and $\widehat{\phi} \circ \phi = [3]$. Our notation for the Selmer groups and Tate–Shafarevich groups follows Silverman [Sil, Chapter X].

**Lemma 2.1** *If $[K : k] = 2$ then the exact sequence*

$$0 \longrightarrow E(K)/\sqrt{-3}E(K) \longrightarrow S^{(\sqrt{-3})}(E/K) \longrightarrow Ш(E/K)[\sqrt{-3}] \longrightarrow 0 \quad (1)$$

*is the direct sum of the exact sequences*

$$0 \longrightarrow \widetilde{E}(k)/\phi E(k) \longrightarrow S^{(\phi)}(E/k) \longrightarrow Ш(E/k)[\phi] \longrightarrow 0 \quad (2)$$

*and*

$$0 \longrightarrow E(k)/\widehat{\phi}\widetilde{E}(k) \longrightarrow S^{(\widehat{\phi})}(\widetilde{E}/k) \longrightarrow Ш(\widetilde{E}/k)[\widehat{\phi}] \longrightarrow 0. \quad (3)$$

**Proof** Since arguments of this type have already appeared in [BF], [N], [SwD] and presumably countless other places in the literature, we will not dwell on the proof. Suffice it to say that we decompose (1) into eigenspaces for the action of $\mathrm{Gal}(K/k)$, and then use the inflation-restriction exact sequence to identify these eigenspaces as (2) and (3). The observation that $[K : k] = 2$ is prime to $\deg \phi = 3$ is crucial throughout the proof. □

**Remark 2.2** Each term of the exact sequence (1) is a $\mathbf{Z}/3\mathbf{Z}$-vector space with an action of $\mathrm{Gal}(K/k)$. Thus each term is a direct sum of the Galois modules $\mathbf{Z}/3\mathbf{Z}$ and $\mu_3$. If we replace $E$ by $\widetilde{E}$ in (1) we obtain the same exact sequence of abelian groups, but as Galois modules the summands $\mathbf{Z}/3\mathbf{Z}$ and $\mu_3$ are interchanged.

# 3 Computation of Selmer groups

Let $k$ be a number field. Let $T[a_0, a_1, a_2]$ be the diagonal plane cubic

$$a_0 x_0^3 + a_1 x_1^3 + a_2 x_2^3 = 0 \tag{4}$$

where $a_0, a_1, a_2 \in k^*/k^{*3}$. Let $E_A$ be the elliptic curve $T[A, 1, 1]$ with identity element $0 = (0 : 1 : -1)$. It is well known [St] that $E_A$ has Weierstrass equation $y^2 = x^3 - 432A^2$. An alternative proof of the following lemma may be found in [CaL, §18].

**Lemma 3.1** *The diagonal plane cubic $T[a_0, a_1, a_2]$ is a smooth curve of genus 1 with Jacobian $E_A$ where $A = a_0 a_1 a_2$.*

**Proof** There is an isomorphism $T[a_0, a_1, a_2] \simeq E_A$ defined over $k(\sqrt[3]{\alpha})$ where $\alpha = a_1 a_2^2$, given by

$$\psi \colon (x_0 : x_1 : x_2) \mapsto (a_2 x_0 : \alpha^{2/3} x_1 : \alpha^{1/3} a_2 x_2).$$

The cocycle $\sigma(\psi)\psi^{-1}$ takes values in the subgroup $\mu_3 \subset \mathrm{Aut}(E_A)$ generated by $x_i \mapsto \omega^i x_i$. But since $\mu_3$ acts on $E_A$ without fixed points, this action belongs to the translation subgroup of $\mathrm{Aut}(E_A)$. It follows that $T[a_0, a_1, a_2]$ is a torsor under $E_A$ and that $E_A$ is the Jacobian of $T[a_0, a_1, a_2]$. □

Temporarily working over $K = k(\omega)$ we note that $E_A$ has complex multiplication by $\mathbf{Z}[\omega]$ where $\omega \colon (x_0 : x_1 : x_2) \mapsto (\omega x_0 : x_1 : x_2)$ and that $E_A[1-\omega] = E_A[\sqrt{-3}]$ is generated by $(0 : \omega : -\omega^2)$. So as in §2 there is a map $\phi$ which gives an exact sequence of Galois modules

$$0 \longrightarrow \mu_3 \longrightarrow E_A \xrightarrow{\ \phi\ } \widetilde{E}_A \longrightarrow 0$$

where $\widetilde{E}_A$ is the $-3$-twist of $E_A$. Taking Galois cohomology we obtain an exact sequence

$$0 \longrightarrow \widetilde{E}_A(k)/\phi E_A(k) \xrightarrow{\ \delta\ } k^*/k^{*3} \longrightarrow H^1(k, E_A)[\phi] \longrightarrow 0. \tag{5}$$

The group $H^1(k, E_A)$ parametrises the torsors under $E_A$. We write $C_{A,\alpha}$ for the torsor under $E_A$ described by $\alpha \in k^*/k^{*3}$. The proof of Lemma 3.1 shows that

$$T[a_0, a_1, a_2] \simeq C_{A,\alpha} \qquad \text{for } A = \prod a_\nu \text{ and } \alpha = \prod a_\nu^\nu \tag{6}$$

where the products are over $\nu \in \mathbf{Z}/3\mathbf{Z}$. Since $T[a_0, a_1, a_2] \simeq T[a_1, a_2, a_0]$ it is clear that $A \in \operatorname{im} \delta$. If $\widetilde{E}_A$ has Weierstrass equation $Y^2 Z = -4AX^3 + Z^3$ then the 3-covering map $T[a_0, a_1, a_2] \to \widetilde{E}_A$ is given by

$$(x_0 : x_1 : x_2) \mapsto (x_0 x_1 x_2 : a_1 x_1^3 - a_2 x_2^3 : a_0 x_0^3).$$

The Selmer group attached to $\phi$ is

$$S^{(\phi)}(E_A/k) = \{\alpha \in k^*/k^{*3} \mid C_{A,\alpha}(k_\mathfrak{p}) \neq \emptyset \text{ for all primes } \mathfrak{p}\}.$$

Since $\deg \phi = 3$ is odd we have ignored the infinite places. We write $\delta_\mathfrak{p}$ for the local connecting map obtained when we apply (5) to the local field $k_\mathfrak{p}$. Then the condition $C_{A,\alpha}(k_\mathfrak{p}) \neq \emptyset$ may also be written $\alpha \in \operatorname{im} \delta_\mathfrak{p}$. Using (6) to give equations for $C_{A,\alpha}$ it is easy to prove

**Lemma 3.2** *Let $k$ be a number field, and $\mathfrak{p}$ a prime not dividing* 3. *Let $\mathfrak{o}_\mathfrak{p}$ denote the ring of integers of $k_\mathfrak{p}$. Then*

$$\operatorname{im} \delta_\mathfrak{p} = \begin{cases} \mathfrak{o}_\mathfrak{p}^*/\mathfrak{o}_\mathfrak{p}^{*3} & \text{if } \operatorname{ord}_\mathfrak{p}(A) \equiv 0 \pmod 3 \\ \langle A \rangle & \text{if } \operatorname{ord}_\mathfrak{p}(A) \not\equiv 0 \pmod 3. \end{cases}$$

If $\mathfrak{p}$ divides 3 the situation is more complicated, although we still have

$$\operatorname{im} \delta_\mathfrak{p} \subset \mathfrak{o}_\mathfrak{p}^*/\mathfrak{o}_\mathfrak{p}^{*3} \quad \text{if } \operatorname{ord}_\mathfrak{p}(A) \equiv 0 \pmod 3. \tag{7}$$

If $\omega \in k_\mathfrak{p}$ Tate local duality tells us that $\operatorname{im} \delta_\mathfrak{p}$ is a maximal isotropic subspace with respect to the Hilbert norm residue symbol

$$k_\mathfrak{p}^*/k_\mathfrak{p}^{*3} \times k_\mathfrak{p}^*/k_\mathfrak{p}^{*3} \longrightarrow \mu_3. \tag{8}$$

The next lemma treats the case $k = \mathbf{Q}(\omega)$. This field has ring of integers $\mathbf{Z}[\omega]$ and class number 1. The unique prime above 3 is $\pi = \omega - \omega^2$.

**Lemma 3.3** *Let $A \in \mathbf{Z}[\omega]$ be nonzero and cube-free. Then*

$$\operatorname{im} \delta_\pi = \begin{cases} \langle A, (1+A)/(1-A) \rangle & \text{if } \operatorname{ord}_\pi(A) \neq 0 \\ \langle A, 1-\pi^3 \rangle & \text{if } \operatorname{ord}_\pi(A) = 0 \text{ and } A \not\equiv \pm 1 \ (\pi^3) \\ \langle \omega(1+3a), 1-\pi^3 \rangle & \text{if } A = \pm(1+a\pi^3) \text{ for some } a \in \mathbf{Z}[\omega]. \end{cases}$$

**Proof** We recall [CF, Exercise 2.13] that $k_\pi^*/k_\pi^{*3}$ has basis $\pi, \omega, 1-\pi^2, 1-\pi^3$ and that these elements define a filtration compatible with the pairing (8). By Tate local duality it follows that $\operatorname{im} \delta_\pi$ has order 9. So to prove the lemma it suffices to prove the inclusions $\supset$. As always $A \in \operatorname{im} \delta_\pi$, whereas (7) and Tate local duality tell us that $1 - \pi^3 \in \operatorname{im} \delta_\pi$. There is at most one more element to find.

(i) Suppose $\operatorname{ord}_\pi(A) \neq 0$. If $\alpha$ satisfies $\alpha - \alpha^{-1} = A$ then $T[A, \alpha, \alpha^{-1}]$ is soluble. Splitting into the cases $\operatorname{ord}_\pi(A) = 1$ or 2 we find

$$4A/(1-A^2) \equiv A \pmod{\pi^4}.$$

Thus $\alpha = (1+A)/(1-A)$ provides a solution mod $\pi^4$.

(ii) Suppose $A = 1 + a\pi^3$ for some $a \in \mathbf{Z}[\omega]$. If $\alpha$ satisfies $A + \alpha + \alpha^{-1} = 0$ then $T[A, \alpha, \alpha^{-1}]$ is soluble. In view of the identity

$$(1 + \pi^3 a) + \omega(1 + 3a) + \omega^2(1 - 3a) = 0$$

we see that $\alpha = \omega(1 + 3a)$ provides a solution mod $\pi^4$. □

# 4 Selmer's conjectures

In this section we take $k = \mathbf{Q}$, so that $K = \mathbf{Q}(\omega)$. We consider the elliptic curves $E_A$ and $\widetilde{E}_A$ over $\mathbf{Q}$ where $A \geq 2$ is a cube-free integer.

**Lemma 4.1** *If $A \geq 3$ then the torsion subgroups are*

$$E_A(\mathbf{Q})_{\text{tors}} = 0 \quad \textit{and} \quad \widetilde{E}_A(\mathbf{Q})_{\text{tors}} \simeq \mathbf{Z}/3\mathbf{Z}.$$

**Proof** See [St, §6] or [K, Chapter 1, Problem 7]. □

Lemma 2.1 gives a decomposition into $\text{Gal}(K/\mathbf{Q})$-eigenspaces

$$S^{(\sqrt{-3})}(E_A/K) \simeq S^{(\phi)}(E_A/\mathbf{Q}) \oplus S^{(\widehat{\phi})}(\widetilde{E}_A/\mathbf{Q}). \tag{9}$$

The following examples were found by Selmer [S1], [S2].

**Example 4.2** Let $A = 60$. Lemmas 3.2 and 3.3 tell us that

$$S^{(\sqrt{-3})}(E_{60}/K) \simeq \langle 2, 3, 5 \rangle \subset K^*/K^{*3}.$$

Then (9) gives $S^{(\phi)}(E_{60}/\mathbf{Q}) \simeq (\mathbf{Z}/3\mathbf{Z})^3$ and $S^{(\widehat{\phi})}(\widetilde{E}_{60}/\mathbf{Q}) = 0$. But a 2-descent [CaL, §15], [Cr] shows that $E_{60}(\mathbf{Q})$ has rank 0. We deduce

$$Ш(E_{60}/\mathbf{Q})[3] \simeq (\mathbf{Z}/3\mathbf{Z})^2.$$

**Example 4.3** Let $A = 473$. Lemmas 3.2 and 3.3 tell us that

$$S^{(\sqrt{-3})}(E_{473}/K) \simeq \langle 11, 1 - 6\omega, 1 - 6\omega^2 \rangle \subset K^*/K^{*3}.$$

Then (9) gives $S^{(\phi)}(E_{473}/\mathbf{Q}) \simeq (\mathbf{Z}/3\mathbf{Z})^2$ and $S^{(\widehat{\phi})}(\widetilde{E}_{473}/\mathbf{Q}) \simeq \mathbf{Z}/3\mathbf{Z}$. But a 2-descent [S2], [Cr] shows that $E_{473}(\mathbf{Q})$ has rank 0. We deduce

$$Ш(E_{473}/\mathbf{Q})[\phi] \simeq \mathbf{Z}/3\mathbf{Z} \quad \text{and} \quad Ш(\widetilde{E}_{473}/\mathbf{Q})[\widehat{\phi}] \simeq \mathbf{Z}/3\mathbf{Z}.$$

**Remark 4.4** According to the formulae and tables of Stephens [St], the above examples have $L(E_A, 1) \neq 0$. So the claims $\text{rank}\, E_A(\mathbf{Q}) = 0$ could equally be deduced from the work of Coates and Wiles [CW].

Example 4.2 tells us that each of the curves

$$\begin{array}{rrcl} T[3,4,5]: & 3x_0^3+4x_1^3+5x_2^3 & = & 0 \\ T[1,3,20]: & x_0^3+3x_1^3+20x_2^3 & = & 0 \\ T[1,4,15]: & x_0^3+4x_1^3+15x_2^3 & = & 0 \\ T[1,5,12]: & x_0^3+5x_1^3+12x_2^3 & = & 0 \end{array} \tag{10}$$

is a counterexample to the Hasse Principle for smooth curves of genus 1 defined over $\mathbf{Q}$. Selmer proves this without the need for a 2-descent. Instead he shows that the equations (10) are insoluble over $\mathbf{Q}$ by writing them as norm equations. As Cassels explains [CaI, §11] this is equivalent to performing a second descent, *i.e.* computing the middle group in

$$\widetilde{E}_A(\mathbf{Q})/\phi E_A(\mathbf{Q}) \subset \widehat{\phi} S^{(3)}(\widetilde{E}_A/\mathbf{Q}) \subset S^{(\phi)}(E_A/\mathbf{Q}). \tag{11}$$

In fact Selmer's calculations suffice to show that $\text{Ш}(E_{60}/\mathbf{Q})(3) \simeq (\mathbf{Z}/3\mathbf{Z})^2$. In other words $\text{Ш}(E_{60}/\mathbf{Q})$ does not contain an element of order 9. More recent work of Rubin [M] improves this to $\text{Ш}(E_{60}/\mathbf{Q}) \simeq (\mathbf{Z}/3\mathbf{Z})^2$.

Selmer also gave practical methods for computing the two right hand groups in

$$E_A(\mathbf{Q})/\widehat{\phi}\widetilde{E}_A(\mathbf{Q}) \subset \phi S^{(3)}(E_A/\mathbf{Q}) \subset S^{(\widehat{\phi})}(\widetilde{E}_A/\mathbf{Q}). \tag{12}$$

Following Stephens [St] we write $g_1+1$, $\lambda_1'+1$, $\lambda_1+1$ for the dimensions of the $\mathbf{Z}/3\mathbf{Z}$-vector spaces (11) and $g_2$, $\lambda_2'$, $\lambda_2$ for the dimensions of the $\mathbf{Z}/3\mathbf{Z}$-vector spaces (12). Trivially we have $0 \le g_1 \le \lambda_1' \le \lambda_1$, $0 \le g_2 \le \lambda_2' \le \lambda_2$ and rank $E_A(\mathbf{Q}) = g_1 + g_2$. Based on a large amount of numerical evidence, Selmer [S3] made the following

**Conjecture 4.5** *Let $A \ge 2$ be a cube-free integer. Let $E_A$ be the elliptic curve $x^3+y^3=Az^3$ defined over $\mathbf{Q}$. Then*

**Weak form** *The second descent excludes an even number of generators, i.e.* $\lambda_1 \equiv \lambda_1' \pmod 2$ *and* $\lambda_2 \equiv \lambda_2' \pmod 2$.

**Strong form** *The number of generators of infinite order for $E_A(\mathbf{Q})$ is an even number less than what is indicated by the first descent, i.e.*

$$\lambda_1+\lambda_2 \equiv g_1+g_2 \pmod 2.$$

For $A=473$, Selmer found $\lambda_1=\lambda_1'=\lambda_2=\lambda_2'=1$ yet $g_1=g_2=0$. He was thus aware of the need to combine the contributions from $\phi$ and $\widehat{\phi}$ in the strong form of his conjecture.

**Remark 4.6** In Heath-Brown's notation [HB] we have

$$r(A) = \operatorname{rank} E_A(\mathbf{Q}) = g_1 + g_2 \quad \text{and} \quad s(A) = \lambda_1 + \lambda_2.$$

By (9) the order of $S^{(\sqrt{-3})}(E_A/K)$ is $3^{s(A)+1}$ and in fact it is this relation that Heath-Brown uses to define $s(A)$. Naturally he writes the strong form of Selmer's conjecture as $r(A) \equiv s(A) \pmod 2$.

Now let $k$ be any number field. Conjecture 4.5 is equivalent to the case $k = \mathbf{Q}$ of the following

**Conjecture 4.7** *Suppose that $A \in k^*$ is not a perfect cube and let $E_A$ be the elliptic curve $x^3 + y^3 = Az^3$ defined over $k$. Then*

**Weak form** *The index of the subgroup $\widehat{\phi}(Ш(\widetilde{E}_A/k)[3]) \subset Ш(E_A/k)[\phi]$ is a perfect square. The same is true for $\phi(Ш(E_A/k)[3]) \subset Ш(\widetilde{E}_A/k)[\widehat{\phi}]$.*

**Strong form** *The order of $Ш(E_A/k)[\phi]$ times that of $Ш(\widetilde{E}_A/k)[\widehat{\phi}]$ is a perfect square.*

In the next section we recall how Conjecture 4.7 follows from the work of Cassels, the strong form being conditional on the finiteness of $Ш(E_A/k)$.

# 5 The Cassels–Tate pairing

Let $E$ be an elliptic curve over a number field $k$. For $\phi \colon E \to E'$ an isogeny of elliptic curves over $k$ we shall write $\widehat{\phi} \colon E' \to E$ for the dual isogeny. Cassels [CaIV] defines an alternating bilinear pairing

$$\langle\ ,\ \rangle \colon Ш(E/k) \times Ш(E/k) \to \mathbf{Q}/\mathbf{Z} \tag{13}$$

with the following nondegeneracy property.

**Theorem 5.1** *Let $\phi \colon E \to E'$ be an isogeny of elliptic curves over $k$. Then $x \in Ш(E/k)$ belongs to the image of $\widehat{\phi} \colon Ш(E'/k) \to Ш(E/k)$ if and only if $\langle x, y \rangle = 0$ for all $y \in Ш(E/k)[\phi]$.*

**Proof** This was proved by Cassels [CaIV] in the case $\phi = [m]$ for $m$ a rational integer. The general case follows by his methods and is explained in [F]. □

The pairing was later generalised to abelian varieties by Tate, and so is known as the Cassels–Tate pairing. The most striking applications in the case of elliptic curves come from the following easy lemma.

**Lemma 5.2** *If a finite abelian group admits a nondegenerate alternating bilinear pairing, then its order must be a perfect square.*

The weak form of Conjecture 4.7 is a special case of

**Corollary 5.3** *Let $\phi\colon E \to E'$ be an $m$-isogeny of elliptic curves over $k$. Then the subgroup $\widehat{\phi}(\text{Ш}(E'/k)[m]) \subset \text{Ш}(E/k)[\phi]$ has index a perfect square.*

**Proof** According to Theorem 5.1 the pairing (13) restricted to $\text{Ш}(E/k)[\phi]$ has kernel $\widehat{\phi}(\text{Ш}(E'/k)[m])$. We are done by Lemma 5.2. □

Let us assume that $\text{Ш}(E/k)$ is finite. So by Theorem 5.1 and Lemma 5.2 the order of $\text{Ш}(E/k)$ is a perfect square. If $\phi\colon E \to E'$ is an isogeny of elliptic curves over $k$ then the same conclusions will hold for $E'$. We define

$$\langle\ ,\ \rangle_\phi \colon \text{Ш}(E/k) \times \text{Ш}(E'/k) \to \mathbf{Q}/\mathbf{Z}\,; \quad (x,y) \mapsto \langle \phi x, y\rangle = \langle x, \widehat{\phi} y\rangle \tag{14}$$

where the equality on the right is [CaVIII, Theorem 1.2]. The strong form of Conjecture 4.7 is a special case of

**Corollary 5.4** *Let $\phi\colon E \to E'$ be an isogeny of elliptic curves over $k$. If $\text{Ш}(E/k)$ is finite then the order of $\text{Ш}(E/k)[\phi]$ times that of $\text{Ш}(E'/k)[\widehat{\phi}]$ is a perfect square.*

**Proof** According to Theorem 5.1 the left and right kernels of $\langle\ ,\ \rangle_\phi$ are $\text{Ш}(E/k)[\phi]$ and $\text{Ш}(E'/k)[\widehat{\phi}]$. We obtain a nondegenerate pairing

$$\text{Ш}(E/k)/\text{Ш}(E/k)[\phi] \times \text{Ш}(E'/k)/\text{Ш}(E'/k)[\widehat{\phi}] \to \mathbf{Q}/\mathbf{Z}.$$

We deduce that these quotients have the same order and are done since $\text{Ш}(E/k)$ and $\text{Ш}(E'/k)$ each have order a perfect square. □

Another well known consequence is

**Corollary 5.5** *Let $E$ be an elliptic curve over $k$ whose Tate–Shafarevich group is finite, and let $m$ be a rational integer. Then the order of $\text{Ш}(E/k)[m]$ is a perfect square.*

**Proof** According to Theorem 5.1 the kernel of $\langle\ ,\ \rangle_m$ is $\text{Ш}(E/k)[m]$. We obtain a nondegenerate alternating pairing

$$\text{Ш}(E/k)/\text{Ш}(E/k)[m] \times \text{Ш}(E/k)/\text{Ш}(E/k)[m] \to \mathbf{Q}/\mathbf{Z}.$$

We apply Lemma 5.2 to this pairing and are done since $\text{Ш}(E/k)$ has order a perfect square. □

**Remark 5.6** We could equally deduce Corollary 5.4 from Corollaries 5.3 and 5.5.

**Proof of Theorem 1.1** Let $E$ be an elliptic curve over $k$ with complex multiplication by $\mathbf{Z}[\omega]$ and suppose that $[K:k]=2$. Lemma 2.1 tells us that

$$Ш(E/K)[\sqrt{-3}] \simeq Ш(E/k)[\phi] \oplus Ш(\widetilde{E}/k)[\widehat{\phi}].$$

Assuming $Ш(E/k)$ is finite, Corollary 5.4 shows that the group on the right has order a perfect square. So the group on the left has order a perfect square, and this is precisely the statement of Theorem 1.1. □

In the first of his celebrated series of papers, Cassels [CaI] defines a pairing $S^{(\sqrt{-3})}(E_A/K) \times S^{(\sqrt{-3})}(E_A/K) \to \mu_3$. It is of course a special case of the pairing (13). He uses it to prove the weak form of Conjecture 4.7 in the case $[K:k]=1$. However in the introduction to the same paper he misquotes the strong form of Selmer's conjecture. The statement he gives is equivalent to

- If $[K:k]=1$ then the order of $Ш(E_A/K)[\sqrt{-3}]$ is a perfect square.

It is this statement to which we have found a counterexample. It is possible that Cassels was misled by earlier work of Selmer at a time when he did not appreciate the need to combine the contributions from $\phi$ and $\widehat{\phi}$ in the strong form of his conjecture.

**Remark 5.7** It is tempting to try and prove Theorem 1.1 also in the case $[K:k]=1$ by imitating the proof of Corollary 5.5. However the isogeny $[\sqrt{-3}]$ has dual $[-\sqrt{-3}]$ and this extra sign means that the pairing $\langle\ ,\ \rangle_{\sqrt{-3}}$ is symmetric rather than alternating. Lemma 5.2 does not apply.

# 6 A new example

In this section we take $K=\mathbf{Q}(\omega)$. Let $E_A$ be the elliptic curve $x^3+y^3=Az^3$. We aim to find $A \in K$ such that the order of $Ш(E_A/K)[\sqrt{-3}]$ is not a perfect square. As in Example 4.3 our method is to compare a 3-descent with a 2-descent. The form of the curves $E_A$ makes the 3-descent easy. We use the results of §3 to compute the Selmer group $S^{(\sqrt{-3})}(E_A/K)$. For the 2-descent we would like to use John Cremona's program `mwrank` [Cr]. But `mwrank` is written specifically for elliptic curves over $\mathbf{Q}$, whereas Theorem 1.1 tells us that there are no examples of the required form with $A^2 \in \mathbf{Q}$. Fortunately we were able to use a program of Denis Simon [Si1], [Si2], written using the computer algebra package `pari` [BBBCO], that extends Cremona's work on 2-descents to general number fields (in practice of degrees 1 up to 5).

We consider all cube-free $A \in \mathbf{Z}[\omega]$ with $A^2 \notin \mathbf{Q}$ and $\mathrm{Norm}(A) \leq 150$. We ignore repeats of the form $\pm\sigma(A)$ for $\sigma \in \mathrm{Gal}(K/\mathbf{Q})$. In all 123 cases a calculation based on Lemmas 3.2 and 3.3 shows that $S^{(\sqrt{-3})}(E_A/K)$ is isomorphic to either $\mathbf{Z}/3\mathbf{Z}$ or $(\mathbf{Z}/3\mathbf{Z})^2$. In the 98 cases where $S^{(\sqrt{-3})}(E_A/K) \simeq \mathbf{Z}/3\mathbf{Z}$ it follows immediately that $\operatorname{rank} E_A(K) = 0$. In the remaining 25 cases we run Simon's program. For 20 of these curves the program exhibits a point of infinite order. Since $E_A(K)$ has the structure of $\mathbf{Z}[\omega]$-module, we are able to deduce that $\operatorname{rank} E_A(K) = 2$. The remaining 5 cases are

$$A = \pm(3+7\omega),\ \pm(9+\omega),\ \pm(12+5\omega),\ \pm(6+13\omega),\ \pm(13+7\omega)$$

and their Galois conjugates. In each of these cases Simon's program reports that $\operatorname{rank} E_A(K) = 0$. Reducing modulo some small primes we find $E_A(K) \simeq \mathbf{Z}/3\mathbf{Z}$. Thus

$$Ш(E_A/K)[\sqrt{-3}] \simeq \mathbf{Z}/3\mathbf{Z}.$$

For the remainder of this article we restrict attention to the first of these examples, namely $A = 3 + 7\omega$, and give further details of the descent calculations involved. In particular we establish the counterexample of the title in a way that is independent of Simon's program.

We begin by checking the above computation of $S^{(\sqrt{-3})}(E_A/K)$ for $A = 3 + 7\omega$. Since $(A)$ is prime, Lemma 3.2 tells us that

$$S^{(\sqrt{-3})}(E_A/K) \subset \langle \omega, 3 + 7\omega \rangle. \tag{15}$$

We check the local conditions at the primes $(\pi)$ and $(A)$ above 3 and 37 respectively.

- Since $37 \equiv 1 \pmod 9$ we know that $\omega$ is a cube locally at $(A)$.
- Lemma 3.3 gives $\operatorname{im} \delta_\pi = \langle A, 1 - \pi^3 \rangle \subset K_\pi^*/K_\pi^{*3}$. Since $A = \omega - \pi^3$ it is clear that $\omega$ belongs to this subgroup.

It follows that equality holds in (15) as required.

Given the provisional nature of Simon's program we have taken the liberty of writing out the 2-descent for $A = 3+7\omega$ in the style of Cassels [CaL, p. 72–73]. The curve $E_A$ has Weierstrass form

$$Y^2 = X^3 - 2^4 3^3 (3 + 7\omega)^2. \tag{16}$$

The 2-descent takes place over the field $L = K(\delta)$ where $\delta^3 = 4(3 + 7\omega)$. According to `pari` [BBBCO][1], $L$ has class number $h = 3$, and fundamental units

$$\begin{aligned} \eta_1 &= (-7 - 3\omega) + (-3 - 2\omega)\delta + (-2 + \omega)\delta^2/2 \\ \eta_2 &= (-7 - 3\omega) + (2 - \omega)\delta + (3 + 2\omega)\delta^2/2. \end{aligned}$$

[1] These calculations were performed using Version 2.0.20 (beta)

Furthermore `pari` is able to certify these results, independent of any conjecture. We have chosen $\eta_1$ and $\eta_2$ to be $K$-conjugates. They have minimal polynomial

$$x^3 + (21 + 9\omega)x^2 + (102 - 165\omega)x - 1.$$

If $(X, Y) = (r/t^2, s/t^3)$ is a solution of (16), with fractions in lowest terms, then a common prime divisor of any two of

$$r - 3\delta^2 t^2, \quad r - 3\omega\delta^2 t^2, \quad r - 3\omega^2\delta^2 t^2$$

must divide $2(1-\omega)(3+7\omega)$. Since $2, (1-\omega), (3+7\omega)$ ramify completely, $r - 3\delta^2 t^2$ must be a perfect ideal square. Since $h$ is odd it follows that $S^{(2)}(E/K)$ is a subgroup of $\langle -1, \eta_1, \eta_2 \rangle \subset L^*/L^{*2}$. We claim that $S^{(2)}(E/K)$ is trivial. By considering norms from $L$ to $K$, it suffices to show that the equation

$$r - 3\delta^2 t^2 = \eta\alpha^2 \quad \text{with } \eta = \eta_1, \eta_2 \text{ or } 1/(\eta_1\eta_2)$$

is insoluble for $r, t \in K$ and $\alpha \in L$. The action of $\mathrm{Gal}(L/K)$ shows that we need only consider the case $\eta = \eta_1$. Put $\alpha = u + v\delta + w\delta^2$. Equating coefficients of powers of $\delta$ we obtain

$$\begin{aligned} 0 &= (-3-2\omega)u^2 + (-14-6\omega)uv + (-26-36\omega)v^2 \\ &\quad + (-52-72\omega)uw + (40-104\omega)vw + (-148\omega)w^2 \\ -3t^2 &= ((-2+\omega)/2)u^2 + (-6-4\omega)uv + (-7-3\omega)v^2 \\ &\quad + (-14-6\omega)uw + (-52-72\omega)vw + (20-52\omega)w^2. \end{aligned}$$

On putting

$$\begin{aligned} u &= (-8+6\omega)e + (-6-34\omega)f + (-20+15\omega)g \\ v &= (-4-4\omega)e + (12+4\omega)f + (-10-11\omega)g \\ w &= (1-\omega)e + (1+4\omega)f + (2-2\omega)g \end{aligned}$$

in the first equation, it becomes

$$0 = (3+7\omega)g^2 - 16ef.$$

Hence there are $m, n$ such that

$$e : f : g = m^2 : (3+7\omega)n^2 : 4mn.$$

On substituting into the second equation, we get

$$\begin{aligned} -3t^2 &= 2(-1-4\omega)m^4 + 8(-4+3\omega)m^3 n + 4(21+12\omega)m^2 n^2 \\ &\quad + 8(4-3\omega)mn^3 + 2(-33-40\omega)n^4. \end{aligned}$$

But this is impossible over the 2-adic completion of $K$. Hence $S^{(2)}(E_A/K)$ is trivial and $\operatorname{rank} E_A(K) = 0$ as claimed.

## Acknowledgements

The author would like to thank Laura Basile, Alexei Skorobogatov and Sir Peter Swinnerton-Dyer for useful conversations.

## References

[BF] C.L. Basile and T.A. Fisher, Diagonal cubic equations in four variables with prime coefficients, *Rational points on algebraic varieties*, 1–12, Progr. Math., **199**, Birkhäuser, Basel, 2001

[BBBCO] C. Batut, K. Belabas, D. Bernardi, H. Cohen and M. Olivier, `pari/gp`, a computer algebra package, `http://www.parigp-home.de`

[CaI] J.W.S. Cassels, Arithmetic on curves of genus 1. I, On a conjecture of Selmer, *J. reine angew. Math.* **202** (1959) 52–99

[CaIV] J.W.S. Cassels, Arithmetic on curves of genus 1. IV, Proof of the Hauptvermutung, *J. reine angew. Math.* **211** (1962) 95–112

[CaVIII] J.W S. Cassels, Arithmetic on curves of genus 1. VIII, On conjectures of Birch and Swinnerton-Dyer, *J. reine angew. Math.* **217** (1965) 180–199

[CaL] J.W.S. Cassels, *Lectures on elliptic curves*, LMSST **24**, Cambridge University Press, Cambridge, 1991

[CF] J.W.S. Cassels and A. Fröhlich (Eds.), *Algebraic number theory*, Academic Press, London, 1967

[CW] J. Coates and A. Wiles, On the conjecture of Birch and Swinnerton-Dyer, *Invent. Math.* **39** (1977) 223–251

[Cr] J.E. Cremona, `mwrank`, a program for performing 2-descent on elliptic curves over **Q**, `http://www.maths.nottingham.ac.uk/personal/jec/ftp/progs`

[F] T.A. Fisher, The Cassels–Tate pairing and the Platonic solids, *J. Number Theory* **98** (2003) 105–155

[HB] D.R. Heath-Brown, The solubility of diagonal cubic Diophantine equations, *Proc. London Math. Soc.* (3) **79** (1999) 241–259

[K] N. Koblitz, *Introduction to elliptic curves and modular forms*, GTM **97**, Springer-Verlag, New York, 1993

[M] B. Mazur, On the passage from local to global in number theory, *Bull. Amer. Math. Soc.* **29** (1993) 14–50

[N] J. Nekovář, Class numbers of quadratic fields and Shimura's correspondence, *Math. Ann.* **287** (1990) 577–594

[S1] E.S. Selmer, The diophantine equation $ax^3+by^3+cz^3=0$, *Acta Math.* **85** (1951) 203–362

[S2] E.S. Selmer, The diophantine equation $ax^3+by^3+cz^3=0$, completion of the tables, *Acta Math.* **92** (1954) 191–197

[S3] E.S. Selmer, A conjecture concerning rational points on cubic curves, *Math. Scand.* **2** (1954) 49–54

[Sil] J.H. Silverman, *The arithmetic of elliptic curves*, GTM **106**, Springer-Verlag, New York, 1986

[Si1] D. Simon, Computing the rank of elliptic curves over number fields, *LMS J. Comput. Math.* **5** (2002) 7–17

[Si2] D. Simon, `ell.gp`, a program for calculating the rank of elliptic curves over number fields, `http://www.math.unicaen.fr/~simon`

[St] N.M. Stephens, The diophantine equation $X^3+Y^3=DZ^3$ and the conjectures of Birch and Swinnerton-Dyer, *J. reine angew. Math.* **231** (1968) 121–162

[SwD] H.P.F. Swinnerton-Dyer, The solubility of diagonal cubic surfaces, *Ann. Sci. École Norm. Sup.* (4) **34** (2001) 891–912

Tom Fisher,
DPMMS, Centre for Mathematical Sciences,
Wilberforce Road, Cambridge CB3 0WB, UK
e-mail: T.A.Fisher@dpmms.cam.ac.uk
web: http://www.dpmms.cam.ac.uk/~taf1000

# Linear relations amongst sums of two squares

D.R. Heath-Brown

## 1 Introduction

It is well known that there are infinitely many sets of three distinct primes in arithmetic progression. This may be proved by an easy adaptation of Vinogradov's treatment of the ternary Goldbach problem. More generally for, any nonzero integers $A, B, C$, not all of the same sign, one can show the existence of infinitely many triples of primes $p_1, p_2, p_3$ satisfying the linear relation

$$Ap_1 + Bp_2 + Cp_3 = 0$$

subject to the natural condition that $A+B+C$ should be even. Balog [1] has made important progress on the question of linear relations involving more than 3 primes, but nonetheless it remains an open problem as to whether there are infinitely many sets of 4 distinct primes in arithmetic progression.

Many open problems involving primes have potentially easier relatives involving sums of two squares. Thus one might ask whether or not there are infinitely many arithmetic progressions of 4 (or more) distinct integers, each of which is a sum of 2 squares. This is trivial. The numbers

$$(n-8)^2+(n-1)^2, \quad (n-7)^2+(n+4)^2, \quad (n+7)^2+(n-4)^2$$
$$\text{and} \quad (n+8)^2+(n+1)^2$$

form an arithmetic progression with common difference $12n$. In this paper we shall address the question of the frequency of such progressions. We shall count the sums of two squares with appropriate multiplicity, so that we shall consider the sum

$$\sum_{\mathbf{x}\in\mathcal{R}} r(L_1(\mathbf{x}))r(L_2(\mathbf{x}))r(L_3(\mathbf{x}))r(L_4(\mathbf{x})), \tag{1.1}$$

where $\mathcal{R}$ is a suitable subset of $\mathbb{R}^2$ and the linear forms $L_i$ are given by

$$\begin{aligned} L_1(\mathbf{x}) &= x_1, & L_2(\mathbf{x}) &= x_1+x_2, \\ L_3(\mathbf{x}) &= x_1+2x_2, & L_4(\mathbf{x}) &= x_1+3x_2, \end{aligned} \tag{1.2}$$

where $\mathbf{x}$ denotes the vector $(x_1, x_2)$. The corresponding problem for arithmetic progressions of length 3 is readily handled by the circle method. However for progressions of length 4 it would appear that one would require a version of the 'Kloosterman refinement' for a double integral

$$\int_0^1 \int_0^1 S(\alpha)^2 S(-2\alpha+\beta)^2 S(\alpha-2\beta)^2 S(\beta)^2 d\alpha d\beta.$$

Since research to date has failed to provide such a technique we shall use a rather different approach.

We shall consider a general set of linear forms $L_1, \ldots, L_4$. However we will find it convenient to work with linear forms which are suitably normalized. Moreover we shall require the region $\mathcal{R}$ in which we work to satisfy certain basic conditions. We therefore introduce the following hypothesis.

**Normalization Condition 1 (NC1)** *We assume:*

*(i) No two of the forms $L_1, \ldots, L_4$ are proportional.*

*(ii) We have*

$$\mathcal{R} = X\mathcal{R}^{(0)} = \{\mathbf{x} \in \mathbb{R}^2 : X^{-1}\mathbf{x} \in \mathcal{R}^{(0)}\},$$

*where $\mathcal{R}^{(0)} \subset \mathbb{R}^2$ is open, bounded and convex, with a piecewise continuously differentiable boundary, and where $X$ is a large positive parameter.*

*(iii) We have $L_i(\mathbf{x}) > 0$ for $1 \leq i \leq 4$ and for all $\mathbf{x} \in \mathcal{R}^{(0)}$.*

*(iv) We have*

$$L_1(x_1, x_2) \equiv L_2(x_1, x_2) \equiv L_3(x_1, x_2) \equiv L_4(x_1, x_2) \equiv x_1 \pmod 4.$$

We have imposed the final condition in order to simplify our analysis. While this may seem a little arbitrary, it can be viewed as an analogue of conditions (ii) and (iii). One can think of (ii) and (iii) as requiring $\mathbf{x}$ to lie in an open neighbourhood of a point $\mathbf{y}$ for which each $L_i(\mathbf{y})$ is a sum of two squares. The 2-adic analogue of this real condition on the domain of summation would involve fixing a 2-adic vector $\mathbf{y}$ such that each value $L_i(\mathbf{y})$ is a sum of two 2-adic squares. We would then require $\mathbf{x}$ to lie in an appropriate 2-adic neighbourhood of $\mathbf{y}$. If one imposes such a condition then it can be shown that there is a suitable change of variables which produces forms satisfying (iv). However we shall not pursue this here.

In view of condition (iv) we shall find it convenient to write

$$\mathcal{R}_4 = \{\mathbf{x} \in \mathcal{R} : x_1 \equiv 1 \pmod 4\},$$

so that our problem is to estimate

$$\sum_{\mathbf{x}\in\mathcal{R}_4} r(L_1(\mathbf{x}))r(L_2(\mathbf{x}))r(L_3(\mathbf{x}))r(L_4(\mathbf{x})) = S, \tag{1.3}$$

say.

From now on, all order constants will be allowed to depend on the set of forms $L_1, \ldots, L_4$, and on the region $\mathcal{R}^{(0)}$. Our first result is then the following.

**Theorem 1** *For a set of forms satisfying* **NC1**, *we have*

$$S = 4\pi^4 \operatorname{meas}\mathcal{R} \prod_{p\geq 3} \sigma_p + O(X^2(\log X)^{-\eta/2}(\log\log X)^{15/4}) \tag{1.4}$$

*where* meas *denotes Lebesgue measure, and*

$$\eta = 1 - \frac{1+\log\log 2}{\log 2} = 0.08607\ldots. \tag{1.5}$$

*Here the product $\prod \sigma_p$ is absolutely convergent and*

$$\sigma_p = E_p\{1-\chi(p)p^{-1}\}^4,$$

*where $\chi$ is the nonprincipal character modulo 4. The factor $E_p$ is given by*

$$E_p = \sum_{a,b,c,d=0}^{\infty} \chi(p)^{a+b+c+d}\rho(p^a,p^b,p^c,p^d)^{-1},$$

*where $\rho(d_1,d_2,d_3,d_4)$ is the determinant of the lattice*

$$\{\mathbf{x}\in\mathbb{Z}^2 : d_i \mid L_i(\mathbf{x}),\ 1\leq i\leq 4\}.$$

*The implied constant in (1.4) may depend on the set of forms $L_1, \ldots, L_4$, and on the region $\mathcal{R}^{(0)}$.*

It may be of interest to note that we can evaluate $E_p$ explicitly in many cases. For $1\leq i<j\leq 4$, let $\Delta_{ij}$ be the determinant of the pair of forms $L_i, L_j$, and let $\Delta$ be the product of the various $\Delta_{ij}$. Then if $p\nmid\Delta$, we can find $E_p$ by a routine, if lengthy, calculation. The result is that

$$E_p = \begin{cases} (1-\frac{1}{p})^{-2}(1-\frac{1}{p^2})^{-2}(1+\frac{2}{p}+\frac{6}{p^2}+\frac{2}{p^3}+\frac{1}{p^4}) & \text{if } \chi(p)=1, \\ (1-\frac{1}{p^2})^{-1}(1-\frac{1}{p^4})^{-1}(1-\frac{1}{p})^4 & \text{if } \chi(p)=-1. \end{cases} \tag{1.6}$$

It follows in particular that $\prod\sigma_p = 0$ if and only if there is some prime $p\mid\Delta$ with $\chi(p)=-1$ for which $E_p=0$.

It is perhaps worth observing that a notional application of the Hardy–Littlewood circle method to the system

$$L_i(x_1, x_2) = u_i^2 + v_i^2, \quad (1 \le i \le 4),$$

consisting of 4 equations in 10 variables predicts exactly the main term given in (1.4). In particular, the singular integral (the density for the real valuation) is $\pi^4 \operatorname{meas} \mathcal{R}$, and the 2-adic density

$$\lim_{n\to\infty} \#\{\mathbf{x}, \mathbf{u}, \mathbf{v} \pmod{2^n} : x_1 \equiv 1 \pmod 4,\ L_i(\mathbf{x}) \equiv u_i^2 + v_i^2 \pmod{2^n}\}$$

is 4.

To apply Theorem 1 to arithmetic progressions of length 4 we note that if 4 integers in arithmetic progression are each a sum of two squares, then the common difference must be a multiple of 4. Take

$$\mathcal{R} = \{(x_1, x_2) \in \mathbb{R}^2 : x_1, x_2 > 0,\ x_1 + 12x_2 < X\}$$

and

$$L_1(\mathbf{x}) = x_1, \quad L_2(\mathbf{x}) = x_1 + 4x_2, \quad L_3(\mathbf{x}) = x_1 + 8x_2, \quad L_4(\mathbf{x}) = x_1 + 12x_2.$$

Since $r(2n) = r(n)$ we see that

$$\sum_{a<b<c<d<X} r(a)r(b)r(c)r(d)$$
$$= \sum_k \sum_{\substack{2^k(x_1,x_2)\in\mathcal{R} \\ 2\nmid x_1}} r(L_1(\mathbf{x}))r(L_2(\mathbf{x}))r(L_3(\mathbf{x}))r(L_4(\mathbf{x})),$$

where the sum over $a, b, c, d$ is restricted to arithmetic progressions of length 4. Now if we set

$$\mathcal{R}_4(k) = \{(x_1, x_2) \in \mathbb{Z}^2 : 2^k(x_1, x_2) \in \mathcal{R},\ x_1 \equiv 1 \pmod 4\},$$

we see that

$$\sum_{a<b<c<d<X} r(a)r(b)r(c)r(d)$$
$$= \sum_k \sum_{(x_1,x_2)\in\mathcal{R}_4(k)} r(L_1(\mathbf{x}))r(L_2(\mathbf{x}))r(L_3(\mathbf{x}))r(L_4(\mathbf{x})).$$

We have sufficient uniformity in Theorem 1 to sum over $k$. Since $\operatorname{meas} \mathcal{R} = X^2/24$ and $\sum_0^\infty 4^{-k} = 4/3$, this therefore yields the asymptotic formula

$$\sum_{a<b<c<d<X} r(a)r(b)r(c)r(d) = CX^2 + O(X^2(\log X)^{-\eta/2}(\log\log X)^{15/4}),$$

Table 1

| $X$ | $S(X)$ | $S(X)/CX^2$ |
|---|---|---|
| 1000 | 21833216 | 21.833 ... |
| 2000 | 91315200 | 22.828 ... |
| 4000 | 381608960 | 23.850 ... |
| 8000 | 1554144256 | 24.283 ... |
| 16000 | 6308194304 | 24.641 ... |
| 32000 | 25428982272 | 24.832 ... |
| 64000 | 102495412736 | 25.023 ... |
| 128000 | 411816625664 | 25.135 ... |

where the sum over $a, b, c, d$ is restricted to arithmetic progressions of length 4. Since $\text{meas}\,\mathcal{R} = X^2/24$, the constant $C$ takes the form

$$C = 4\pi^4 \frac{1}{24}\frac{4}{3}\prod_{p\geq 3} E_p\{1-\chi(p)p^{-1}\}^4,$$

with $E_p$ given by (1.6) for $p \geq 5$. Moreover one may compute that

$$E_3 = \frac{27}{80}.$$

Since progressions with $d = X$ clearly contribute $O(X^{1+\varepsilon})$ for any $\varepsilon > 0$ we may summarize our conclusion as follows.

**Corollary 1** *There is a positive constant $C$ such that*

$$\sum_{a<b<c<d\leq X} r(a)r(b)r(c)r(d) = CX^2 + O(X^2(\log X)^{-\eta/2}(\log\log X)^{15/4}),$$

*where the sum over $a, b, c, d$ is restricted to arithmetic progressions of length 4. The constant $C$ has the approximate value* $25.3039\ldots$.

The corollary is illustrated by Table 1, in which

$$S(X) = \sum_{a<b<c<d<X} r(a)r(b)r(c)r(d).$$

The general problem as formulated above is relevant to a very different question. The simultaneous equations

$$V: \quad \begin{cases} L_1(x_1,x_2)L_2(x_1,x_2) = x_3^2 + x_4^2 \\ L_3(x_1,x_2)L_4(x_1,x_2) = x_5^2 + x_6^2 \end{cases} \tag{1.7}$$

will, in general, define a 3-fold in $\mathbb{P}^5$. We can estimate the number of rational points on this variety as $\mathbf{x}$ runs over a region $\mathcal{R}$ by examining the sum

$$\sum_{\mathbf{x}\in\mathcal{R}} r(L_1(\mathbf{x})L_2(\mathbf{x}))r(L_3(\mathbf{x})L_4(\mathbf{x})).$$

Varieties of the type (1.7) are of considerable interest, since they may fail to satisfy the Hasse Principle. Thus they may have no nontrivial rational points even though they have nonsingular points over $\mathbb{R}$ and each of the $p$-adic fields $\mathbb{Q}_p$. For general pairs of quadratic forms this observation is due to Iskovskih [6]. For varieties of the particular shape (1.7) the phenomenon is illustrated by the example

$$x_1x_2 = x_3^2 + x_4^2, \qquad (3x_1+4x_2)(8x_1+11x_2) = x_5^2 + x_6^2, \tag{1.8}$$

as we proceed to show. There are nonsingular points with $x_1 = x_2 = 1$ in $\mathbb{R}$ and in $\mathbb{Q}_p$ for every prime $p$ other than $p = 7$ and $p = 19$. Similarly for these two exceptional fields there are nonsingular points with $x_1 = 2$ and $x_2 = 1$. We proceed to assume that the equations (1.7) have a nonzero integral solution $x_1, \ldots, x_6$. In particular it follows that $x_1$ and $x_2$ cannot both be zero. For any $d \in \mathbb{N}$, if $nd^2$ is a sum of two squares, then $n$ is also a sum of two squares. Thus we may assume, without loss of generality, that $x_1$ and $x_2$ are coprime. Moreover, we may change the signs if necessary, so as to suppose that at least one of $x_1$ and $x_2$ is positive. Then, since their product is a sum of two squares, we see that the other must be nonnegative. It follows firstly that each of $x_1$ and $x_2$ is a sum of two squares, and secondly that each of $3x_1 + 4x_2$ and $8x_1 + 11x_2$ is strictly positive. Now

$$\begin{vmatrix} 3 & 4 \\ 8 & 11 \end{vmatrix} = 1,$$

so that $3x_1 + 4x_2$ and $8x_1 + 11x_2$ must be coprime. Thus both $3x_1 + 4x_2$ and $8x_1 + 11x_2$ will be sums of two squares.

Now if $x_1$ is odd, then $x_1 = a^2 + b^2 \equiv 1 \pmod 4$, so that we must have $3x_1 + 4x_2 \equiv 3 \pmod 4$. Thus $3x_1 + 4x_2$ cannot be a sum of two squares. Similarly if $x_1$ is even, then $x_2$ must be odd, and hence $x_2 \equiv 1 \pmod 4$, since $x_2$ is a sum of two squares. However this means that $8x_1 + 11x_2 \equiv 3 \pmod 4$ so that $8x_1 + 11x_2$ cannot be a sum of two squares. This completes the proof.

Even when the variety does possess rational points, it may fail to satisfy the weak approximation principle. In general, a variety $V$ is said to satisfy the *weak approximation principle* if its rational points are dense in the adélic points. To put this in concrete terms, for our variety (1.7), suppose we are given a real point $(x_1^{(\mathbb{R})}, \ldots, x_6^{(\mathbb{R})})$ and $p$-adic points $(x_1^{(p)}, \ldots, x_6^{(p)})$ for a finite number of distinct primes $p$, all lying on the variety (1.7). The weak approximation principle then asserts that, for any $\varepsilon > 0$, we can find a rational point $(x_1, \ldots, x_6)$ on (1.7) satisfying the simultaneous conditions

$$|x_i - x_i^{(\mathbb{R})}| < \varepsilon \quad \text{and} \quad |x_i - x_i^{(p)}|_p < \varepsilon, \quad (1 \le i \le 6)$$

for each of the primes $p$.

However it can happen that $V$ fails to satisfy even the real condition. In particular the variety may have two real components, on one of which the rational points are dense, and on the other of which there are no rational points. This is demonstrated by the example

$$x_1 x_2 = x_3^2 + x_4^2, \qquad (x_1 - x_2)(3x_1 - 8x_2) = x_5^2 + x_6^2, \tag{1.9}$$

due to Colliot-Thélène, Coray and Sansuc [2]. There is clearly a rational point with $x_1 = 1$ and $x_2 = 2$. Moreover the real points belong to two components, namely those with $x_2/x_1 \ge 1$ and $0 \le x_2/x_1 \le 3/8$. (We regard points with $x_1 = 0$ as being of the first type.) The special feature of this example is that all rational points lie on the first of these components. To prove this we shall suppose we have an integer point for which $0 \le x_2/x_1 \le 3/8$, and derive a contradiction. As with (1.8) we may assume that $x_1$ and $x_2$ are coprime and nonnegative, so that they must both be sums of two squares. Our assumption on the size of $x_2/x_1$ implies that $x_1 - x_2$ and $3x_1 - 8x_2$ are both nonnegative. Since

$$\begin{vmatrix} 1 & -1 \\ 3 & -8 \end{vmatrix} = -5,$$

the highest common factor of $x_1 - x_2$ and $3x_1 - 8x_2$ must be either 1 or 5. Thus, since the product of the linear forms $x_1 - x_2$ and $3x_1 - 8x_2$ is a sum of two squares, they must each be a sum of two squares.

Now if $x_1$ is odd, then $x_1 = a^2 + b^2 \equiv 1 \pmod 4$, so that we must have $3x_1 - 8x_2 \equiv 3 \pmod 4$. Thus $3x_1 - 8x_2$ cannot be a sum of two squares. Similarly if $2 \,\|\, x_1$ we will have $x_1 \equiv 2 \pmod 8$ and $3x_1 - 8x_2 \equiv 6 \pmod 8$, so that $3x_1 - 8x_2$ is not a sum of two squares. Finally, if $4 \mid x_1$, then $x_2$ is odd, and we will have $x_2 = c^2 + d^2 \equiv 1 \pmod 4$. In this case $x_1 - x_2 \equiv 3 \pmod 4$ and $x_1 - x_2$ cannot be a sum of two squares. This establishes our claim.

In general there is a heuristic expectation that the number of rational points on a given variety which lie in a large region should be given by a product of local densities. This is indeed the type of asymptotic formula that the Hardy–Littlewood circle method provides, in those cases for which the error terms can be successfully estimated. However when the rational points on a variety are not evenly distributed amongst the admissible adélic points, the entire rationale for this heuristic expectation breaks down. It is thus of considerable interest to estimate the number of points on such a variety, and to compare the result with that predicted from the product of local densities. This is what we shall do for the varieties (1.7).

We shall introduce the same type of normalization condition as before. Specifically, we require the following:

**Normalization Condition 2 (NC2)** *We assume:*

*(i) No two of the forms $L_1, \ldots, L_4$ are proportional.*

*(ii) We have*

$$\mathcal{R} = X\mathcal{R}^{(0)} = \{\mathbf{x} \in \mathbb{R}^2 : X^{-1}\mathbf{x} \in \mathcal{R}^{(0)}\},$$

*where $\mathcal{R}^{(0)} \subset \mathbb{R}^2$ is open, bounded and convex, with a piecewise continuously differentiable boundary, and where $X$ is a large positive parameter.*

*(iii) We have $L_i(\mathbf{x}) > 0$ for $1 \leq i \leq 4$ and for all $\mathbf{x} \in \mathcal{R}^{(0)}$.*

*(iv) We have*

$$L_1(x_1, x_2) \equiv L_2(x_1, x_2) \equiv \nu x_1 \pmod 4$$

*and*

$$L_3(x_1, x_2) \equiv L_4(x_1, x_2) \equiv \nu' x_1 \pmod 4,$$

*for appropriate $\nu, \nu' = \pm 1$.*

In connection with condition (iii) we note that the equations (1.7) do not require that $L_i(\mathbf{x}) > 0$. However, apart from $O(X)$ points where some $L_i$ vanishes, the solutions may be subdivided into regions in which each $L_i$ is one signed. On each such region we can then replace $L_i$ by $\pm L_i$ as necessary, so as to ensure that we have points with $L_i(\mathbf{x}) > 0$.

As with **NC1**, condition (iv) is imposed in order to simplify the exposition. However it may be viewed, as before, as being the result of restricting $\mathbf{x}$ to a suitable 2-adic region.

As an example, we note that the variety defined by (1.8) has a 2-adic point $x_1^{(0)}, \ldots, x_6^{(0)}$ with $x_1^{(0)} = x_2^{(0)} = 1$. The region given by $x_1 - x_2 \equiv x_1^{(0)} - x_2^{(0)} \equiv 0 \pmod 4$ is a 2-adic neighbourhood of the point $x_1^{(0)}, \ldots, x_6^{(0)}$. For any point in this neighbourhood we may write $x_1 = y_1$ and $x_2 = y_1 + 4y_2$ to produce the equations

$$y_1(y_1 + 4y_2) = x_3^2 + x_4^2, \qquad (7y_1 + 16y_2)(19y_1 + 44y_2) = x_5^2 + x_6^2. \qquad (1.10)$$

The linear forms now satisfy part (iv) of **NC2**.

Similarly for the example (1.9) we have a 2-adic point with $x_1^{(0)} = 1$ and $x_2^{(0)} = 2$, and we use the 2-adic region

$$x_2 - 2x_1 \equiv x_2^{(0)} - 2x_1^{(0)} \equiv 0 \pmod{8}.$$

We thus write $x_1 = y_1$ and $x_2 = 2y_1 + 8y_2$ to produce the equations

$$y_1(y_1 + 4y_2) = y_3^2 + y_4^2, \qquad (y_1 + 8y_2)(13y_1 + 64y_2) = x_5^2 + x_6^2, \tag{1.11}$$

all of whose rational points we have shown to satisfy $y_2/y_1 \geq -1/8$. Again the linear forms satisfy part (iv) of **NC2**.

In view of part (iv) of **NC2** it is natural to restrict consideration to the case in which $(x_1, x_2)$ lies in the set

$$\mathcal{R}_2 = \{\mathbf{x} \in \mathcal{R} : x_1 \equiv 1 \pmod{2}\}.$$

Our principal result describing the number of rational points on the general variety (1.7) is now as follows.

**Theorem 2** *Suppose* **NC2** *holds. The local densities for the variety $V$ with equations (1.7), for the set $\mathcal{R}_2$, are given by*

$$\sigma_\infty = \pi^2 \operatorname{meas} \mathcal{R}, \qquad \sigma_2 = 2$$

*and*

$$\sigma_p = (1 - \chi(p)/p)^2 T_\chi(p), \qquad (p \geq 3), \tag{1.12}$$

*where*

$$T_\chi(p) = E_p^{(0,0)} - \chi(p) E_p^{(0,1)} - \chi(p) E_p^{(1,0)} + E_p^{(1,1)} \tag{1.13}$$

*and*

$$E_p^{(u,v)} = \sum_{\alpha,\beta,\gamma,\delta=0}^{\infty} \chi(p)^{\alpha+\beta+\gamma+\delta} \rho(p^{\alpha+u}, p^{\beta+u}, p^{\gamma+v}, p^{\delta+v})^{-1}. \tag{1.14}$$

*Here $\rho(d_1, d_2, d_3, d_4)$ is as in Theorem 1. Moreover, when $p \nmid \Delta$ we have*

$$\sigma_p = (1 + \chi(p)/p)^2. \tag{1.15}$$

*If $\sigma_p = 0$ for any prime $p$ then $V$ has no rational point with $(x_1, x_2) \in \mathcal{R}_2$. If $\sigma_p \neq 0$ for every prime $p$, then*

$$\sum_{\mathbf{x} \in \mathcal{R}_2} r(L_1(\mathbf{x}) L_2(\mathbf{x})) r(L_3(\mathbf{x}) L_4(\mathbf{x})) = \{1 + \varepsilon\} \sigma_\infty \prod_p \sigma_p + o(X^2).$$

*where*

$$\varepsilon = \chi(\nu\nu') \prod_{p|\Delta,\, \chi(p)=-1} T_-(p)/T_+(p), \tag{1.16}$$

*with*

$$T_\pm(p) = E_p^{(0,0)} \pm E_p^{(0,1)} \pm E_p^{(1,0)} + E_p^{(1,1)}. \tag{1.17}$$

*Moreover, when* $p \equiv -1 \pmod 4$ *we have* $E_p^{(u,v)} \geq 0$*, so that*

$$|T_-(p)| \leq T_+(p).$$

*We also have* $E_p^{(1,0)} = E_p^{(0,1)} = 0$ *for any prime* $p \equiv -1 \pmod 4$ *not dividing* $\Delta_{12}\Delta_{34}$.

*If* $\varepsilon = -1$ *then* $V$ *has no rational point with* $(x_1, x_2) \in \mathcal{R}_2$.

Thus the factor $1+\varepsilon$ measures the discrepancy between the true asymptotic formula and the Hardy–Littlewood prediction. Although we shall not prove it here, we may remark that the sums $T_\pm(p)$ are always rational numbers, so that the factor $1+\varepsilon$ is a rational number in the range $[0,2]$.

We see that Theorem 2 establishes a local to global principle in the shape of the assertion that if $\sigma_p > 0$ for every $p$, then there exist rational points on $V$, providing that $1+\varepsilon \neq 0$. Moreover it is a standard fact that we will have $\sigma_p > 0$ for any prime for which $V$ has a nonsingular $p$-adic point. In contrast, our result does not give a full solution to the weak approximation problem, since we are unable to restrict the variables $x_3, x_4, x_5, x_6$ in (1.7). However, we are able to control the variables $x_1, x_2$ by our method.

In fact it is known that the Brauer–Manin obstruction is the only obstruction to both the Hasse Principle and Weak Approximation, for varieties of the form (1.7). Although this is not formally stated in the literature, it is possible to use a descent argument to reduce the problem to one involving a certain intersection of two quadrics in $\mathbb{P}^6$, to which Theorem 6.7 of Colliot-Thélène, Sansuc and Swinnerton-Dyer [3] may be applied. In particular it follows that our condition $1+\varepsilon > 0$ must be equivalent to the emptiness of the Brauer–Manin obstruction for the Hasse Principle.

In the final section of the paper we shall investigate the examples (1.8) and (1.9) more fully, as well as the variety

$$x_1(x_1+12x_2) = x_3^2 + x_4^2, \qquad (x_1+4x_2)(x_1+16x_2) = x_5^2 + x_6^2, \tag{1.18}$$

for which we shall show that $0 < 1+\varepsilon < 2$.

We conclude this introduction by remarking that it should be possible to replace the character $\chi$ by any other nonprincipal real character. Indeed one should be able to use different characters for each of the four linear forms in Theorem 1. In the same way, in Theorem 2 one would take any two nonprincipal real characters $\chi_1, \chi_2$. One would then hope to be able to replace the original expression $r(L_1(\mathbf{x})L_2(\mathbf{x}))r(L_3(\mathbf{x})L_4(\mathbf{x}))$ by $r_1(L_1(\mathbf{x})L_2(\mathbf{x}))r_2(L_3(\mathbf{x})L_4(\mathbf{x}))$, where

$$r_i(m) = 4\sum_{d|m} \chi_i(m) \quad (i = 1, 2).$$

If one also imposed congruence restrictions on the values of the forms $L_j(\mathbf{x})$, one would then be able to count the representations of $L_1(\mathbf{x})L_2(\mathbf{x})$ and $L_3(\mathbf{x})L_4(\mathbf{x})$ by individual genera of quadratic forms. However, while these generalizations look plausible, we have checked none of the details, and make no claim as to the results one might obtain.

# 2 The level of distribution

In this section we shall investigate the distribution of points $\mathbf{x}$ in subsets of $\mathcal{R}_4$, subject to a set of simultaneous divisibility conditions $d_i \mid L_i(\mathbf{x})$ for $1 \leq i \leq 4$. Naturally, we shall only be interested in odd values of $d_i$. If we write $\mathbf{d} = (d_1, d_2, d_3, d_4)$, it is clear that

$$\left\{\mathbf{x} \in \mathbb{Z}^2 : d_i \mid L_i(\mathbf{x}),\ 1 \leq i \leq 4\right\} = \Lambda_{\mathbf{d}},$$

say, is a lattice in $\mathbb{Z}^2$. We set

$$\rho(\mathbf{d}) = \det(\Lambda_{\mathbf{d}})$$

as in the statement of Theorem 1. We note that

$$\rho(\mathbf{d}) = [\mathbb{Z}^2 : \Lambda_{\mathbf{d}}] \mid d_1 d_2 d_3 d_4. \tag{2.1}$$

We shall consider convex regions $\mathcal{R}(\mathbf{d}) \subseteq \mathcal{R}$ for which $\mathcal{R}(\mathbf{d})$ is also the interior of a simple, piecewise continuously differentiable closed curve. We will write $\partial\mathcal{R}(\mathbf{d})$ for the length of the boundary curve defining $\mathcal{R}(\mathbf{d})$ and we set

$$\mathcal{R}_4(\mathbf{d}) = \left\{\mathbf{x} \in \mathcal{R}(\mathbf{d}) : x_1 \equiv 1 \pmod 4\right\}.$$

Since $\mathcal{R}(\mathbf{d}) \subseteq \mathcal{R} \subseteq [-cX, cX]^2$ for some constant $c$, by part (ii) of **NC1**, we deduce that

$$\partial\mathcal{R}(\mathbf{d}) \leq 8cX,$$

since $\mathcal{R}(\mathbf{d})$ is convex. We may now state our basic result on the level of distribution of a set of linear forms $L_i$.

**Lemma 2.1** *Let $Q_1, Q_2, Q_3, Q_4 \geq 2$, and write*

$$Q = \max Q_i \quad \textit{and} \quad V = Q_1 Q_2 Q_3 Q_4.$$

*Then there is an absolute constant $A$ such that*

$$\sum_{d_i \leq Q_i} \left| \#(\Lambda_{\mathbf{d}} \cap \mathcal{R}_4(\mathbf{d})) - \frac{\operatorname{meas}(\mathcal{R}(\mathbf{d}))}{4\rho(d)} \right| \ll (XV^{1/2} + XQ + V)(\log Q)^A,$$

*where the $d_i$ run over odd integers.*

A very similar result is proved by Daniel [4, Lemma 3.2], to which we refer the reader for details. As in [4, (3.11)] we find that

$$\left| \#(\Lambda_{\mathbf{d}} \cap \mathcal{R}(\mathbf{d})) - \frac{\operatorname{meas}(\mathcal{R}(\mathbf{d}))}{\rho(d)} \right| \ll \frac{\partial \mathcal{R}(\mathbf{d})}{|\mathbf{v}|} + 1 \ll \frac{X}{|\mathbf{v}|} + 1,$$

for some nonzero vector $\mathbf{v} \in \Lambda_{\mathbf{d}}$ with coprime coordinates, satisfying

$$|\mathbf{v}| \ll \det(\Lambda_{\mathbf{d}})^{1/2}.$$

By (2.1) we then deduce that $|\mathbf{v}| \ll V^{1/2}$. A trivial modification of Daniel's argument yields

$$\left| \#(\Lambda_{\mathbf{d}} \cap \mathcal{R}_4(\mathbf{d})) - \frac{\operatorname{meas}(\mathcal{R}(\mathbf{d}))}{4\rho(d)} \right| \ll \frac{X}{|\mathbf{v}|} + 1.$$

When none of the forms $L_i(\mathbf{v})$ vanish, we may estimate

$$\sum_{d_1, d_2, d_3, d_4 \leq Q} |\mathbf{v}|^{-1} \tag{2.2}$$

exactly as in [4, §3], giving a bound $O(V^{1/2}(\log Q)^A)$. However if $L_i(\mathbf{v}) = 0$ for some $i$ we must argue differently. (This situation does not arise in Daniel's work since he has an irreducible form $f$ of degree $k > 1$, so that $f(\mathbf{v})$ cannot vanish.) Since $\mathbf{v}$ has coprime coordinates, there can be only two possibilities for $\mathbf{v}$ for each value of $i$. Thus we will have $|\mathbf{v}| \ll 1$, with a constant depending only on the forms $L_i$. Moreover, if $L_i(\mathbf{v}) = 0$ we then have $0 \neq L_j(\mathbf{v}) \ll 1$ for $j \neq i$. Thus $d_i$ may take any value up to $Q$, while for $j \neq i$ there are only $O(1)$ available values for $d_j$. It follows that vectors $\mathbf{v}$ for which some $L_i(\mathbf{v})$ vanishes will contribute $O(Q_i)$ to (2.2). This is sufficient for Lemma 2.1.

# 3 The leading term

In this section we shall examine the dominant contribution to the sum $S$ given by (1.3). We shall use the fact that

$$r(n) = 4\sum_{d|n} \chi(d)$$

for any positive integer $n$, where

$$\chi(d) = \begin{cases} +1 & \text{if } d \equiv 1 \pmod 4, \\ -1 & \text{if } d \equiv 3 \pmod 4, \\ 0 & \text{if } d \equiv 0 \pmod 2. \end{cases}$$

Since $L_i(\mathbf{x}) > 0$ and $L_i(\mathbf{x}) \equiv 1 \pmod 4$ in our situation, we have

$$\begin{aligned} r(L_i(\mathbf{x})) = 4\sum_{d|L_i(\mathbf{x})} \chi(d) &= 4\sum_{\substack{d|L_i(\mathbf{x}) \\ d \le X^{1/2}}} \chi(d) + 4\sum_{\substack{d|L_i(\mathbf{x}) \\ d > X^{1/2}}} \chi(d) \\ &= 4\sum_{\substack{d|L_i(\mathbf{x}) \\ d \le X^{1/2}}} \chi(d) + 4\sum_{\substack{L_i(\mathbf{x})=ed \\ d > X^{1/2}}} \chi(d) \\ &= 4\sum_{\substack{d|L_i(\mathbf{x}) \\ d \le X^{1/2}}} \chi(d) + 4\sum_{\substack{L_i(\mathbf{x})=ed \\ d > X^{1/2}}} \chi(e) \\ &= 4\sum_{\substack{d|L_i(\mathbf{x}) \\ d \le X^{1/2}}} \chi(d) + 4\sum_{\substack{e|L_i(\mathbf{x}) \\ L_i(\mathbf{x}) > eX^{1/2}}} \chi(e) \\ &= 4A_+(L_i(\mathbf{x})) + 4A_-(L_i(\mathbf{x})), \end{aligned} \tag{3.1}$$

say. We shall use this decomposition for the terms corresponding to $L_1, L_2, L_3$, and for $L_4$ we shall write similarly

$$r(L_4(\mathbf{x})) = 4B_+(L_4(\mathbf{x})) + 4C(L_4(\mathbf{x})) + 4B_-(L_4(\mathbf{x})),$$

where

$$B_+(m) = \sum_{\substack{d|m \\ d \le Y}} \chi(d), \quad C(m) = \sum_{\substack{d|m \\ Y < d \le X/Y}} \chi(d), \quad \text{and} \quad B_-(m) = \sum_{\substack{e|m \\ m > eX/Y}} \chi(e). \tag{3.2}$$

Here $Y \le X^{1/2}$ is a parameter to be specified in due course. For the sums $A_-$ and $B_-$ we note that if $\mathbf{x}$ is confined to a region $\mathcal{R}$ satisfying part (iii) of **NC1**, then the variables $e$ which occur in the defining sums will satisfy $e \ll X^{1/2}$ and $e \ll Y$ in the two cases respectively.

We now write

$$S = \sum_{\mathbf{x} \in \mathcal{R}_4} r(L_1(\mathbf{x}))r(L_2(\mathbf{x}))r(L_3(\mathbf{x}))r(L_4(\mathbf{x}))$$

in the form

$$4S_+ + 4S_- + 4S_0,$$

where

$$S_\pm = \sum_{\mathbf{x} \in \mathcal{R}_4} r(L_1(\mathbf{x}))r(L_2(\mathbf{x}))r(L_3(\mathbf{x}))B_\pm(L_4(\mathbf{x}))$$

$$\text{and} \quad S_0 = \sum_{\mathbf{x} \in \mathcal{R}_4} r(L_1(\mathbf{x}))r(L_2(\mathbf{x}))r(L_3(\mathbf{x}))C(L_4(\mathbf{x})). \tag{3.3}$$

For the sums $S_\pm$ we shall use the decomposition (3.1) to produce a total of 8 subsums

$$S_{\pm,\pm,\pm,\pm} = \sum_{\mathbf{x} \in \mathcal{R}_4} A_\pm(L_1(\mathbf{x}))A_\pm(L_2(\mathbf{x}))A_\pm(L_3(\mathbf{x}))B_\pm(L_4(\mathbf{x})),$$

so that

$$S = 4S_0 + 4^4 \sum S_{\pm,\pm,\pm,\pm}. \tag{3.4}$$

We shall see later that $S_0$ is negligible. In this section we consider the remaining terms. Each of the sums $S_{\pm,\pm,\pm,\pm}$ is treated in the same way, so we shall consider the case of $S_{+,+,-,-}$, which is typical. We shall write $Q_1 = Q_2 = X^{1/2}$, and take

$$Q_3 = c_3 X^{1/2} \quad \text{and} \quad Q_4 = c_4 Y,$$

with suitable constants $c_3$ and $c_4$, so that the variables $e$ in the sums for $A_-(L_3(\mathbf{x}))$ and $B_-(L_4(\mathbf{x}))$ will satisfy $e \le Q_3$ and $e \le Q_4$ respectively. With this convention, the definitions of $A_\pm$ and $B_\pm$ show that

$$S_{+,+,-,-} = \sum_{d_i \le Q_i} \chi(d_1 d_2 d_3 d_4) \#(\Lambda_{\mathbf{d}} \cap \mathcal{R}_4(\mathbf{d})),$$

where

$$\mathcal{R}(\mathbf{d}) = \{\mathbf{x} \in \mathcal{R} : L_3(\mathbf{x}) > d_3 X^{1/2},\ L_4(\mathbf{x}) > d_4 X/Y\}. \tag{3.5}$$

Since these sets are convex, we conclude from Lemma 2.1 that

$$S_{+,+,-,-} = \frac{1}{4}\sum_{d_i\le Q_i} \chi(d_1d_2d_3d_4)\rho^{-1}(\mathbf{d})\,\mathrm{meas}(\mathcal{R}(\mathbf{d})) + O(\{X^{7/4}Y^{1/2}+X^{3/2}+X^{3/2}Y\}(\log X)^A).$$

Since $Y \le X^{1/2}$, the error term is $O(X^{7/4}Y^{1/2}(\log X)^A)$, which will be acceptable if we take

$$Y = X^{1/2}(\log X)^{-2A-2}, \tag{3.6}$$

as we now do. Thus for the general sum we have

$$S_{\pm,\pm,\pm,\pm} = \frac{1}{4}\sum_{d_i\le Q_i} \chi(d_1d_2d_3d_4)\rho^{-1}(\mathbf{d})\,\mathrm{meas}(\mathcal{R}(\mathbf{d})) + O(X^2(\log X)^{-1}). \tag{3.7}$$

We now consider the sum

$$\sum_{A_i<d_i\le B_i} \chi(d_1d_2d_3d_4)\rho^{-1}(\mathbf{d}), \tag{3.8}$$

where $B_i \le 2A_i$ for $1 \le i \le 4$. We may suppose without loss of generality that

$$A_4 \ge A_1, A_2, A_3. \tag{3.9}$$

We shall require some information on the function $\rho(\mathbf{d})$. By the Chinese Remainder Theorem there is a multiplicative property

$$\rho(d_1e_1,\ldots,d_4e_4) = \rho(d_1,\ldots,d_4)\rho(e_1,\ldots,e_4), \tag{3.10}$$

whenever

$$\mathrm{hcf}(d_1d_2d_3d_4, e_1e_2e_3e_4) = 1.$$

For most primes it is easy to handle the function $\rho$ explicitly. As in the introduction, we write $\Delta$ for the product of the 6 possible $2\times 2$ determinants $\Delta_{ij}$ formed from the various pairs $L_i, L_j$ of forms. Thus if $p$ is a prime which does not divide $\Delta$, then for any pair $i \ne j$, we see that $p \mid L_i(\mathbf{x}), L_j(\mathbf{x})$ implies $p \mid \mathbf{x}$. Hence if

$$p^{e_i} \mid L_i(\mathbf{x}) \quad (1 \le i \le 4) \tag{3.11}$$

for a prime $p \nmid \Delta$, and $e_{\sigma(1)} \ge e_{\sigma(2)} \ge e_{\sigma(3)} \ge e_{\sigma(4)}$ for some permutation $\sigma$, then (3.11) is equivalent to

$$p^{e_{\sigma(2)}} \mid \mathbf{x} \quad \text{and} \quad p^{e_{\sigma(1)}-e_{\sigma(2)}} \mid L_{\sigma(1)}(p^{-e_{\sigma(2)}}\mathbf{x}).$$

Thus

$$\rho(p^{e_1}, \dots, p^{e_4}) = p^{e_{\sigma(1)}+e_{\sigma(2)}}, \quad p \nmid \Delta. \tag{3.12}$$

For primes $p \mid \Delta$ we conclude similarly that

$$\rho(p^{e_1}, \dots, p^{e_4}) \gg_\Delta p^{e_{\sigma(1)}+e_{\sigma(2)}}. \tag{3.13}$$

Turning to (3.8) we set $f = d_1 d_2 d_3 \Delta$, and we write $d_4 = gh$, where

$$g = \prod_{p^e \,\|\, d_4,\, p|f} p^e, \quad \text{and} \quad (h, f) = 1.$$

Then

$$\sum_{A_4 < d_4 \le B_4} \chi(d_4)\rho^{-1}(\mathbf{d}) = \sum_{g \le B_4} \chi(g)\rho^{-1}(d_1, d_2, d_3, g) \sum_{\substack{A_4/g < h \le B_4/g \\ (h,f)=1}} \chi(h)\rho^{-1}(1, 1, 1, h).$$

In view of (3.12) we see that the inner sum is

$$\begin{aligned} \sum_{\substack{A_4/g < h \le B_4/g \\ (h,f)=1}} \chi(h)/h &= \sum_{d|f} \mu(d) \sum_{\substack{A_4/g < h \le B_4/g \\ d|h}} \chi(h)/h \\ &= \sum_{d|f} \mu(d)\chi(d)/d \sum_{A_4/gd < j \le B_4/gd} \chi(j)/j. \end{aligned}$$

However

$$\sum_{J < j \le K} \chi(j)/j \ll J^{-1},$$

so the sum above is $O(gf^\varepsilon A_4^{-1})$, for any $\varepsilon > 0$.

It follows that (3.8) is

$$\ll A_4^{-1} \sum_{d_1, d_2, d_3} \sum_{g \le B_4} (d_1 d_2 d_3)^\varepsilon g \rho^{-1}(d_1, d_2, d_3, g). \tag{3.14}$$

We shall estimate this sum by Rankin's method. For any fixed $\delta > 0$ we have

$$d_i^{\varepsilon} \ll d_i^{\delta} \ll A_i^{2\delta} d_i^{-\delta}$$

providing that $\varepsilon$ is small enough. Similarly we have

$$1 \ll A_4^{\delta} g^{-\delta}.$$

It follows that

$$\begin{aligned}
&\sum_{d_1,d_2,d_3} \sum_{g \le B_4} g(d_1 d_2 d_3)^{\varepsilon} \rho(d_1, d_2, d_3, g)^{-1} \\
&\quad \ll (A_1 A_2 A_3 A_4)^{2\delta} \sum_{d_1,d_2,d_3} \sum_{g \le B_4} g^{1-\delta} (d_1 d_2 d_3)^{-\delta} \rho(d_1, d_2, d_3, g)^{-1} \\
&\quad \ll (A_1 A_2 A_3 A_4)^{2\delta} \sum_{d_1,d_2,d_3=1}^{\infty} \sum_{g=1}^{\infty} g^{1-\delta} (d_1 d_2 d_3)^{-\delta} \rho(d_1, d_2, d_3, g)^{-1}, \quad (3.15)
\end{aligned}$$

where $g$ is still restricted to integers composed solely of prime factors $p$ dividing $f = \Delta d_1 d_2 d_3$. In view of the multiplicative property (3.10) we can factorize the 4-fold sum on the right. For each prime $p$ we write $d_1 = p^a$, $d_2 = p^b$, $d_3 = p^c$ and $g = p^d$, so that the corresponding factor is

$$\sum_{a,b,c,d=0}^{\infty} p^{d-(a+b+c+d)\delta} \rho(p^a, p^b, p^c, p^d)^{-1}, \quad (3.16)$$

subject to the condition that if $p \nmid \Delta$ then there are no terms with $a = b = c = 0$ and $d > 0$. For those primes $p$ which do not divide $\Delta$ the above sum is $1 + O(\Sigma_p)$, where $\Sigma_p$ is a sum of the form

$$\begin{aligned}
&\sum_{a=1}^{\infty} \sum_{0 \le b,c \le a} \sum_{d=0}^{\infty} p^{d-(a+b+c+d)\delta} \rho(p^a, p^b, p^c, p^d)^{-1} \\
&\qquad \le \sum_{a=1}^{\infty} \sum_{0 \le b,c \le a} \sum_{d=0}^{\infty} p^{d-(a+b+c+d)\delta} p^{-a-d} \\
&\qquad \le p^{-1-\delta} \left\{ \sum_{e=0}^{\infty} p^{-e\delta} \right\}^4 = O_{\delta}(p^{-1-\delta}),
\end{aligned}$$

by (3.12). The product of all such factors (3.16) is therefore $O_{\delta}(1)$. For the remaining primes we use (3.13) to show similarly that (3.16) is $O_{\delta,\Delta}(1)$. The 4-fold sum in (3.16) is therefore bounded, and on choosing $\delta = 1/10$, say, we see from (3.9) that (3.15) is $O(A_4^{4/5})$, and hence, from (3.14) that

$$\sum_{A_i < d_i \le B_i} \chi(d_1 d_2 d_3 d_4) \rho^{-1}(\mathbf{d}) \ll (A_1 A_2 A_3 A_4)^{-1/20}.$$

We may now use repeated summation by parts to show that

$$\sum_{d_i \le A_i} \chi(d_1 d_2 d_3 d_4) \rho^{-1}(\mathbf{d})(d_1 d_2 d_3 d_4)^{-\delta} = S(\delta) + O((\min A_i)^{-1/20}) \qquad (3.17)$$

uniformly for $\delta > 0$, with

$$S(\delta) = \sum_{d_1,d_2,d_3,d_4=1}^{\infty} \chi(d_1 d_2 d_3 d_4) \rho^{-1}(\mathbf{d})(d_1 d_2 d_3 d_4)^{-\delta}.$$

The sum $S(\delta)$ is absolutely convergent for such $\delta$. Indeed by (3.10) it suffices to consider the behaviour of the various Euler factors. For each prime the corresponding factor is

$$\sum_{a,b,c,d=0}^{\infty} \chi(p)^{a+b+c+d} p^{-(a+b+c+d)\delta} \rho(p^a, p^b, p^c, p^d)^{-1} = E_p(\delta), \qquad (3.18)$$

say. We write this in the form $1 + \Sigma$ where

$$\Sigma \ll \sum_{a=1}^{\infty} \sum_{b,c,d=0}^{\infty} p^{-a-(a+b+c+d)\delta} \ll p^{-1-\delta},$$

by (3.12) and (3.13). This suffices to ensure absolute convergence for $\delta > 0$. Similarly, when $p \nmid \Delta$ we have $\rho(p, 1, 1, 1) = p$ by (3.12), whence

$$E_p(\delta) = 1 + 4\chi(p)/p^{-1-\delta} + O(p^{-2}) = \{1 - \chi(p)/p^{1+\delta}\}^{-4}\{1 + O(p^{-2})\},$$

uniformly for $\delta > 0$. It follows that we can write $S(\delta) = L(1+\delta, \chi)^4 F(1+\delta)$ where

$$F(s) = \prod_p E_p(s-1)\{1 - \chi(p)p^{-s}\}^4 \qquad (3.19)$$

is absolutely and uniformly convergent for $\mathrm{Re}(s) \ge 1$. This allows us to take the limit in (3.17) as $\delta$ tends to zero, so that

$$\sum_{d_i \le A_i} \chi(d_1 d_2 d_3 d_4) \rho^{-1}(\mathbf{d}) = \left(\frac{\pi}{4}\right)^4 F(1) + O((\min A_i)^{-1/20}).$$

It remains to introduce the factor $\mathrm{meas}(\mathcal{R}(\mathbf{d}))$ into this sum, which we proceed to do via partial summation. Recall that we are working with the example (3.5). For ease of notation we shall set $d_3 = x, d_4 = y$ and $f(x, y) =$

meas$(\mathcal{R}(\mathbf{d}))$. Then

$$\sum_{d_i\le Q_i}\chi(d_1d_2d_3d_4)\rho^{-1}(\mathbf{d})\,\mathrm{meas}(\mathcal{R}(\mathbf{d}))=$$

$$\int_0^{Q_3}\int_0^{Q_4}f_{xy}(x,y)\sum_{\substack{d_1\le Q_1,\,d_2\le Q_2\\ d_3\le x,\,d_4\le y}}\chi(d_1d_2d_3d_4)\rho^{-1}(\mathbf{d})dxdy$$

by partial summation, on noting that $f(Q_3,y)=f(x,Q_4)=0$ for all $x,y$. We therefore obtain

$$S_{+,+,-,-}=\frac{1}{4}\left(\frac{\pi}{4}\right)^4F(1)\,\mathrm{meas}\,\mathcal{R}+O\Big(\int_0^{Q_3}\int_0^{Q_4}|f_{xy}(x,y)|(\min(x,y))^{-1/20}\Big).$$

However $F_{xy}(x,y)\ll X^2/Q_3Q_4$, as one sees from (3.15). Hence the error term above is $O(X^2(\min Q_i)^{-1/20})$. We therefore deduce that

$$S_{+,+,-,-}=\frac{1}{4}\left(\frac{\pi}{4}\right)^4F(1)\,\mathrm{meas}\,\mathcal{R}+O(X^{79/40}(\log X)^A),$$

and similarly for each of the sums $S_{\pm,\pm,\pm,\pm}$. If we now refer to (3.4) and (3.7), we may conclude as follows.

**Lemma 3.1** *We have*

$$S=4\pi^4F(1)\,\mathrm{meas}\,\mathcal{R}+4S_0+O(X^2(\log X)^{-1}),$$

*where $F(1)$ is given by (3.18) and (3.19), and $S_0$ is given by (3.3).*

# 4 The sum $S_0$—first steps

Clearly we have

$$\begin{aligned} S_0 &\ll \sum_{\mathbf{x}\in\mathcal{R}_4} r(L_1(\mathbf{x}))r(L_2(\mathbf{x}))r(L_3(\mathbf{x}))|C(L_4(\mathbf{x}))| \\ &= \sum_{m\in\mathcal{B}} S_0(m)|C(m)|, \end{aligned} \tag{4.1}$$

where

$$\mathcal{B}=\left\{m\in\mathbb{Z}:\begin{matrix}\exists d\mid m\text{ s.t.}\\ Y<d\le X/Y\end{matrix}\right\}\cap\left\{m\in\mathbb{Z}:\begin{matrix}\exists\mathbf{x}\in\mathcal{R}_4\text{ s.t.}\\ L_4(\mathbf{x})=m\end{matrix}\right\} \tag{4.2}$$

and

$$S_0(m) = \sum_{\mathbf{x} \in \mathcal{A}(m)} r(L_1(\mathbf{x}))r(L_2(\mathbf{x}))r(L_3(\mathbf{x}))$$

with

$$\mathcal{A}(m) = \{\mathbf{x} \in \mathcal{R}_4 : L_4(\mathbf{x}) = m\}.$$

Suppose that the forms $L_i$ are given by

$$L_i(x_1, x_2) = A_i x_1 + B_i x_2, \quad (1 \le i \le 4). \tag{4.3}$$

We have arranged that $L_i(x_1, x_2) \equiv 1 \pmod 4$ whenever we have $x_1, x_2 \in \mathbb{Z}$ and $x_1 \equiv 1 \pmod 4$. It follows that $A_i \equiv 1 \pmod 4$ and $B_i \equiv 0 \pmod 4$. In particular $A_i \ne 0$. If we now substitute $m = L_4(\mathbf{x})$ for $x_1$, so that $x_1 = (m - B_4 x_2)/A_4$, and write $x_2 = n$ for ease of notation, we find that

$$L_i(\mathbf{x}) = \frac{a_i m + b_i n}{A_4} = L_i'(m, n),$$

say, where

$$a_i = A_i, \quad b_i = A_4 B_i - B_4 A_i, \quad (1 \le i \le 3).$$

Thus we have

$$a_i \equiv 1 \pmod 4, \quad b_i \equiv 0 \pmod 4, \quad (1 \le i \le 3). \tag{4.4}$$

Note that, as $\mathbf{x}$ runs over $\mathbb{Z}^2$, not every value $m \in \mathbb{Z}$ need occur. Indeed, since $x_1 \equiv 1 \pmod 4$ we will have $m \equiv 1 \pmod 4$. We also observe that if $\mathbf{x}$ runs over $\mathcal{R}$, then the corresponding values of $m$ and $n$ will satisfy $m, n \ll X$. Finally we note that we can clear the denominator in $L_i'$, so that $r(L_i'(m,n)) \le r(A_4(a_i m + b_i n))$.

We now write

$$H = 6\Delta A_4^3 \prod_{1 \le i \le 3} b_i.$$

This will be nonzero since no two of the original forms $L_1, \ldots, L_4$ were proportional. We also define a multiplicative function $r_1(n)$ by setting

$$r_1(p^e) = \begin{cases} (e+1)^3, & \text{if } p \mid H \text{ or } e \ge 2, \\ 1 + \chi(p), & \text{otherwise.} \end{cases}$$

Using the multiplicative property of the function $r(n)$ one can then verify that

$$r(L_1'(m,n))r(L_2'(m,n))r(L_3'(m,n)) \le 64 r_1(F(n)),$$

where

$$F(n) = A_4^3 \prod_{i=1}^{3} (a_i m + b_i n). \tag{4.5}$$

Our principal tool in handling $S_0(m)$ will be a theorem of Nair [7], which will provide an upper bound of the correct order of magnitude. In order to apply Nair's result we must remove fixed prime factors from $F$. Thus we first write $F(X) = cG(X)$, where $G(X)$ is a primitive integer polynomial, and $c \mid H$. It follows that $r_1(F(n)) \ll r_1(G(n))$. The only fixed prime factors that a primitive cubic polynomial can have are $p = 2$ and $p = 3$. However since $m \equiv 1 \pmod 4$ we see from (4.5) that $p = 2$ can never divide $G(n)$. If $G(X)$ has $p = 3$ as a fixed prime divisor then $G(X) \equiv \pm(X^3 - X) \pmod 3$. Thus if we split the integers $n$ into the three possible congruence classes $n \equiv n_0 \pmod 3$, and write $n = 3\widehat{n} + n_0$ we see that

$$\frac{G(n)}{3} = 9\frac{G'''(n_0)}{6}\widehat{n}^3 + 3\frac{G''(n_0)}{2}\widehat{n}^2 + G'(n_0)\widehat{n} + \frac{1}{3}G(n_0) = \widehat{G}(\widehat{n}),$$

say. Since $G'(n_0) \equiv \mp 1 \pmod 3$ we see that $\widehat{G}$ does not have $p = 3$ as a fixed prime divisor. Thus, by splitting the range for $n$ into three congruence classes if necessary, we can produce a polynomial with no fixed prime divisor.

We now state the following special case of Nair's theorem [7].

**Lemma 4.1** *Let $f(n)$ be a nonnegative multiplicative function satisfying the bound $f(p^e) \leq (e+1)^4$ for every prime power $p^e$. Let $G(X) \in \mathbb{Z}[X]$ be a polynomial of degree at most 4, without repeated roots, and with no fixed prime factor. Write $\rho(p)$ for the number of roots of $G$ modulo $p$, and $||G||$ for the sum of the moduli of the coefficients of $G$. Then for any $\delta > 0$ there is a constant $c_\delta$ such that*

$$\sum_{n \leq N,\, G(n) > 0} f(G(n)) \ll_\delta N \prod_{p \leq N} \left(1 - \frac{\rho(p)}{p}\right) \exp\left(\sum_{p \leq N} \frac{f(p)\rho(p)}{p}\right),$$

*for $N \geq c_\delta ||G||^\delta$.*

For our application the range for $n$ will be an interval of length $N \ll X$, which will have to be translated by a distance $O(X)$ in order to produce the interval $(0, N]$. This has the effect of modifying the coefficients of the original polynomial $G$. However even after this translation we will have $||G|| \ll X^3$. Given the form (4.5) of $F$ we see that $G$ will have three linear factors. Moreover we have $\rho(p) = 1$ for $p \mid m$, while if $p \nmid mH$ we will have $\rho(p) = 3$, since $p \mid a_ib_j - a_jb_i$ would imply $p \mid \Delta$. We will therefore have

$$\begin{aligned}
S_0(m) &\ll \sum_{n \le N,\, G(n)>0} r_1(G(n)) \\
&\ll N \prod_{p \le N}\left(1 - \frac{\rho(p)}{p}\right) \exp\left(\sum_{p \le N} \frac{r_1(p)\rho(p)}{p}\right) \\
&\ll N \prod_{3<p \le N}\left(1 - \frac{\rho(p)}{p}\right) \exp\left(\sum_{p \le N} \frac{3r_1(p)}{p}\right) \\
&\ll N \prod_{p \mid m,\, p>3} \frac{1-1/p}{1-3/p} \prod_{3<p \le N}\left(1 - \frac{3}{p}\right) \exp\left(\sum_{p \le N} \frac{3r_1(p)}{p}\right) \\
&\ll N\left(\frac{\sigma(m)}{m}\right)^2 \\
&\ll N(\log\log N)^2, \qquad (4.6)
\end{aligned}$$

providing that $N \gg_\delta X^{3\delta}$. (Here $\sigma(m)$ is the usual sum of divisors function.) Since we trivially have $r_1(G(n)) \ll X^{1/2}$ we see on taking $\delta = 1/6$ that $S_0(m) \ll X(\log\log X)^2$ whether $N \gg X^{1/2}$ or not. We therefore deduce the following result from (4.1).

**Lemma 4.2** *We have*

$$S_0 \ll X(\log\log X)^2 \sum_{m \in \mathcal{B}} |C(m)|,$$

*where $\mathcal{B}$ and $C(m)$ are given by (4.2) and (3.2) respectively.*

# 5 Completion of the proof of Theorem 1

Cauchy's inequality shows that

$$\sum_{m \in \mathcal{B}} |C(m)| \le (\#\mathcal{B})^{1/2}\Big(\sum_{1 \le m \ll X} |C(m)|^2\Big)^{1/2}. \qquad (5.1)$$

However it is clear that if we let $M$ and $D$ run over powers of 2, then

$$\begin{aligned}
\#\mathcal{B} &\le \#\{m \ll X : \exists d \mid m,\ Y < d \le X/Y\} \\
&\ll \log(X/Y^2) \sum_M \#\{M < m \le 2M : \exists d \mid m,\ D < d \le 2D\} \qquad (5.2)
\end{aligned}$$

for some $D$ in the range $Y \ll D \ll X/Y$. Clearly we may replace $d$ by $m/d$, so that

$$\#\{M < m \le 2M : \exists d \mid m,\ D < d \le 2D\} \\ \le \#\{M < m \le 2M : \exists d \mid m,\ M/2D < d < 2M/D\}.$$

Now we may apply the following result.

**Lemma 5.1** *We have*

$$\#\{n \le x : \exists d \mid n,\ y < d \le 2y\} \ll \frac{x}{(\log y)^{\eta}(\log\log y)^{1/2}}$$

*uniformly for* $3 \le y \le x$*, where* $\eta$ *is given by (1.5).*

This is the case $u = 1$, $\beta = 0$ of Theorem 21, part (ii) in Hall and Tenenbaum, see [5, (2.2) and (2.3)].

Lemma 5.1 yields

$$\#\{M < m \le 2M : \exists d \mid m,\ D < d \le 2D\} \ll \frac{M}{(\log X)^{\eta}(\log\log X)^{1/2}}$$

whenever $M \ge X^{3/4}$. For smaller values of $M$ we merely use the trivial bound $O(M)$. Then (5.2) and (3.6) imply that

$$\#\mathcal{B} \ll X(\log X)^{-\eta}(\log\log X)^{1/2}. \tag{5.3}$$

It remains to consider

$$\sum_{1 \le m \le cX} |C(m)|^2,$$

for a suitable constant $c$. We expand the term $|C(m)|^2$ and write $(d_1, d_2) = h$ and $d_i = hk_i$ to produce

$$\begin{aligned} \sum_{1 \le m \le cX} |C(m)|^2 &= \sum_{d_1, d_2 \in (Y, X/Y]} \chi(d_1 d_2)\#\{m \le cX : [d_1, d_2] \mid m\} \\ &= \sum_{h \le X/Y} \sum_{\substack{k_1, k_2 \in (Y/h, X/Yh] \\ (k_1, k_2) = 1}} \chi(h^2 k_1 k_2)\#\{m \le cX : hk_1k_2 \mid m\} \\ &= \sum_{h \le X/Y} \sum_{\substack{k_1, k_2 \in (Y/h, X/Yh] \\ (k_1, k_2) = 1}} \chi(h^2 k_1 k_2)\#\{n \le cX/hk_1k_2\} \\ &= \sum_{h \le X/Y} \sum_{k_1 \in (Y/h, X/Yh]} \chi(h^2 k_1) \sum_{n \le \min(cX/Yk_1, cX/hk_1)} \sum_{k_2} \chi(k_2), \end{aligned} \tag{5.4}$$

where the innermost sum in the final expression is subject to the conditions $Y/h < k_2 \le \min(X/Yh, cX/hk_1n)$ and $(k_2, k_1) = 1$.

In general we have

$$\begin{aligned}
\sum_{k\le K,\,(k,s)=1} \chi(k) &= \sum_{d|s} \mu(d) \sum_{k\le K,\, d|k} \chi(k) \\
&= \sum_{d|s} \mu(d)\chi(d) \sum_{j\le K/d} \chi(j) \\
&\ll \sum_{d|s} |\mu(d)\chi(d)| \\
&\ll \tau(s),
\end{aligned}$$

where $\tau$ is the usual divisor function. Inserting this bound into (5.4) we deduce that

$$\begin{aligned}
\sum_{1\le m\le cX} |C(m)|^2 &\ll \sum_{h\le X/Y} \sum_{k_1\in(Y/h,X/Yh]} \sum_{n\le\min(cX/Yk_1,cX/hk_1)} \tau(k_1) \\
&\ll \sum_{h\le X/Y} \sum_{k_1\in(Y/h,X/Yh]} \min\Big(\frac{X}{Yk_1}, \frac{X}{hk_1}\Big)\tau(k_1) \\
&= \sum_{h\le X/Y} \min\Big(\frac{X}{Y}, \frac{X}{h}\Big) \sum_{k_1\in(Y/h,X/Yh]} \tau(k_1)/k_1 \\
&\ll \sum_{h\le X/Y} \min\Big(\frac{X}{Y}, \frac{X}{h}\Big) \log^2(X/Yh) \\
&\ll XY^{-1}\sum_{h\le Y} \log^2(X/Yh) + X \sum_{Y<h\le X/Y} h^{-1}\log^2(X/Yh) \\
&\ll X\log^2(XY^{-2}) + X\log^3(XY^{-2}).
\end{aligned}$$

Our choice (3.6) of $Y$ then ensures that

$$\sum_{1\le m\le cX} |C(m)|^2 \ll X(\log\log X)^3,$$

so that (5.1), (5.2) and Lemma 4.2 produce the bound

$$S_0 \ll X^2(\log X)^{-\eta/2}(\log\log X)^{15/4}.$$

This suffices, in conjunction with Lemma 3.1, for Theorem 1.

## 6 Proof of Theorem 2—preliminaries

Our starting point for the proof of Theorem 2 is the identity

$$r(mn) = \frac{1}{4}\sum_{d|m,n} \mu(d)\chi(d)r(m/d)r(n/d),$$

valid for any positive integers $m, n$. This identity allows us to pass from a problem about solutions of a single equation $mn = r^2 + s^2$ to one which involves a series of systems $m = d(t^2 + u^2)$, $n = d(v^2 + w^2)$ for varying $d$. One can think of this as corresponding to a simple 'descent' process.

In view of part (iii) of **NC2**, we may take $m = L_1, n = L_2$, or alternatively $m = L_3, n = L_4$ in the above identity. Thus, if

$$S = \sum_{\mathbf{x}\in\mathcal{R}_2} r(L_1(\mathbf{x})L_2(\mathbf{x}))r(L_3(\mathbf{x})L_4(\mathbf{x})),$$

we have

$$S = \frac{1}{16}\sum_{d,d'} \mu(d)\mu(d')\chi(dd') \times$$
$$\sum_{\mathbf{x}\in\mathcal{R}_2} r(L_1(\mathbf{x})/d)r(L_2(\mathbf{x})/d)r(L_3(\mathbf{x})/d')r(L_4(\mathbf{x})/d'),$$

where we set $r(q) = 0$ if $q$ is not an integer. Since $L_i$ is always odd for $\mathbf{x} \in \mathcal{R}_2$, part (iv) of **NC2** shows that we must have $x_1 \equiv \nu d \pmod 4$ if $r(L_1/d) \neq 0$, and similarly $x_1 \equiv \nu' d' \pmod 4$ if $r(L_3/d) \neq 0$. In particular, only terms for which $dd' \equiv \nu\nu' \pmod 4$ make a nonzero contribution, so that

$$S = \frac{\chi(\nu\nu')}{16} \sum_{dd'\equiv\nu\nu' \pmod 4} \mu(d)\mu(d')S(d,d'), \tag{6.1}$$

where

$$S(d,d') = \sum_{\mathbf{x}\in\mathcal{R},\, x_1\equiv\nu d \pmod 4} r(L_1(\mathbf{x})/d)r(L_2(\mathbf{x})/d)r(L_3(\mathbf{x})/d')r(L_4(\mathbf{x})/d').$$

Henceforth we shall assume, as we clearly may, that $d$ and $d'$ are both odd.

We now show that it suffices to establish an asymptotic formula for each individual sum $S(d, d')$.

**Lemma 6.1** *Suppose that*

$$S(d,d') \ll X^2\tau(d)^5\tau(d')^5[d,d']^{-2} \tag{6.2}$$

*uniformly for all square-free $d, d'$, where $[d, d']$ denotes the least common multiple of $d$ and $d'$. Assume further that*

$$S(d,d') = C(d,d')\,\mathrm{meas}\,\mathcal{R} + o(X^2) \tag{6.3}$$

*for all fixed square-free $d, d'$, and that*

$$C(d,d') \ll \tau(d)^5\tau(d')^5[d,d']^{-2} \tag{6.4}$$

*for square-free $d, d'$. Then, under* **NC2**, *we have*

$$S = C \operatorname{meas} \mathcal{R} + o(X^2), \tag{6.5}$$

*with*

$$C = \frac{\chi(\nu\nu')}{16} \sum_{dd' \equiv \nu\nu' \pmod 4} \mu(d)\mu(d')C(d, d'). \tag{6.6}$$

Notice that we do not require any uniformity in $d, d'$ for (6.3). It suffices that (6.3) should hold for each fixed pair $d, d'$.

To prove the lemma we set

$$E(d, d'; X) = X^{-2}|S(d, d') - C(d, d') \operatorname{meas} \mathcal{R}|,$$

so that (6.2) and (6.4) yield

$$E(d, d'; X) \ll \tau(d)^5 \tau(d')^5 [d, d']^{-2}$$

uniformly in $X$. On the other hand, for fixed $d, d'$ we will have $E(d, d'; X) \to 0$ as $X \to \infty$. The required result will therefore follow from the dominated convergence of the double sum

$$\sum_{d,d'=1}^{\infty} E(d, d'; X),$$

providing that we can show that

$$\sum_{d,d'=1}^{\infty} \tau(d)^5 \tau(d')^5 [d, d']^{-2}$$

converges. However if we set $(d, d') = h$ and $d = hk, d' = hk'$ we will have

$$\sum_{d,d'=1}^{\infty} \tau(d)^5 \tau(d')^5 [d, d']^{-2} \le \sum_{h,k,k'=1}^{\infty} \tau(k)^5 \tau(k')^5 \tau(h)^{10} (hkk')^{-2},$$

and the required result follows.

We now establish the bound (6.2), using Nair's result, Lemma 4.1. We begin by writing $\Delta$ for the product of the 6 possible $2 \times 2$ determinants formed from the various pairs $L_i, L_j$ of forms, as previously. Thus if $p$ is a prime which does not divide $\Delta$, then $p \mid L_i(\mathbf{x}), L_j(\mathbf{x})$ implies $p \mid \mathbf{x}$, providing that $i \neq j$. We shall put $e = (d, \Delta), e' = (d', \Delta)$ and $f = d/e, f' = d'/e'$. If $d, d'$ are square-free, we see that $e$ and $f$ are square-free and that $(f, \Delta) = 1$. Similarly $e'$ and $f'$ are square-free and $(f', \Delta) = 1$. The condition $d \mid L_1(\mathbf{x}), L_2(\mathbf{x})$ now implies

$f \mid \mathbf{x}$, while $d' \mid L_3(\mathbf{x}), L_4(\mathbf{x})$ implies $f' \mid \mathbf{x}$. We therefore set $\mathbf{x} = g\mathbf{y}$, where $g = [f, f']$ is the lowest common multiple of $f$ and $f'$. We shall henceforth assume that $g \ll X$, as we clearly may. It now follows that

$$S(d,d') \leq \sum_{\mathbf{y}} r(gL_1(\mathbf{y})/d)r(gL_2(\mathbf{y})/d)r(gL_3(\mathbf{y})/d')r(gL_4(\mathbf{y})/d'),$$

where the sum is for vectors $\mathbf{y}$ such that $g\mathbf{y} \in \mathcal{R}$ and $y_1 \equiv g\nu d \pmod 4$. If the forms $L_i$ are given by (4.3), we conclude, using part (iv) of **NC2**, that $A_i \neq 0$ for $1 \leq i \leq 4$. We proceed to define a multiplicative function $r_2(n)$ by setting

$$r_2(p^e) = \begin{cases} 1 + \chi(p) & \text{if } p \nmid 3dd' \prod A_i \text{ and } e = 1, \\ (1+e)^4 & \text{otherwise.} \end{cases}$$

Then

$$r(gL_1(\mathbf{y})/d)r(gL_2(\mathbf{y})/d)r(gL_3(\mathbf{y})/d')r(gL_4(\mathbf{y})/d') \\ \leq 4^4\tau(g)^4 r_2\big(L_1(\mathbf{y})L_2(\mathbf{y})L_3(\mathbf{y})L_4(\mathbf{y})\big).$$

Moreover, if we regard $y_2$ as fixed and set $F(X) = \prod L_i(X, y_2)$, we will have $F(X) = cG(X)$ for some primitive quartic polynomial $G(X)$, with $c \mid \prod A_i$. Since we are taking the forms $L_i$ to be fixed, it follows that

$$r(gL_1(\mathbf{y})/d)r(gL_2(\mathbf{y})/d)r(gL_3(\mathbf{y})/d')r(gL_4(\mathbf{y})/d') \ll \tau(g)^4 r_2(G(y_1)).$$

We intend to apply Lemma 4.1, and we therefore investigate possible fixed prime factors $p$ of $H(X) = G(2X+1)$. Since $G$ is quartic and primitive we must have $p = 2$ or $p = 3$. However, for $y_1 \equiv g\nu d \pmod 4$, we see from part (iv) of **NC2** that $F(y_1)$, and hence also $G(y_1)$, must be odd. Thus $H(0) = G(1)$ is odd. There remains the case $p = 3$. Suppose that $3 \mid H(n)$ for all $n \in \mathbb{Z}$. We split the available $y$ into congruence classes modulo 3 and consider the three polynomials

$$H_j(X) = \frac{H(3X+j)}{3}, \quad (j = 0, 1, 2).$$

Clearly the only possible fixed prime factor of $H_j$ is $p = 3$. We claim that if $H_j$ does have 3 as a fixed prime factor, then $H_j$ is divisible by 3 as a polynomial. Moreover, if we then put $H_j(X) = 3K_j(X)$ we claim that $H_j$ does not have 3 as a fixed prime factor. To prove these assertions, suppose that there is some $j$ such that $3 \mid H_j(n)$ for all $n \in \mathbb{Z}$. Then $9 \mid H(3n+j)$, whence $9 \mid H(j) + 3nH'(j)$ for every $n$. It follows that $9 \mid H(j)$ and $3 \mid H'(j)$ so that 9 divides the polynomial $H(3X+j)$. Thus $3 \mid H_j(X)$ as claimed. Moreover, if $9 \mid H_j(n)$ for every $n$, then $27 \mid H(3n+j)$, whence $27 \mid H(j) + 3nH'(j) + 9n^2H''(j)/2$. From this we deduce that $3 \mid H''(j)$. However we then see that

$$\begin{aligned} H(m+j) &= H(j) + mH'(j) + m^2\frac{H''(j)}{2} + m^3\frac{H^{(3)}(j)}{6} + m^4\frac{H^{(4)}(j)}{24} \\ &\equiv m^3\frac{H^{(3)}(j)}{6} + m^4\frac{H^{(4)}(j)}{24} \pmod 3. \end{aligned}$$

This produces a contradiction, since we are supposing that $H(X)$ is primitive and has 3 as a fixed prime factor.

It therefore follows that we may replace $H(X)$ if necessary by a set of 3 polynomials $H_j(X)$ or $K_j(X)$ which have no fixed prime divisor. Moreover $r_2(H(3n+j)) \le r_2(3)r_2(H_j(n))$ and $r_2(H(3n+j)) \le r_2(9)r_2(K_j(n))$, so that only a factor $O(1)$ is lost. Now, if

$$S(y_2) = \sum_y r(gL_1(y,y_2)/d)r(gL_2(y,y_2)/d)r(gL_3(y,y_2)/d')r(gL_4(y,y_2)/d'),$$

where the sum over $y$ is subject to $g(y,y_2) \in \mathcal{R}$ and $y \equiv g\nu d \pmod 4$, we find from Lemma 4.1 that if $y_2 \neq 0$, then

$$\begin{aligned} S(y_2) &\ll \frac{X}{g}\tau(g)^4 \prod_{p\le X}\Big(1-\frac{\rho(p)}{p}\Big)\exp\Big(\sum_{p\le X}\frac{r_2(p)\rho(p)}{p}\Big) \\ &\ll \frac{X}{g}\tau(g)^4 \prod_{5<p\le N}\Big(1-\frac{4}{p}\Big)\exp\Big(\sum_{p\le X}\frac{4r_2(p)}{p}\Big)\exp\Big(\sum_{p\mid dd'y_2}\frac{64}{p}\Big) \\ &\ll \frac{X}{g}\tau(g)^4\Big(\frac{\sigma(dd')}{dd'}\Big)^{64}\Big(\frac{\sigma(|y_2|)}{|y_2|}\Big)^{64}, \end{aligned}$$

as in (4.6). We trivially have

$$S(0) \ll \sum_{y\ll X/g} \tau(y)^4 \ll X^2g^{-2}.$$

We therefore deduce that

$$\begin{aligned} S(d,d') &\ll X^2g^{-2} + Xg^{-1}\tau(g)^4\tau(dd') \sum_{1\le y_2\ll X/g} \Big(\frac{\sigma(|y_2|)}{|y_2|}\Big)^{64} \\ &\ll X^2g^{-2}\tau(g)^4\tau(dd'). \end{aligned}$$

Since $g|dd'$ and $[d,d'] \mid \Delta g$, the bound (6.2) then follows.

# 7 Proof of Theorem 2—the asymptotic formula

We must now establish the asymptotic formula (6.5), and analyse its main term, with a view to proving the bound (6.4). We begin by showing how Theorem 1 may be applied.

The conditions $d \mid L_1(\mathbf{x}), L_2(\mathbf{x})$ and $d' \mid L_3(\mathbf{x}), L_4(\mathbf{x})$ will hold if and only if $\mathbf{x} \in \Lambda_{(d,d,d',d')}$. We therefore take $\mathbf{a}, \mathbf{b}$ as a basis for $\Lambda_{(d,d,d',d')}$ and write $\mathbf{a} = (a_1, a_2)$ and $\mathbf{b} = (b_1, b_2)$. Since $(dd', dd')$ is clearly in $\Lambda_{(d,d,d',d')}$, we see that at least one of $a_1$ and $b_1$ must be odd, and we can therefore take $a_1$ to be odd. By changing the sign of $a_1$ if necessary we can then assume that we have $a_1 \equiv \nu d \pmod{4}$, and finally, replacing $\mathbf{b}$ by $\mathbf{b} - k\mathbf{a}$ for a suitable integer $k$, we can assume that $4 \mid b_1$. Having normalized the basis $\mathbf{a}, \mathbf{b}$ of $\Lambda_{(d,d,d',d')}$ in this way we set $\mathbf{x} = y_1\mathbf{a} + y_2\mathbf{b}$. Moreover we write $L_i'(\mathbf{y}) = d^{-1}L_i(y_1\mathbf{a} + y_2\mathbf{b})$ for $i = 1, 2$ and similarly $L_i'(\mathbf{y}) = d'^{-1}L_i(y_1\mathbf{a} + y_2\mathbf{b})$ for $i = 3, 4$, and we set

$$\mathcal{R}'^{(0)} = \{\mathbf{y} \in \mathbb{R}^2 : y_1\mathbf{a} + y_2\mathbf{b} \in \mathcal{R}^{(0)}\}.$$

It now follows that

$$x_1 = y_1 a_1 + y_2 b_1 \equiv y_1 \nu d \pmod{4},$$

so that for $i = 1, 2$ the condition $L_i(\mathbf{x}) \equiv \nu x_1 \pmod{4}$ becomes

$$L_i'(\mathbf{y}) \equiv d^{-1}L_i(\mathbf{x}) \equiv d^{-1}\nu x_1 \equiv y_1 \pmod{4}.$$

Similarly for $i = 3, 4$ we have

$$L_i'(\mathbf{y}) \equiv d'^{-1}L_i(\mathbf{x}) \equiv d'^{-1}\nu' x_1 \equiv y_1 \pmod{4},$$

since $d\nu \equiv d'\nu' \pmod{4}$ in (6.1).

It is now apparent that, for fixed $d, d'$, the forms $L_i'(\mathbf{y})$, and the region $\mathcal{R}'^{(0)}$ satisfy **NC1**. Evidently we have $\operatorname{meas}(\mathcal{R}') = \operatorname{meas}\mathcal{R}/\rho(d,d,d',d')$. For fixed $d, d'$ we therefore deduce that

$$S(d, d') = \frac{4\pi^4 \prod_p \sigma_p(d, d')}{\rho(d, d, d', d')} \operatorname{meas}\mathcal{R} + o(X^2)$$

for each fixed pair $d, d'$. Here we have

$$\sigma_p(d, d') = E_p(d, d')\{1 - \chi(p)/p\}^4$$

where

$$E_p(d, d') = \sum_{\alpha,\beta,\gamma,\delta=0}^{\infty} \chi(p)^{\alpha+\beta+\gamma+\delta} \rho_0(p^\alpha, p^\beta, p^\gamma, p^\delta)^{-1}$$

and $\rho_0(d_1, d_2, d_3, d_4)$ is the determinant of the lattice

$$\Lambda_1 = \{\mathbf{y} \in \mathbb{Z}^2 : d_i \mid L_i'(\mathbf{y}),\ 1 \le i \le 4\}.$$

We now observe that $\rho_0(d_1, d_2, d_3, d_4)$ is also the index of the lattice $\Lambda_1$ in $\mathbb{Z}^2$, and hence can equally be identified as the index of

$$\Lambda_2 = \{\mathbf{x} = y_1\mathbf{a} + y_2\mathbf{b} : \mathbf{y} \in \mathbb{Z}^2,\ d_i \mid L_i'(\mathbf{y}),\ 1 \le i \le 4\}$$

in

$$\Lambda_3 = \{\mathbf{x} = y_1\mathbf{a} + y_2\mathbf{b} : \mathbf{y} \in \mathbb{Z}^2\}.$$

However we have

$$\Lambda_2 = \{\mathbf{x} \in \mathbb{Z}^2 : dd_1 \mid L_1(\mathbf{x}),\ dd_2 \mid L_2(\mathbf{x}),\ d'd_3 \mid L_3(\mathbf{x}),\ d'd_4 \mid L_4(\mathbf{x})\},$$

and

$$\Lambda_3 = \{\mathbf{x} \in \mathbb{Z}^2 : d \mid L_1(\mathbf{x}),\ d \mid L_2(\mathbf{x}),\ d' \mid L_3(\mathbf{x}),\ d' \mid L_4(\mathbf{x})\}.$$

It therefore follows that the index of $\Lambda_2$ in $\Lambda_3$ is

$$\frac{\rho(dd_1, dd_2, d'd_3, d'd_4)}{\rho(d, d, d', d')},$$

and hence that

$$\rho_0(d_1, d_2, d_3, d_4) = \frac{\rho(dd_1, dd_2, d'd_3, d'd_4)}{\rho(d, d, d', d')}.$$

We now see that $\rho_0(p^\alpha, p^\beta, p^\gamma, p^\delta) = \rho(p^\alpha, p^\beta, p^\gamma, p^\delta)$ if $p \nmid dd'$, by the multiplicative property (3.10). It therefore follows that $E_p(d, d') = E_p$ for $p \nmid dd'$, with $E_p$ as in Theorem 1.

We now define

$$N = \prod_{E_p = 0} p$$

so that we must have $\prod_p \sigma_p(d, d') = 0$ unless $N \mid dd'$. For a typical prime factor $p$ of $dd'$ let $p^u \,\|\, d$ and $p^v \,\|\, d'$, so that

$$\rho(d, d, d', d') = \prod_{p \mid dd'} \rho(p^u, p^u, p^v, p^v).$$

Assuming now that $N \mid dd'$ we set

$$F_N = \prod_{p \nmid N} E_p(1 - \chi(p)/p)^4.$$

Moreover we define $E_p^{(u,v)}$ by (1.14), so that $E_p = E_p^{(0,0)}$. We then see that

$$\frac{\prod_p \sigma_p(d,d')}{\rho(d,d,d',d')} = F_N \prod_{p|dd'} g(p^u,p^v),$$

where

$$g(p^u,p^v) = \begin{cases} E_p^{(u,v)}(1-\chi(p)/p)^4 & \text{if } p \mid N, \\ E_p^{(u,v)}/E_p^{(0,0)} & \text{if } p \nmid 2N. \end{cases}$$

If we extend $g(m,n)$ by the multiplicativity condition

$$g(ef,e'f') = g(e,e')g(f,f') \quad \text{if } \mathrm{hcf}(ee',ff') = 1,$$

we then deduce that (6.3) holds with

$$C(d,d') = 4\pi^4 F_N g(d,d')$$

when $N \mid dd'$, and $C(d,d') = 0$ otherwise. Although we have defined $g(p^u,p^v)$ for all nonnegative integer exponents $u,v$ the reader should note that only the values $u,v = 0,1$ are relevant for us, since $d$ and $d'$ may be taken to be square-free in (6.1).

When $p \nmid \Delta$ we see from (3.12) that $E_p = 1 + O(p^{-1})$ and

$$E_p^{(u,v)} \leq \frac{(u+1)^2}{p^{2u}}\{1+O(p^{-1})\},$$

for $u \geq v$. Thus

$$g(p^u,p^v) \leq \tau(p^u)^3\tau(p^v)^3[p^u,p^v]^{-2}$$

for $p \gg_{\Delta,N} 1$. For the remaining primes $p \ll_{\Delta,N} 1$, and in particular those primes which divide $\Delta$, we automatically have

$$g(p^u,p^v) \ll_\Delta \tau(p^u)^3\tau(p^v)^3[p^u,p^v]^{-2}, \quad (0 \leq u,v \leq 1).$$

We may now deduce the required bound (6.4), with an implied constant depending on $\Delta$, using the multiplicative property of the function $g(d,d')$.

We have now established the asymptotic formula (6.5) and the bound (6.4), and it remains to consider the constant $C$ given by (6.6). Our work thus far shows that

$$C = \frac{\chi(\nu\nu')}{16} 4\pi^4 F_N \sum_{\substack{dd' \equiv \nu\nu' \pmod 4 \\ N|dd'}} \mu(d)\mu(d')g(d,d').$$

We shall rewrite this as

$$\frac{\pi^4 F_N}{4} \sum_{2\nmid dd',\, N\mid dd'} \frac{\chi(\nu\nu') + \chi(dd')}{2} \mu(d)\mu(d')g(d,d') = \frac{\pi^4 F_N}{8}\{\chi(\nu\nu')\Sigma_1 + \Sigma_2\},$$

where

$$\Sigma_1 = \sum_{2\nmid dd',\, N\mid dd'} \mu(d)\mu(d')g(d,d') \quad \text{and} \quad \Sigma_2 = \sum_{N\mid dd'} \chi(dd')\mu(d)\mu(d')g(d,d').$$

To evaluate $\Sigma_1$ we set $d = ef$ where $e \mid N$ and $(f, N) = 1$, and similarly $d' = e'f'$. Then

$$\Sigma_1 = \Big\{ \sum_{e,e'\mid N,\, N\mid ee'} \mu(e)\mu(e')g(e,e') \Big\}\Big\{ \sum_{(ff',2N)=1} \mu(f)\mu(f')g(f,f') \Big\},$$

so that we may use the multiplicative property to deduce that

$$\Sigma_1 = \prod_{p\mid N}\{-g(1,p) - g(p,1) + g(p,p)\} \prod_{p\nmid 2N}\{1 - g(1,p) - g(p,1) + g(p,p)\},$$

whence

$$\begin{aligned} F_N\Sigma_1 &= F_N \prod_{p\mid N}\{E_p^{(0,0)} - E_p^{(0,1)} - E_p^{(1,0)} + E_p^{(1,1)}\}(1-\chi(p)/p)^4 \\ &\qquad \times \prod_{p\nmid 2N}\{E_p^{(0,0)} - E_p^{(0,1)} - E_p^{(1,0)} + E_p^{(1,1)}\}/E_p^{(0,0)} \\ &= \prod_{p\neq 2}\{E_p^{(0,0)} - E_p^{(0,1)} - E_p^{(1,0)} + E_p^{(1,1)}\}(1-\chi(p)/p)^4, \end{aligned} \tag{7.1}$$

since $E_p^{(0,0)} = 0$ when $p \mid N$. In exactly the same way we find that

$$F_N\Sigma_2 = \prod_{p\neq 2}\{E_p^{(0,0)} - \chi(p)E_p^{(0,1)} - \chi(p)E_p^{(1,0)} + E_p^{(1,1)}\}(1-\chi(p)/p)^4. \tag{7.2}$$

Using the functions $T_\chi(p)$ and $T_\pm(p)$ given by (1.13) and (1.17) we therefore deduce that

$$C = \frac{\pi^4}{8}\left\{ \chi(\nu\nu') \prod_{p\neq 2} T_-(p)(1-\chi(p)/p)^4 + \prod_{p\neq 2} T_\chi(p)(1-\chi(p)/p)^4 \right\}.$$

This suffices for Theorem 2, providing that we can confirm the evaluation of $\sigma_2$ and $\sigma_\infty$, and verify that $E_p^{(1,0)} = E_p^{(0,1)} = 0$ for any prime $p \equiv -1 \pmod 4$ that does not divide $\Delta_{12}\Delta_{34}$.

# 8 Proof of Theorem 2—local densities

We begin this section by defining and then computing the local densities for the variety given by (1.7), subject to the condition $\mathbf{x} \in \mathcal{R}_2$. For a prime $p > 2$ the $p$-adic density $\sigma_p$ is merely

$$\sigma_p = \lim_{e\to\infty} p^{-4e} N(p^e), \tag{8.1}$$

where

$$N(p^e) = \#\left\{ \begin{matrix} x_1, \ldots, x_6 \\ \pmod{p^e} \end{matrix} : \begin{matrix} L_1(x_1,x_2)L_2(x_1,x_2) \equiv x_3^2 + x_4^2 \pmod{p^e}, \\ L_3(x_1,x_2)L_4(x_1,x_2) \equiv x_5^2 + x_6^2 \pmod{p^e} \end{matrix} \right\}.$$

Similarly, for $p = 2$ the 2-adic density in $\mathcal{R}_2$ will be given by (8.1), for $p = 2$, but with

$$N(2^e) = \#\left\{ \begin{matrix} x_1, \ldots, x_6 \\ \pmod{2^e} \end{matrix} : \begin{matrix} 2 \nmid x_1, \\ L_1(x_1,x_2)L_2(x_1,x_2) \equiv x_3^2 + x_4^2 \pmod{2^e}, \\ L_3(x_1,x_2)L_4(x_1,x_2) \equiv x_5^2 + x_6^2 \pmod{2^e} \end{matrix} \right\}. \tag{8.2}$$

Finally, the real density is given by

$$\sigma_\infty = \int_{-\infty}^{\infty} \int_{-\infty}^{\infty} \int_{x_1,\ldots,x_6} e(\alpha Q_1 + \beta Q_2) dx_1 \ldots dx_6 \, d\beta d\alpha,$$

where

$$Q_1 = L_1(x_1,x_2)L_2(x_1,x_2) - x_3^2 - x_4^2, \quad Q_2 = L_3(x_1,x_2)L_4(x_1,x_2) - x_5^2 - x_6^2.$$

Here $(x_1, x_2)$ runs over $\mathcal{R}$, and $x_3, x_4, x_5, x_6$ each run over an interval of the form $[-cX, cX]$, with $c$ a suitably large constant. According to part (iii) of **NC2**, this is sufficient.

For a prime $p \equiv 1 \pmod 4$ one easily finds that

$$\#\{x, y \pmod{p^e} : x^2 + y^2 \equiv A \pmod{p^e}\}$$
$$= \begin{cases} p^e + ep^{e-1}(p-1) & \text{if } p^e \mid A, \\ (1 + \nu_p(A))p^{e-1}(p-1) & \text{if } \nu_p(A) < e, \end{cases}$$

for any integer $A$, where $\nu_p(A)$ is the value of $\nu$ for which $p^\nu \parallel A$. Similarly, when $p \equiv -1 \pmod 4$ we have

$$\#\{x, y \pmod{p^e} : x^2 + y^2 \equiv A \pmod{p^e}\} =$$
$$\begin{cases} p^{2[e/2]} & \text{if } p^e \mid A, \\ p^{e-1}(p+1) & \text{if } \nu_p(A) < e,\ 2 \mid \nu_p(A), \\ 0 & \text{if } \nu_p(A) < e,\ 2 \nmid \nu_p(A). \end{cases} \tag{8.3}$$

Finally, for $p = 2$ we have

$$\#\{x, y \pmod{2^e} : x^2 + y^2 \equiv A \pmod{2^e}\} = 2^{e+1}, \tag{8.4}$$

providing that $e \geq 2$ and $A \equiv 1 \pmod 4$.

It follows that, for a fixed prime $p \equiv 1 \pmod 4$, we have

$$N(p^e) = \sum_{x_1, x_2} p^{2e-2}(p-1)^2\{1 + \nu_p(L_1(\mathbf{x})L_2(\mathbf{x}))\}\{1 + \nu_p(L_3(\mathbf{x})L_4(\mathbf{x}))\} + O(e^2 p^{3e})$$

as $e \to \infty$, where the summation is for $\mathbf{x} \pmod{p^e}$, subject to the condition that $p^e \nmid L_1(\mathbf{x})L_2(\mathbf{x})$ and $p^e \nmid L_3(\mathbf{x})L_4(\mathbf{x})$. Now, if $\nu_1, \nu_2, \nu_3, \nu_4 < e$, then we see that

$$\begin{aligned}
&\#\{\mathbf{x} \pmod{p^e} : \nu_p(L_i(\mathbf{x})) = \nu_i,\ (1 \leq i \leq 4)\} \\
&= \sum_{f_1, f_2, f_3, f_4 = 0,1} (-1)^{f_1+f_2+f_3+f_4} \#\{\mathbf{x} \pmod{p^e} : p^{\nu_i + f_i} \mid L_i(\mathbf{x}),\ (1 \leq i \leq 4)\} \\
&= \sum_{f_1, f_2, f_3, f_4 = 0,1} (-1)^{f_1+f_2+f_3+f_4} p^{2e} \rho(p^{\nu_1+f_1}, p^{\nu_2+f_2}, p^{\nu_3+f_3}, p^{\nu_4+f_4})^{-1}.
\end{aligned} \tag{8.5}$$

It therefore follows that

$$\begin{aligned}
N(p^e) &= p^{4e-2}(p-1)^2 \sum_{\nu_1+\nu_2<e,\, \nu_3+\nu_4<e} (1 + \nu_1 + \nu_2)(1 + \nu_3 + \nu_4) \times \\
&\quad \sum_{f_1, f_2, f_3, f_4 = 0,1} (-1)^{f_1+f_2+f_3+f_4} \rho(p^{\nu_1+f_1}, p^{\nu_2+f_2}, p^{\nu_3+f_3}, p^{\nu_4+f_4})^{-1} \\
&\quad + O(e^2 p^{3e}) \\
&= p^{4e-2}(p-1)^2 \sum_{\nu_1, \nu_2, \nu_3, \nu_4 = 0}^{\infty} (1 + \nu_1 + \nu_2)(1 + \nu_3 + \nu_4) \times \\
&\quad \sum_{f_1, f_2, f_3, f_4 = 0,1} (-1)^{f_1+f_2+f_3+f_4} \rho(p^{\nu_1+f_1}, p^{\nu_2+f_2}, p^{\nu_3+f_3}, p^{\nu_4+f_4})^{-1} \\
&\quad + O(e^2 p^{3e}) \\
&= p^{4e-2}(p-1)^2 \sum_{\mu_1, \mu_2, \mu_3, \mu_4 = 0}^{\infty} \rho(p^{\mu_1}, p^{\mu_2}, p^{\mu_3}, p^{\mu_4})^{-1} \times \\
&\quad \sum_{0 \leq f_i \leq \min(1, \mu_i)} (-1)^{f_1+f_2+f_3+f_4} (1 + \mu_1 + \mu_2 - f_1 - f_2)(1 + \mu_3 + \mu_4 - f_3 - f_4) \\
&\quad + O(e^2 p^{3e}).
\end{aligned}$$

The sum over the $f_i$ vanishes unless $\min(\mu_1, \mu_2) = \min(\mu_3, \mu_4) = 0$, in which case it is 1. We now conclude that

$$\begin{aligned}\sigma_p &= (1-1/p)^2 \sum_{\min(\mu_1,\mu_2)=\min(\mu_3,\mu_4)=0} \rho(p^{\mu_1}, p^{\mu_2}, p^{\mu_3}, p^{\mu_4})^{-1} \\ &= (1-1/p)^2 T_-(p).\end{aligned} \tag{8.6}$$

We proceed to investigate the case $p \equiv -1 \pmod 4$ in much the same way. Using (8.3) and (8.5) we deduce that

$$N(p^e) = p^{4e-2}(p+1)^2 \sum_{\mu_1,\mu_2,\mu_3,\mu_4=0}^{\infty} (-1)^{\mu_1+\mu_2+\mu_3+\mu_4} \rho(p^{\mu_1}, p^{\mu_2}, p^{\mu_3}, p^{\mu_4})^{-1} F + O(e^2 p^{3e}),$$

where $F$ is the number of integers $f_1, f_2, f_3, f_4$ in the range $0 \le f_i \le \min(1, \mu_i)$ such that $f_1 + f_2 \equiv \mu_1 + \mu_2 \pmod 2$ and $f_3 + f_4 \equiv \mu_3 + \mu_4 \pmod 2$. The sum over the $f_i$ therefore equals 4 if $\mu_i \ge 1$ for every $i$, and equals 1 when $\min(\mu_1, \mu_2) = \min(\mu_3, \mu_4) = 0$. In the remaining case the sum is equal to 2. From this we deduce that

$$\sigma_p = (1+1/p)^2 T_+(p) \quad (p \equiv -1 \pmod 4)). \tag{8.7}$$

The formula (1.12) therefore follows.

We turn next to the case of $p = 2$. In view of part (iv) of **NC2**, we will have $L_1(\mathbf{x})L_2(\mathbf{x}) \equiv L_3(\mathbf{x})L_4(\mathbf{x}) \equiv 1 \pmod 4$, providing that $2 \nmid x_1$. According to (8.2) and (8.4) we deduce that

$$N(2^e) = 2^{2e+2}\#\{\mathbf{x} \pmod{2^e} : 2 \nmid x_1\} = 2^{4e+1},$$

whence

$$\sigma_2 = 2.$$

Finally, to evaluate $\sigma_\infty$, we restrict $x_3, x_4, x_5, x_6$ to be nonnegative, and substitute $q_1 = L_1(x_1, x_2)L_2(x_1, x_2) - x_3^2 - x_4^2$ for $x_4$, and similarly $q_2 = L_3(x_1, x_2)L_4(x_1, x_2) - x_5^2 - x_6^2$ for $x_6$. We write

$$G_1 = L_1(x_1, x_2)L_2(x_1, x_2) - x_3^2 - q_1, \quad G_2 = L_3(x_1, x_2)L_4(x_1, x_2) - x_5^2 - q_2,$$

and we set

$$F(q_1, q_2) = \frac{1}{4}\int_{x_1,x_2,x_3,x_5} G_1^{-1/2} G_2^{-1/2} dx_1 dx_2 dx_3 dx_5,$$

where the integral is subject to $(x_1, x_2) \in \mathcal{R}$ and $0 \leq x_3, x_5 \leq cX$, together with the constraints

$$L_1(x_1, x_2)L_2(x_1, x_2) - x_3^2 \geq q_1, \quad L_3(x_1, x_2)L_4(x_1, x_2) - x_5^2 \geq q_2.$$

Then we have

$$\sigma_\infty = 16 \int_{-\infty}^{\infty} \int_{-\infty}^{\infty} \int_{q_1, q_2} F(q_1, q_2) e(\alpha q_1 + \beta q_2) dq_1 dq_2 \, d\beta d\alpha,$$

and by the Fourier inversion theorem this reduces to $16F(0,0)$. To evaluate $F(0,0)$ we observe that

$$\int_0^{\sqrt{A}} \{A - x^2\}^{-1/2} dx = \frac{\pi}{2},$$

whence $F(0,0) = \pi^2 \operatorname{meas} \mathcal{R}/16$ and

$$\sigma_\infty = \pi^2 \operatorname{meas} \mathcal{R}.$$

Suppose next that the equations (1.7) has an integer solution $x_1, \ldots, x_6$ with $(x_1, x_2) \in \mathcal{R}_2$. It follows from part (iv) of **NC2** that $x_3^2 + x_4^2$ and $x_5^2 + x_6^2$ are nonzero integers, so that the solution is nonsingular. A standard argument now shows that this solution can be lifted via Hensel's Lemma to a positive $p$-adic density of points, for any prime $p$. Thus we must have $\sigma_p > 0$ for every $p$.

We now evaluate $\sigma_p$ when $p \nmid \Delta$. For such primes, (3.12) gives

$$\rho(p^{\mu_1}, p^{\mu_2}, p^{\mu_3}, p^{\mu_4}) = p^{a+b}$$

where $a$ is the maximum of the $\mu_i$, and if $a = \mu_j$, say, then $b$ is the maximum of the set $\{\mu_1, \mu_2, \mu_3, \mu_4\} \setminus \{\mu_j\}$. When $\min(\mu_1, \mu_2) = \min(\mu_3, \mu_4) = 0$ we therefore have

$$\rho(p^{\mu_1}, p^{\mu_2}, p^{\mu_3}, p^{\mu_4}) = p^{\mu_1 + \mu_2 + \mu_3 + \mu_4}, \tag{8.8}$$

so that (8.6) yields

$$\begin{aligned} \sigma_p &= (1 - 1/p)^2 \sum_{\min(\mu_1, \mu_2) = \min(\mu_3, \mu_4) = 0} p^{-\mu_1 - \mu_2 - \mu_3 - \mu_4} \\ &= (1 - 1/p)^2 \Big\{ \sum_{\min(m,n) = 0} p^{-m-n} \Big\}^2 \\ &= (1 + 1/p)^2, \end{aligned}$$

when $p \equiv 1 \pmod 4$. This proves (1.15) for such primes.

The computation for the case $p \equiv -1 \pmod 4$ is somewhat more involved. We first evaluate

$$S_1 = \sum_{\min(\mu_1,\mu_2)=0,\ \min(\mu_3,\mu_4)=0} (-1)^{\mu_1+\mu_2+\mu_3+\mu_4}\rho(p^{\mu_1},p^{\mu_2},p^{\mu_3},p^{\mu_4})^{-1}.$$

Using the argument of the previous paragraph we find that

$$\begin{aligned} S_1 &= \sum_{\min(\mu_1,\mu_2)=0,\ \min(\mu_3,\mu_4)=0} (-1)^{\mu_1+\mu_2+\mu_3+\mu_4}p^{-\mu_1-\mu_2-\mu_3-\mu_4} \\ &= \Big\{\sum_{\min(m,n)=0} (-1)^{m+n}p^{-m-n}\Big\}^2 \\ &= (p-1)^2(p+1)^{-2}. \end{aligned}$$

Next we consider

$$S_2 = \sum_{\mu_1,\mu_2,\mu_3\geq 1} (-1)^{\mu_1+\mu_2+\mu_3}\rho(p^{\mu_1},p^{\mu_2},p^{\mu_3},1)^{-1}.$$

We may write this as

$$\begin{aligned} S_2 &= \sum_{a,b,c\geq 1} (-1)^{a+b+c}p^{\min(a,b,c)}p^{-a-b-c} \\ &= \sum_{k=1}^{\infty} p^k \sum_{\min(a,b,c)=k} (-1)^{a+b+c}p^{-a-b-c} \\ &= \sum_{k=1}^{\infty} p^k\Big\{\sum_{a,b,c=k}^{\infty} (-1)^{a+b+c}p^{-a-b-c} - \sum_{a,b,c=k+1}^{\infty} (-1)^{a+b+c}p^{-a-b-c}\Big\} \\ &= \sum_{k=1}^{\infty} p^k\left\{(\frac{(-p^{-1})^k}{1+p^{-1}})^3 - (\frac{(-p^{-1})^{k+1}}{1+p^{-1}})^3\right\} \\ &= \frac{1+p^{-3}}{(1+p^{-1})^3}\sum_{k=1}^{\infty} p^k(-p^{-1})^{3k} \\ &= -\frac{1+p^{-3}}{(1+p^{-1})^3}\frac{1}{p^2+1}. \end{aligned}$$

Of course we get the same result for any sum in which three of the $\mu_i$ are at least 1 and the fourth is 0. The next sum to compute is

$$S_3 = \sum_{\mu_1,\mu_2\geq 1} (-1)^{\mu_1+\mu_2}\rho(p^{\mu_1},p^{\mu_2},1,1)^{-1}.$$

This is easily found to be

$$S_3 = \sum_{a,b\geq 1}(-1)^{a+b}p^{-a-b} = (p+1)^{-2}.$$

Now if

$$S_4 = \sum_{\mu_1,\mu_2 \geq 1,\ \min(\mu_3,\mu_4)=0} (-1)^{\mu_1+\mu_2+\mu_3+\mu_4} \rho(p^{\mu_1}, p^{\mu_2}, p^{\mu_3}, p^{\mu_4})^{-1},$$

then

$$S_4 = 2S_2 + S_3 = -\frac{(1-p^{-1})^2}{(1+p^{-1})^2} \frac{1}{p^2+1}.$$

Clearly we have the same result if the rôles of $\mu_1, \mu_2$ and $\mu_3, \mu_4$ are interchanged. Finally we examine

$$S_5 = \sum_{\mu_1,\mu_2,\mu_3,\mu_4=1}^{\infty} (-1)^{\mu_1+\mu_2+\mu_3+\mu_4} \rho(p^{\mu_1}, p^{\mu_2}, p^{\mu_3}, p^{\mu_4})^{-1}.$$

Now, according to (3.12) we have

$$\begin{aligned} S_5 &= p^{-2} \sum_{\mu_1,\mu_2,\mu_3,\mu_4=0}^{\infty} (-1)^{\mu_1+\mu_2+\mu_3+\mu_4} \rho(p^{\mu_1}, p^{\mu_2}, p^{\mu_3}, p^{\mu_4})^{-1} \\ &= p^{-2}\{S_1 + 2S_4 + S_5\}, \end{aligned}$$

whence

$$S_5 = \frac{S_1 + 2S_4}{p^2 - 1} = -S_4.$$

Then, as in the proof of (8.7), we have

$$\sigma_p = (1+1/p)^2\{S_1 + 4S_4 + 4S_5\} = (1+p^{-1})^2 S_1 = (1-p^{-1})^2.$$

This establishes (1.15) when $p \equiv -1 \pmod 4$.

Having dealt with the evaluation of the densities $\sigma_p$, our next task is to interpret the sums $E_p^{(u,v)}$ given by (1.14). Only primes $p \equiv -1 \pmod 4$ need concern us. We claim that whenever $p \equiv -1 \pmod 4$ we have

$$E_p^{(u,v)} = p^{-2u-2v}(1+1/p)^{-4} \lim_{e \to \infty} p^{-6e} N^{(u,v)}(p^e), \tag{8.9}$$

where

$$N^{(u,v)}(p^e) = \# \left\{ \begin{array}{c} x_1, \ldots, x_{10} \\ \pmod{p^e} \end{array} : \begin{array}{r} L_1(x_1,x_2) \equiv p^u(x_3^2+x_4^2) \pmod{p^e}, \\ L_2(x_1,x_2) \equiv p^u(x_5^2+x_6^2) \pmod{p^e}, \\ L_3(x_1,x_2) \equiv p^v(x_7^2+x_8^2) \pmod{p^e}, \\ L_4(x_1,x_2) \equiv p^v(x_9^2+x_{10}^2) \pmod{p^e} \end{array} \right\}.$$

If $p^u \mid L_1(x_1,x_2)$, then the number of pairs $x_3, x_4$ modulo $p^e$ for which

$$p^{-u}L_1(x_1,x_2) \equiv x_3^2 + x_4^2 \pmod{p^{e-u}}$$

will be given by (8.3). Thus if $p^e \mid L_1(x_1, x_2)$ there are $O(p^e)$ such pairs. Otherwise suppose that $p^f \parallel L_1(x_1, x_2)$. Then if $f-u$ is even there are $p^{e+u-1}(p+1)$ pairs, and if $f-u$ is odd there are no such pairs. If we set $u_1 = u_2 = u$ and $u_3 = u_4 = v$ we then find that

$$N^{(u,v)}(p^e) = p^{4e+2u+2v-4}(p+1)^4 \sum_{\substack{0 \le \nu_i < e \\ \nu_i \equiv u_i \pmod 2}} N(p^e; \nu_1, \nu_2, \nu_3, \nu_4) + O(p^{5e}),$$

where

$$N(p^e; \nu_1, \nu_2, \nu_3, \nu_4) = \#\{x_1, x_2 \pmod{p^e} : \nu_p(L_i(x_1, x_2)) = \nu_i, \ \ (1 \le i \le 4)\}.$$

The sum over the $\nu_i$ may be re-written as

$$\sum_{\substack{0 \le \nu_i < e \\ \nu_i \equiv u_i \pmod 2}} \ \sum_{f_1, f_2, f_3, f_4 = 0,1} (-1)^{f_1+f_2+f_3+f_4} \frac{p^{2e}}{\rho(p^{\nu_1+f_1}, p^{\nu_2+f_2}, p^{\nu_3+f_3}, p^{\nu_4+f_4})},$$

whence

$$\lim_{e \to \infty} p^{-6e} N^{(u,v)}(p^e) = p^{2u+2v}(1+1/p)^4 \Sigma,$$

with

$$\begin{aligned} \Sigma &= \sum_{\nu_i \equiv u_i \pmod 2} \ \sum_{f_1, f_2, f_3, f_4 = 0,1} \frac{(-1)^{f_1+f_2+f_3+f_4}}{\rho(p^{\nu_1+f_1}, p^{\nu_2+f_2}, p^{\nu_3+f_3}, p^{\nu_4+f_4})} \\ &= \sum_{g_i \ge u_i} (-1)^{g_1+g_2+g_3+g_4} \rho(p^{g_1}, p^{g_2}, p^{g_3}, p^{g_4})^{-1} \end{aligned}$$

as in our treatment of (8.7). This suffices for the proof of (8.9).

It is now clear that $E_p^{(u,v)} \ge 0$ for $p \equiv -1 \pmod 4$. Now let $p \nmid \Delta_{12}\Delta_{34}$ for some prime $p \equiv -1 \pmod 4$, and let $u = u_1 = u_2 = 1$ and $v = u_3 = u_4 = 0$, say. Suppose we have a solution to the congruences

$$\begin{gathered} L_1(x_1, x_2) \equiv p(x_3^2 + x_4^2), \quad L_2(x_1, x_2) \equiv p(x_5^2 + x_6^2) \pmod{p^e}, \\ L_3(x_1, x_2) \equiv x_7^2 + x_8^2, \quad L_4(x_1, x_2) \equiv x_9^2 + x_{10}^2 \pmod{p^e} \end{gathered}$$

in which $p^{2f} \mid x_1, x_2$ for some exponent $2f \le e-2$. Then $p^f$ must divide each of $x_3, \ldots, x_{10}$ and therefore

$$\begin{gathered} L_1(y_1, y_2) \equiv p(y_3^2 + y_4^2), \quad L_2(y_1, y_2) \equiv p(y_5^2 + y_6^2) \pmod{p^{e-2f}}, \\ L_3(y_1, y_2) \equiv y_7^2 + y_8^2, \quad L_4(y_1, y_2) \equiv y_9^2 + y_{10}^2 \pmod{p^{e-2f}} \end{gathered}$$

where $x_i = p^{2f} y_i$ for $i = 1, 2$ and $x_i = p^f y_i$ for $3 \le i \le 10$. Since the first two of these congruences imply that $p \mid L_1(y_1, y_2), L_2(y_1, y_2)$ we deduce that $p \mid y_1, y_2$, since $p \nmid \Delta_{12}$. It follows that $p \mid L_3(y_1, y_2), L_4(y_1, y_2)$, and hence that $p$ divides both $y_7^2 + y_8^2$ and $y_9^2 + y_{10}^2$. Thus $p^2 \mid y_7^2 + y_8^2, y_9^2 + y_{10}^2$, so that $p^2 \mid L_3(y_1, y_2), L_4(y_1, y_2)$. Since $p \nmid \Delta_{34}$ this requires $p^2 \mid y_1, y_2$, whence, finally, $p^{2f-2} \mid x_1, x_2$. We therefore conclude that any solution of the original congruences must have $p^{e-1} \mid x_1, x_2$. In view of (8.3) we deduce that $N^{(1,0)}(p^e) = O(p^{4e})$, whence $E_p^{(1,0)} = 0$, by (8.9). Similarly we will have $E_p^{(0,1)} = 0$.

It remains to show that if $\varepsilon = -1$ then the variety (1.7) has no points with $(x_1, x_2) \in \mathcal{R}_2$. Clearly, if $\varepsilon = -1$ then we must have $T_-(p) = \pm T_+(p)$ for every prime $p \mid \Delta$ with $p \equiv -1 \pmod 4$. Let

$$\mathcal{P} = \{p \mid \Delta : p \equiv -1 \pmod 4,\ T_-(p) = -T_+(p)\}.$$

We now argue by contradiction, assuming that we have a point $(x_1, x_2) \in \mathcal{R}_2$ on the variety (1.7). Then, since $L_i(x_1, x_2) \neq 0$ by part (iv) of **NC2**, we see that the equations (1.7) entail

$$\nu_p(L_1(x_1, x_2)) \equiv \nu_p(L_2(x_1, x_2)) \pmod 2,$$
$$\nu_p(L_3(x_1, x_2)) \equiv \nu_p(L_4(x_1, x_2)) \pmod 2,$$

for any prime $p \equiv -1 \pmod 4$. We now suppose that

$$2 \mid \nu_p(L_1(x_1, x_2)) - u \quad \text{and} \quad 2 \mid \nu_p(L_3(x_1, x_2)) - v$$

with $0 \le u, v \le 1$. Then we can find a nonsingular $p$-adic solution to the equations

$$L_1(x_1, x_2) = p^u(y_3^2 + y_4^2), \quad L_2(x_1, x_2) = p^u(y_5^2 + y_6^2),$$
$$L_3(x_1, x_2) = p^v(y_7^2 + y_8^2), \quad L_4(x_1, x_2) = p^v(y_9^2 + y_{10}^2).$$

This can then be lifted by the standard procedure to show, via (8.9), that $E_p^{(u,v)} > 0$. Thus

$$E_p^{(u,v)} > 0 \quad \text{if} \quad 2 \mid \nu_p(L_1(x_1, x_2)) - u \quad \text{and} \quad 2 \mid \nu_p(L_3(x_1, x_2)) - v. \tag{8.10}$$

We now show that $\nu_p(L_1(x_1, x_2))$ and $\nu_p(L_3(x_1, x_2))$ have opposite parities whenever $p \in \mathcal{P}$. Since $T_-(p) = -T_+(p)$ for such a prime, and $E_p^{(u,v)} \ge 0$ for all $u, v$, we will have $E_p^{(0,0)} = E_p^{(1,1)} = 0$. The claim then follows from (8.10).

Conversely we now show that if $\nu_p(L_1(x_1, x_2))$ and $\nu_p(L_3(x_1, x_2))$ have opposite parities, and $p \equiv -1 \pmod 4$, then $p \in \mathcal{P}$. For such a prime, it follows from (8.10) that either $E_p^{(1,0)} > 0$ or $E_p^{(0,1)} > 0$ . However we have already seen that $E_p^{(1,0)} = E_p^{(0,1)} = 0$ unless $p \mid \Delta_{12}\Delta_{34}$. Thus if $\nu_p(L_1(x_1, x_2))$ and $\nu_p(L_3(x_1, x_2))$ have opposite parities, and $p \equiv -1 \pmod 4$, then $p \mid \Delta$. Thus $p$ must occur in the product for $\varepsilon$, whence $T_-(p) = \pm T_+(p)$. Since either $E_p^{(1,0)} > 0$ or $E_p^{(0,1)} > 0$ we cannot have $T_-(p) = T_+(p)$, so that we must indeed have $p \in \mathcal{P}$.

We have therefore shown that the set $\mathcal{P}$ consists precisely of those primes $p \equiv -1 \pmod 4$ which divide $L_1(x_1, x_2)L_3(x_1, x_2)$ to an odd power. Since part (iii) of **NC2** implies that $L_1(x_1, x_2)L_3(x_1, x_2)$ is positive, we conclude from part (iv) of **NC2** that

$$\chi(\nu\nu') \equiv L_1(x_1, x_2)L_3(x_1, x_2) \equiv (-1)^{\#\mathcal{P}} \pmod 4. \tag{8.11}$$

On the other hand we have

$$\prod_{p|\Delta,\, \chi(p)=-1} T_-(p)/T_+(p) = (-1)^{\#\mathcal{P}},$$

and since $\varepsilon = -1$ we deduce that

$$(-1)^{\#\mathcal{P}} = -\chi(\nu\nu').$$

This contradicts (8.11), and therefore completes the proof of Theorem 2.

# 9 Examples

In this section we shall discuss Theorem 2 in the context of the examples (1.10), (1.11) and (1.18). We begin with (1.10), which we repeat here as

$$y_1(y_1 + 4y_2) = x_3^2 + x_4^2, \quad (7y_1 + 16y_2)(19y_1 + 44y_2) = x_5^2 + x_6^2.$$

This has been shown to have no nontrivial rational points, even though it has nonsingular points in every completion of $\mathbb{Q}$. We take the region $\mathcal{R}^{(0)}$ to be the square $(0,1)^2$, so that parts (i), (ii) and (iii) of **NC2** will be satisfied. Moreover part (iv) is clearly satisfied with $\nu = 1$ and $\nu' = -1$.

The existence of nonsingular local points is sufficient to ensure that $\sigma_p > 0$ for every prime $p$. However for the forms in (1.10) we find that $\Delta_{12}\Delta_{34} = 2^4$, so that $E_p^{(1,0)} = E_p^{(0,1)} = 0$ for any primes entering into the product in (1.16). It follows that $T_-(p) = T_+(p)$ for such primes, so that $\varepsilon = \chi(\nu\nu') = \chi(-1) = -1$. Thus the failure of the Hasse Principle is fully explained by Theorem 2, at least as far as points with $(y_1, y_2) \in \mathcal{R}_2$ are concerned.

We turn now to the example (1.11), namely

$$y_1(y_1 + 4y_2) = x_3^2 + x_4^2, \quad (y_1 + 8y_2)(13y_1 + 64y_2) = x_5^2 + x_6^2.$$

Although there are rational points in this example, we showed in §1 that all such points have $y_2/y_1 \geq -1/8$. We shall therefore consider the application of Theorem 2 to two different regions. We begin by examining the case

$$y_1, y_1 + 4y_2 > 0, \quad y_1 + 8y_2 < 0, \quad 13y_1 + 64y_2 < 0,$$

for which there are no rational points. Here we must replace $L_3$ and $L_4$ by $-L_3$ and $-L_4$ respectively, to produce linear forms which will all be positive. Having made this change we then take $\mathcal{R}^{(0)} = (0,1)^2$. Then parts (i), (ii) and (iii) of **NC2** will hold. We also see that part (iv) holds, with $\nu = 1$ and $\nu' = -1$. We may now proceed as in the previous example, noting that $\Delta_{12}\Delta_{34} = 2^5 \cdot 5$. Once again it follows that $\varepsilon = -1$, so that $\mathcal{R}_2$ produces no solutions.

On the other hand, if we look at the case

$$y_1, y_1 + 4y_2 > 0, \quad y_1 + 8y_2 > 0, \quad 13y_1 + 64y_2 > 0,$$

we may again work with $\mathcal{R}^{(0)} = (0,1)^2$. This time we have $\nu = \nu' = 1$ in part (iv) of Normalization Condition 2. The value $\Delta_{12}\Delta_{34} = 2^5 \cdot 5$ is the same as before, so that (1.16) yields $\varepsilon = \chi(\nu\nu') = \chi(1) = 1$. It therefore follows that the density of rational points in $\mathcal{R}_2$ is twice the product of local densities, while the density of rational points in the first case was of course zero.

The examples we have looked at so far all have $\varepsilon = \pm 1$. However other values may occur, as the example (1.18)

$$x_1(x_1 + 12x_2) = x_3^2 + x_4^2, \quad (x_1 + 4x_2)(x_1 + 16x_2) = x_5^2 + x_6^2,$$

will demonstrate. We shall use the region

$$\mathcal{R} = \{0 < x_1, x_1 + 16x_2 < X\}$$

so that

$$\sigma_\infty = \pi^2 \operatorname{meas} \mathcal{R} = \frac{\pi^2}{16} X^2.$$

There is a nonsingular rational point with $(x_1, x_2) = (1, 0)$, and this is enough to ensure that all the local densities are positive. Since $\Delta_{12}\Delta_{34} = 2^4 \cdot 3^2$ and $\nu = \nu' = 1$, we now find that $\varepsilon = T_-(3)/T_+(3)$. In order to show that $\varepsilon \neq \pm 1$ it will suffice to demonstrate that $E_3^{(0,0)}$ and $E_3^{(1,0)}$ are positive. To do this we shall use (8.9). When $u = v = 0$ the congruences

$$x_1 \equiv x_3^2 + x_4^2 \pmod 3, \quad x_1 + 12x_2 \equiv x_5^2 + x_6^2 \pmod 3,$$
$$x_1 + 4x_2 \equiv x_7^2 + x_8^2 \pmod 3, \quad x_1 + 16x_2 \equiv x_9^2 + x_{10}^2 \pmod 3$$

have a nonsingular solution with $x_1 = 1$ and $x_2 = 0$, which is sufficient to ensure that $E_3^{(0,0)} > 0$. Similarly, for $u = 1$, $v = 0$, the congruences

$$x_1 \equiv 3(x_3^2 + x_4^2) \pmod{3^e}, \quad x_1 + 12x_2 \equiv 3(x_5^2 + x_6^2) \pmod{3^e},$$
$$x_1 + 4x_2 \equiv x_7^2 + x_8^2 \pmod{3^e}, \quad x_1 + 16x_2 \equiv x_9^2 + x_{10}^2 \pmod{3^e}$$

require $x_1 = 3x_1'$, say, so that they are equivalent to

$$x_1' \equiv x_3^2 + x_4^2 \pmod{3^{e-1}}, \quad x_1' + 4x_2 \equiv x_5^2 + x_6^2 \pmod{3^{e-1}},$$
$$3x_1 + 4x_2 \equiv x_7^2 + x_8^2 \pmod{3^e}, \quad 3x_1 + 16x_2 \equiv x_9^2 + x_{10}^2 \pmod{3^e}.$$

There is now a nonsingular solution with $x_1' = x_2 = 1$, so that $E_3^{(1,0)} > 0$, as required.

Thus (1.8) provides an example with $0 < 1 + \varepsilon < 2$. We illustrate this example numerically. Since $\sigma_2 = 2$, we see that (1.15) yields

$$\prod_p \sigma_p = \frac{2\sigma_3}{(1-1/3)^2} \prod_p (1 + \chi(p)/p)^2 = \frac{18}{\pi^2}\sigma_3.$$

Moreover one finds from (1.12) that

$$\sigma_3\Big(1 + \frac{T_-(3)}{T_+(3)}\Big) = \frac{16}{9}(T_+(3) + T_-(3)) = \frac{32}{9}(E_3^{(0,0)} + E_3^{(1,1)}).$$

One may now evaluate $E_3^{(0,0)}$ and $E_3^{(1,1)}$ by a somewhat tedious calculation along the lines of that given in the previous section to prove (1.15). The starting point is the fact that (3.12) remains true for $p = 3$, except when $\min(e_1, e_2) > \max(e_3, e_4)$, in which case

$$\rho(3^{e_1}, 3^{e_2}, 3^{e_3}, 3^{e_4}) = 3^{e_1+e_2-1},$$

or $\min(e_3, e_4) > \max(e_1, e_2)$, in which case

$$\rho(3^{e_1}, 3^{e_2}, 3^{e_3}, 3^{e_4}) = 3^{e_3+e_4-1}.$$

The conclusion is that

$$E_3^{(0,0)} = \frac{9}{20} \quad \text{and} \quad E_3^{(1,1)} = \frac{1}{20}.$$

It follows that we will have asymptotically $2X^2$ solutions to (1.18) in $\mathcal{R}_2$. This is illustrated by Table 2, in which

$$S(X) = \sum_{\mathbf{x} \in \mathcal{R}_2} r(L_1(\mathbf{x})L_2(\mathbf{x}))r(L_3(\mathbf{x})L_4(\mathbf{x})).$$

# 10 Acknowledgements

Parts of this work were undertaken while the author was visiting the University of Hong Kong, and the Max-Planck Institute for Mathematics in Bonn. Their hospitality and financial support is gratefully acknowledged.

The author is also extremely grateful to Professors Colliot-Thélène and Salberger for their helpful remarks on the proof of the Hasse Principle and Weak Approximation for the variety (1.7).

Table 2

| $X$ | $S(X)$ | $S(X)/2X^2$ |
|---|---|---|
| 1000 | 1993472 | 0.9967 ... |
| 2000 | 8030592 | 1.0038 ... |
| 4000 | 32057728 | 1.0018 ... |
| 8000 | 1276046726 | 0.9969 ... |
| 16000 | 511437824 | 0.9989 ... |
| 32000 | 2043518720 | 0.9978 ... |

# References

[1] A. Balog, Linear equations in primes, *Mathematika* **39** (1992) 367–378

[2] J.-L. Colliot-Thélène, D. Coray, and J.-J. Sansuc, Descente et principe de Hasse pour certaines variétés rationnelles, *J. reine angew. Math.* **320** (1980) 150–191

[3] J.-L. Colliot-Thélène, J.-J. Sansuc, and H.P.F. Swinnerton-Dyer, Intersections of two quadrics and Châtelet surfaces. I, *J. reine angew. Math.* **373** (1987) 37–107

[4] S. Daniel, On the divisor-sum problem for binary forms, *J. reine. angew. Math.* **507** (1999) 107–129

[5] R.R. Hall and G. Tenenbaum, *Divisors,* Cambridge Tracts in Mathematics **90**, Cambridge University Press, Cambridge, 1988

[6] V.A. Iskovskih, A counterexample to the Hasse principle for systems of two quadratic forms in five variables, *Mat. Zametki* **10** (1971) 253–257

[7] M. Nair, Multiplicative functions of polynomial values in short intervals, *Acta Arith.* **62** (1992) 257-269

D. R. Heath-Brown,
Mathematical Institute,
24-29 St Giles',
Oxford OX1 3LB, England
e-mail: rhb@maths.ox.ac.uk

# Kronecker double series and the dilogarithm

Andrey Levin

**Abstract**

In this article we give an explicit expression for the value of a certain Kronecker double series at any point of complex multiplication as a sum of dilogarithms whose arguments are values of some modular unit of higher level at the corresponding points. This result can be interpreted in the spirit of the Zagier conjecture. The special value of the Kronecker double series is equal to the value of the partial $\zeta$-function of an ideal class for an order in an imaginary quadratic field. The values of the above mentioned modular unit belong to ray class field corresponding to this order. Thus we get an explicit formula for the value of a partial $\zeta$-function at $s = 2$ as a combination of dilogarithms of algebraic numbers.

# 1 Introduction

## 1.1 Modular part

**1.1.1** We start by fixing notation and recalling some standard facts about modular curves. We set $\mathcal{H} = \{\tau \in \mathbb{C} \mid \Im(\tau) > 0\}$. Then a matrix $M = \left(\begin{smallmatrix} a & b \\ c & d \end{smallmatrix}\right) \in \mathrm{GL}_2(\mathbb{Z})$ acts on $\mathcal{H}$ by $M(\tau) = \frac{a\tau+b}{c\tau+d}$. Write $L = L_\tau$ for the lattice in $\mathbb{C}$ generated by $\tau$ and 1, and $E = E_\tau$ for the corresponding elliptic curve $\mathbb{C}/L_\tau$. The matrix $M$ defines a map of lattices $M^\tau \colon L_{M(\tau)} \to L_\tau$ given by $w \to (c\tau+d)w$ and an isogeny $M_\tau \colon E_\tau \to E_{M(\tau)}$ given by $\xi \to (ad-bc)(c\tau+d)^{-1}\xi$. If $\tau$ is a fixed point of $M$, then $M_\tau$ is a map of the curve $E_\tau$ onto itself. In this case we omit the subscript $\tau$ in our notation.

A point $\xi \in E_\tau$ defines a character $\chi_\xi$ of the lattice $L_\tau$ given by $\chi_\xi(w) = \exp\left(\frac{2\pi i(w\bar{\xi}-\bar{w}\xi)}{\tau-\bar{\tau}}\right)$. This pairing is $\mathrm{GL}_2(\mathbb{Z})$-invariant: $\chi_{M(\xi)}(w) = \chi_\xi(M(w))$.

**Definition 1.1.2** The *second Kronecker double series* $\mathcal{K}_2(\xi;\tau)$ is the $C^\infty$-function on $\mathbb{C} \times \mathcal{H}$ defined by the convergent series

$$\mathcal{K}_2(\xi;\tau) = \left(\frac{\tau-\bar{\tau}}{2\pi i}\right)^2 {\sum_{w\in L}}' \frac{\chi_\xi(w)}{|w|^4}, \tag{1}$$

where, as usual, $\sum'$ denotes the sum over $L \setminus \{0\}$.

One checks that this function is invariant under the action of $\mathrm{SL}_2(\mathbb{Z})$ on $\mathbb{C} \times \mathcal{H}$ defined above.

**1.1.3** The Weierstrass $\wp$-function is the elliptic function defined by the convergent series

$$\wp(\xi;\tau) = \frac{1}{\xi^2} + \sum_{w \in L_\tau}{}' \Big( \frac{1}{(w+\xi)^2} - \frac{1}{w^2} \Big).$$

## 1.2 Dilogarithms

**Definition 1.2.1** The *Euler dilogarithm* $Li_2(z)$ is the multivalued analytic function on $\mathbb{P}^1 \setminus \{0, 1, \infty\}$ defined as the analytic continuation of the series $\sum_{j \geq 1} \frac{z^j}{j^2}$ (which converges for $|z| < 1$).

**Definition 1.2.2** The formula

$$D_2(z) = \Im(Li_2(z)) + \arg(1-z) \cdot \log|z| \quad \text{for } z \notin \{0,1,\infty\},$$
$$D_2(0) = D_2(1) = D(\infty) = 0.$$

defines a single-valued real function on $\mathbb{P}^1$ that we call the *Bloch–Wigner dilogarithm.* It is continuous on $\mathbb{P}^1$ and smooth on $\mathbb{P}^1 \setminus \{0, 1, \infty\}$

We can extend the function $D_2$ byt linearity to a function on the $\mathbb{Q}$-vector space $\mathbb{Q}[\mathbb{C}]$.

**1.2.3** Define a map

$$\delta\colon \mathbb{Q}[\mathbb{C} \setminus \{0,1\}] \to \textstyle\bigwedge^2 \mathbb{C}^* \quad \text{by the formula} \quad [x] \to x \wedge (1-x).$$

## 1.3 Results

**Main Theorem 1.3.1** *Let $\tau$ be a fixed point of $M = \left(\begin{smallmatrix} a & b \\ c & d \end{smallmatrix}\right) \neq 0, 1$. Set $m = \det M$ and $n = \det(M-1)$. Then*

$$\begin{aligned} &-(m+1)(n+1)\frac{c(\tau-\overline{\tau})}{i}\mathcal{K}_2(0;\tau) \\ &= 4mnD_2\Big(\frac{ad-bc}{c\tau+d}\Big) + \sum_{\substack{\alpha \in \mathrm{Ker}(M)\setminus 0 \\ \beta \in \mathrm{Ker}(M-1)\setminus 0}} \sum_{k=1}^{m}\sum_{l=1}^{n} D_2\Big(\frac{\wp(k\alpha+l\beta)-\wp(\alpha)}{\wp(\beta)-\wp(\alpha)}\Big); \end{aligned}$$

*the arguments of the dilogarithms are in the kernel of $\delta$ (see 1.2.3):*

$$\delta\Big(4mn \times [c\tau+d] + \sum_{\alpha,\beta,k,l} \Big(\frac{\wp(k\alpha+l\beta)-\wp(\alpha)}{\wp(\beta)-\wp(\alpha)}\Big)\Big) = 0 \in \textstyle\bigwedge^2(\mathbb{C}^* \otimes_{\mathbb{Z}} \mathbb{Q}).$$

**1.3.2** Our proof is based on introducing a new function $\mathcal{L}_{1,1}$, which we call the *elliptic* $(1,1)$*-logarithm* and define in Section 2. The Main Theorem follows from Theorems A and B below.

**Theorem A 1.3.3** *Let $\tau, M, m$ and $n$ be as in the Theorem 1.3.1. Then*

$$\frac{c(\tau-\bar{\tau})}{4i}\frac{(m+1)(n+1)}{mn}\mathcal{K}_2(0;\tau) = -D_2\left(\frac{ad-bc}{c\tau+d}\right) + \sum_{\substack{\alpha\in\operatorname{Ker} M\backslash 0\\ \beta\in\operatorname{Ker}(M-1)\backslash 0}} \mathcal{L}_{1,1}(\alpha,\beta,0). \quad (2)$$

**Theorem B 1.3.4** *Let $\alpha$ and $\beta$ be two distinct nontrivial torsion points on an elliptic curve $E_\tau$, say $m\alpha = n\beta = 0$ for some $m, n \in \mathbb{N}$. Then*

$$\sum_{k=1}^{m}\sum_{l=1}^{n} D_2\Big(\frac{\wp(k\alpha+l\beta)-\wp(\alpha)}{\wp(\beta)-\wp(\alpha)}\Big) = -2mn\big(\mathcal{L}_{1,1}(\alpha,\beta,0)+\mathcal{L}_{1,1}(-\alpha,\beta,0)\big). \quad (3)$$

## 1.4 Generalities and structure of the article

**1.4.1** The fact that the value of the Kronecker double series $\mathcal{K}_2$ at a CM point can be expressed as a combination of dilogarithms is not new. It can be derived from a result of Deninger as follows. A CM point $\tau$ defines a ring $R$ of endomorphisms of the correspondent lattice $L_\tau$; this ring is an order in some imaginary quadratic extension $F = R\otimes_{\mathbb{Z}}\mathbb{Q}$ of $\mathbb{Q}$. Extend the field $F$ by the value of the $\jmath$-invariant at the point $\tau$. Deninger [D1] constructed an element in the third algebraic $K$-group of the field $F(\jmath(\tau))$ that the regulator maps to the value of $\mathcal{K}_2$ at $\tau$. We know by Suslin and Bloch that the regulator on $K_3$ of a number field is given by the Bloch–Wigner dilogarithm. Hence we can conclude that the value of the $\mathcal{K}_2$ equals a combination of values of the dilogarithm at numbers in $F(\jmath(\tau))$. The arguments of the dilogarithm in our formula belong to some extension of the field $F(\jmath(\tau))$, but the set of them is Galois invariant.

Our proof is in some sense parallel to Deninger's construction.

**1.4.2** We can also interpret the Main Theorem independently of algebraic $K$-theory. The value of $\mathcal{K}_2(0;\tau)$ for a CM point $\tau$ is (a rational multiple of) the value at $s=2$ of the partial zeta function $\zeta_{F,\mathcal{A}}(s) = \sum_{\mathfrak{a}\in\mathcal{A}} N(\mathfrak{a})^{-s}$ for some ideal class $\mathcal{A}$. The fact that this value can be written as a combination of dilogarithms with arguments belonging to the associated class field over $F$ is a special case of the refined version of Zagier's conjecture that the values of all partial zeta functions at arbitrary integer argument $s = m$ can be expressed in terms of $m$-logarithms.

For the case at hand, the same result, of course, also follows from the theorems of Deninger (existence of elements in $K_2$ with required values of regulator) and Bloch–Suslin (structure of $K_2$ and description of the regulator from $K_2$). Our proof, as well as giving an explicit formula, also has the advantage of avoiding algebraic $K$-theory. It is possible in principle that this method could be applied for higher values of $m$ where it is not known that the regulator may be expressed in terms of polylogarithms.

**1.4.3** Theorem A reflects a general phenomenon. Another reflection of this phenomenon is the following fact. For any elliptic modular curve over $\mathbb{Q}$ the value of its $L$-function at $s = 2$ can be expressed as a combination of the values of a special function (Goncharov's elliptic $(1,2)$-logarithm [G2]), a "relative" of our elliptic $(1,1)$-logarithm. On the other hand, for a CM-curve the value of the $L$-function is equal to a combination of the values of a certain Kronecker double series [D1]. Therefore this Kronecker double series must be equal to a combination of values of the elliptic $(1,2)$-logarithm.

**1.4.4** The paper is organized as follows. Section 2 defines the elliptic $(1,1)$-logarithm for an arbitrary elliptic curve and studies its elementary properties. In Section 3 we realize the Kronecker series as an integral over the square of the elliptic curve, and, for a curve with complex multiplication, reduce this integral to an integral over the elliptic curve itself. In Section 4 we compare the elliptic $(1,1)$-logarithm and the dilogarithm. In Section 5 we check that $\delta$ vanishes on the arguments of the dilogarithms on the right-hand side of the Main Theorem, thus completing its proof. In the final Section 6 we prove a more general formula, relating values of $\mathcal{K}_2$ at torsion points to the dilogarithm.

**1.4.5 Acknowledgments** I wish to thank Sasha Goncharov for explaining his ideas about Chow polylogarithms. I am also grateful to Don Zagier and Herbert Gangl for very stimulating discussions and computer experiments at a crucial moment for this work during my stay at the Max-Planck Institut für Mathematik, Bonn in 1997. I wish to thank the MPI for hospitality.

# 2 The elliptic $(1,1)$-logarithm

In this section we define and study properties of the elliptic $(1,1)$-logarithm. Some motivation for considering this function is Goncharov's integral representation of the Bloch–Wigner dilogarithm. This representation uses a special case of a very general differential operator, that we call $A_*$.

## 2.1 The operations $A_n$

**Definition 2.1.1** Let $\varphi_1, \dots, \varphi_n$ be smooth functions on a complex variety $X$. We set

$$A_n(\varphi_1, \dots, \varphi_n) = \mathrm{Alt}_n \left( \sum_{j=0}^{n-1} (-1)^j \varphi_1 \, \partial\varphi_2 \wedge \cdots \wedge \partial\varphi_{n-j} \wedge \bar\partial\varphi_{n-j+1} \wedge \cdots \wedge \bar\partial\varphi_n \right), \quad (4)$$

where $\mathrm{Alt}_n$ denotes alternation under the symmetric group $S_n$, with a factor of $1/(n!)$:

$$\mathrm{Alt}_n(F(x_1, \dots, x_n) = \frac{1}{n!} \sum_{\sigma \in S_n} \mathrm{sign}(\sigma) F(x_{\sigma(1)}, \dots, x_{\sigma(n)}).$$

**Remark 2.1.2** If $\varphi_j = \log |f_j|^2$ for *analytic* functions $f_j$, then $A_n$ is the so-called Beilinson–Deligne product of the $f_j$, up to a factor 2 for odd $n$ and $2i$ for even $n$.

**Remark 2.1.3** The $(p,q)$-component of $A_n$ equals $(-1)^q \frac{p!q!}{(p+q)!}$ multiplied by the $(p,q)$-component of $\mathrm{Alt}_n(\varphi_1 \, \mathrm{d}\varphi_2 \wedge \cdots \wedge \mathrm{d}\varphi_n)$.

An important property of the operations $A_n$ is the following:

**Proposition 2.1.4** *For $n > 1$*

$$\mathrm{d}A_n(\varphi_1, \dots, \varphi_n) = \partial\varphi_1 \wedge \partial\varphi_2 \wedge \cdots \wedge \partial\varphi_n + (-1)^{n-1} \bar\partial\varphi_1 \wedge \bar\partial\varphi_2 \wedge \cdots \wedge \bar\partial\varphi_n + \sum_{j=1}^{n} (-1)^j \bar\partial\partial \, \varphi_j \wedge A_{n-1}(\varphi_1, \dots, \hat\varphi_j, \dots, \varphi_n). \quad (5)$$

The proof is a straightforward computation.

## 2.2 Goncharov's integral representation of the Bloch–Wigner dilogarithm

**Lemma 2.2.1 (Goncharov [G1])** *The value of the Bloch–Wigner dilogarithm at a point $a \neq 0, 1, \infty$ is equal to the following convergent integral:*

$$\frac{1}{4\pi} \int_{\mathbb{P}^1} A_3(\log |z|^2, \log |1-z|^2, \log |a-z|^2).$$

**Proof** Since we are integrating over a curve, only the $(1,1)$-component contributes. Thus we can replace $A_3$ by

$$-\frac{1}{6}\Big(\log|z|^2 \mathrm{d}\log|1-z|^2 \wedge \mathrm{d}\log|a-z|^2 \\ - \log|1-z|^2 \mathrm{d}\log|z|^2 \wedge \mathrm{d}\log|a-z|^2 \\ + \log|a-z|^2 \mathrm{d}\log|z|^2 \wedge \mathrm{d}\log|1-z|^2\Big).$$

We first prove that the integral converges. The integrand is smooth outside 0, 1, $a$ and $\infty$. Let $(r,\varphi)$ be a polar coordinate system near one of the first three points; any term of the integrand is asymptotic to one of $\log|r| r dr \wedge \mathrm{d}\varphi$ or $r^{-1} r dr \wedge \mathrm{d}\varphi$, and is integrable. As $A_3$ is trilinear and totally antisymmetric, we can replace the three arguments $(\log|z|^2, \log|1-z|^2, \log|a-z|^2)$ by

$$(\log|z|^2, \log|1-z|^2 - \log|z|^2, \log|a-z|^2 - \log|z|^2) = \\ (\log|z|^2, \log|1-z^{-1}|^2, \log|az^{-1}-1|^2);$$

the convergence at $\infty$ can be checked for these by the same considerations.

By Stokes's formula we reduce the integral to

$$-\frac{1}{16\pi}\int\Big(\log|z|^2 \mathrm{d}\log|1-z|^2 \wedge \mathrm{d}\log|a-z|^2 \\ - \log|1-z|^2 \mathrm{d}\log|z|^2 \wedge \mathrm{d}\log|a-z|^2\Big).$$

The $(1,1)$-component of $\mathrm{d}\varphi_1 \wedge \mathrm{d}\varphi_2$ is the negative of the $(1,1)$-component of $(\partial - \overline{\partial})\varphi_1 \wedge (\partial - \overline{\partial})\varphi_2$. Hence the integral is equal to

$$\frac{1}{16\pi}\int\Big(\log|z|^2(\partial-\overline{\partial})\log|1-z|^2 - \log|1-z|^2(\partial-\overline{\partial})\log|z|^2\Big) \\ \wedge (\partial-\overline{\partial})\log|a-z|^2.$$

An easy calculation shows that

$$dD_2(z) = -i\Big(\log|z|(\partial-\overline{\partial})\log|1-z| - \log|1-z|(\partial-\overline{\partial})\log|z|\Big).$$

Therefore the integral is equal to

$$\frac{i}{4\pi}\int \mathrm{d}D_2(z)\wedge(\partial-\overline{\partial})\log|a-z|^2 = \frac{1}{4\pi i}\int D_2(z)\mathrm{d}(\partial-\overline{\partial})\log|a-z|^2.$$

As $\mathrm{d}(\partial-\overline{\partial})\log|a-z|^2 = 2\overline{\partial}\partial\log|a-z|^2 = 4\pi i(\delta_a(z)-\delta_\infty(z))$ and $D_2(\infty)=0$, this completes the proof. □

**Remark 2.2.2** It follows from the proof that the integral representation

$$\sum_{p\in X} \mathrm{ord}_p(g) D_2(f(p)) = \frac{1}{4\pi} \int_X A_3(\log|f|^2, \log|1-f|^2, \log|g|^2) \tag{6}$$

holds for any two meromorphic functions $f$ and $g$ on a compact curve $X$.

**Remark 2.2.3** For any curve $C$

$$\int_C A_3(\varphi_1, \varphi_2, \varphi_3) = \int_C \frac{1}{2}\varphi_1 \, \mathrm{d}\varphi_2 \wedge \mathrm{d}\varphi_3,$$

so that this integral is zero if $\varphi_j = \mathrm{const}$ for some $j$.

**Lemma 2.2.4** *For any three distinct points $a, b, c \in \mathbb{C}$, we have*

$$\int_{\mathbb{P}^1} A_3\big(\log|z-a|^2, \log|z-b|^2, \log|z-c|^2\big) = 4\pi D_2\Big(\frac{c-a}{b-a}\Big). \tag{7}$$

**Proof** We write $r$ for the ratio $\frac{c-a}{b-a}$ and make the change of variable $x = \frac{z-a}{b-a}$. Then

$$\log|z-a|^2 = \log|x|^2 + \log|b-a|^2, \quad \log|z-b|^2 = \log|x-1|^2 + \log|b-a|^2$$
$$\text{and} \quad \log|z-c|^2 = \log|x-r|^2 + \log|b-a|^2.$$

By the preceding remark, any terms that include the constant $\log|b-a|^2$ vanish. This reduces the integral to that of Lemma 2.2.1. □

**Remark 2.2.5** The integrand in (7) vanishes formally by antisymmetry if any of $a$, $b$ and $c$ are equal (we say formally because we are not really allowed to multiply 1-forms if their singularities coincide). If only two of $a$, $b$, $c$ are equal, then the r.-h.s. of (7) is also zero, since $D_2(0) = D_2(1) = D_2(\infty) = 0$. Along the triple diagonal the expression in (7) has a discontinuity. However, it is continuous on the blowup of $\mathbb{C}^3$ along the triple diagonal.

## 2.3 The elliptic $(1,1)$-logarithm

In this section we define a real-valued function $\mathcal{L}_{1,1}(\alpha, \beta, \gamma; \tau)$, called the *elliptic* $(1,1)$-*logarithm*, on the third power of an elliptic curve (more precisely, on the fibered third power of the universal elliptic curve over the modular curve), which is invariant under the diagonal action of the elliptic curve by translations and antisymmetric under permutations of the variables.

**2.3.1** A natural generalization of the function $\log|x-a|^2$ on $\mathbb{P}^1$ to an elliptic curve $E_\tau$ is the *Kronecker double series*

$$\mathcal{L}_1(\xi;\tau) := \log\left|\widetilde{\theta}(\xi,\tau)\right|^2 + \frac{1}{2}\frac{2\pi i}{\tau-\overline{\tau}}(\xi-\overline{\xi})^2 = -\frac{\tau-\overline{\tau}}{2\pi i}\sum_{w\in L}{}_e' \frac{\chi_\xi(w)}{|w|^2}$$

where $\sum_e$ denotes Eisenstein summation (see Weil [W]), and

$$\widetilde{\theta}(\xi,\tau) = \frac{\theta(\xi,\tau)}{\eta(\tau)} = q^{1/12}(z^{\frac{1}{2}} - z^{-\frac{1}{2}})\prod_{j=1}^{\infty}(1-q^j z)(1-q^j z^{-1}).$$

$q = \exp(2\pi i\tau)$ (or, more precisely, $q^{1/12} = \exp(\frac{1}{6}\pi i\tau)$), and $z = \exp(2\pi i\xi)$ (or, more precisely, $z^{\pm\frac{1}{2}} = \exp(\pm\pi i\xi)$). The notation $\mathcal{L}_1$ for this function is not standard. It is meant to emphasize that this function is the elliptic 1-logarithm.

**2.3.2** The function $\mathcal{L}_1$ is the Green function for the operator $\overline{\partial}\partial$ on an elliptic curve:

$$\overline{\partial}\partial(\mathcal{L}_1(\xi;\tau)) = 2\pi i\delta_0 + \frac{2\pi i}{\tau-\overline{\tau}}d\xi\wedge d\overline{\xi}.$$

Here $\delta_0$ denotes the delta function at zero.

**2.3.3** The function $\mathcal{L}_1$ satisfies the distribution relation

$$\sum_{\alpha:M_\tau(\alpha)=\beta}\mathcal{L}_1(\alpha;\tau) = \mathcal{L}_1(\beta;M(\tau)).$$

**2.3.4** Any elliptic function $f$ with $\mathrm{ord}_\beta(f)=0$ has a "theta decomposition"

$$\log|f(\xi)|^2 - \log|f(\beta)|^2 = \sum \mathrm{ord}_\alpha(f)\Big(\mathcal{L}_1(\xi-\alpha) - \mathcal{L}_1(\beta-\alpha)\Big). \qquad (8)$$

Thus the natural elliptic generalization of the function $D_2\left(\frac{c-a}{b-a}\right)$ is the following

**Definition 2.3.5** The *elliptic* (1, 1)*-logarithm* $\mathcal{L}_{1,1}(\alpha,\beta,\gamma;\tau)$ is the *convergent* integral:

$$\frac{1}{4\pi}\int_{E_\tau} A_3\Big(\mathcal{L}_1(\xi-\alpha),\mathcal{L}_1(\xi-\beta),\mathcal{L}_1(\xi-\gamma)\Big).$$

The convergence of the integral can be checked by the same consideration as for $D_2(a)$.

**Remark 2.3.6** As above, only the $(1,1)$-component of the integrand gives a nontrivial contribution, so

$$\mathcal{L}_{1,1}(\alpha,\beta,\gamma;\tau) = -\frac{1}{8\pi}\int_{E_\tau} \mathcal{L}_1(\xi-\alpha)\,\mathrm{d}\mathcal{L}_1(\xi-\beta)\wedge \mathrm{d}\mathcal{L}_1(\xi-\gamma).$$

We give another definition of this function based on Fourier expansions:

**Lemma 2.3.7** *Considered as a distribution, the function $\mathcal{L}_{1,1}(\alpha,\beta,\gamma;\tau)$ is equal to the following series*

$$\frac{(\tau-\overline{\tau})^2}{16\pi^2 i}\times \sum_{\substack{w_1+w_2+w_3=0\\ w_i\neq 0}} \frac{\chi_\alpha(w_1)\chi_\beta(w_2)\chi_\gamma(w_3)(w_2\overline{w}_3-\overline{w}_2 w_3)}{|w_1|^2|w_2|^2|w_3|^2}.$$

A series of this kind was introduced by Deninger [D2]. In contrast to his case, however, our series is not absolutely convergent.

**Proof** By the preceding remark, we can compute the following integral

$$-\frac{1}{8\pi}\int -\frac{\tau-\overline{\tau}}{2\pi i}\times\sum_{w_1\neq 0}\frac{\chi_{\xi-\alpha}(w_1)}{|w_1|^2}\times$$

$$\Big(\sum_{w_2\neq 0}\frac{\chi_{\xi-\beta}(w_2)(\overline{w}_2\mathrm{d}\xi-w_2\mathrm{d}\overline{\xi})}{|w_2|^2}\Big)\wedge\Big(\sum_{w_3\neq 0}\frac{\chi_{\xi-\gamma}(w_3)(\overline{w}_3\mathrm{d}\xi-w_3\mathrm{d}\overline{\xi})}{|w_3|^2}\Big)$$

$$=-\frac{1}{8\pi}\frac{\tau-\overline{\tau}}{2\pi i}\int\sum_{w_i\neq 0}\frac{\chi_{\xi-\alpha}(w_1)\chi_{\xi-\beta}(w_2)\chi_{\xi-\gamma}(w_3)(w_2\overline{w}_3-\overline{w}_2 w_3)}{|w_1|^2|w_2|^2|w_3|^2}\mathrm{d}\xi\wedge\mathrm{d}\overline{\xi}.$$

The integrals of the terms with $w_1+w_2+w_3\neq 0$ vanish, as integrals of nontrivial harmonics over a torus, so we get

$$-\frac{1}{8\pi}\frac{\tau-\overline{\tau}}{2\pi i}\sum_{\substack{w_1+w_2+w_3=0\\ w_i\neq 0}}\frac{\chi_\alpha(w_1)\chi_\beta(w_2)\chi_\gamma(w_3)(w_2\overline{w}_3-\overline{w}_2 w_3)}{|w_1|^2|w_2|^2|w_3|^2}\int\mathrm{d}\xi\wedge\mathrm{d}\overline{\xi}$$

$$=-\frac{1}{8\pi}\frac{\tau-\overline{\tau}}{2\pi i}\sum_{\substack{w_1+w_2+w_3=0\\ w_i\neq 0}}\frac{\chi_\alpha(w_1)\chi_\beta(w_2)\chi_\gamma(w_3)(w_2\overline{w}_3-\overline{w}_2 w_3)}{|w_1|^2|w_2|^2|w_3|^2}\times(-(\tau-\overline{\tau})).$$

□

**Remark 2.3.8** This Fourier series is antisymmetric under permutations of $\alpha$, $\beta$ and $\gamma$, since

$$w_2\overline{w}_3-\overline{w}_2 w_3=\frac{1}{3}\begin{vmatrix}1&1&1\\ w_1&w_2&w_3\\ \overline{w}_1&\overline{w}_2&\overline{w}_3\end{vmatrix}\qquad\text{for } w_1+w_2+w_3=0.$$

**2.3.9** The function $\mathcal{L}_{1,1}$ is smooth outside the diagonals; this follows from the general formula for differentiation of an integral with respect to a parameter. $\mathcal{L}_{1,1}$ is zero on the diagonals, because $A_3$ is antisymmetric.

**Lemma 2.3.10** *$\mathcal{L}_{1,1}$ is continuous on the complement of the triple diagonal $\alpha = \beta = \gamma$.*

**Proof** We prove that

$$\lim_{t \to 0} \mathcal{L}_{1,1}(t\alpha, t\beta, \gamma; \tau) = 0 \quad \text{for } \gamma \neq 0.$$

We choose some rather small $\varepsilon$ and represent the integral as the sum of the integral over the disk of radius $\varepsilon$ around 0 and the integral over the complement of this disk inside the elliptic curve. The second integral tends to zero, as the integral of any term of $A_3$ converges and $A_3$ is antisymmetric. Inside the disk, $\mathcal{L}_1(\xi - t\alpha)$ equals $\log|\xi - t\alpha|^2 + \varphi(\xi, t\alpha)$, where $\varphi(\xi, t\alpha)$, is a smooth function, the same is true for $\mathcal{L}_1(\xi - t\beta)$. Substitute this decomposition into $A_3$; we get several types of summands: 1) all the arguments of $A_3$ are smooth, 2) one argument is singular and 3) two arguments are singular. In the first two cases the integral tends to zero for the same reason as above. To estimate the last summand, perform the change of variable $z = t^{-1}\xi$. We get

$$\begin{aligned}
&\int_{|\xi|<\varepsilon} A_3\Big(\log|\xi - t\alpha|^2, \log|\xi - t\beta|^2, \mathcal{L}_1(\xi - \gamma)\Big) \\
&\quad = \int_{|\xi|<t^{-1}\varepsilon} A_3\Big(\log|z - \alpha|^2, \log|z - t\beta|^2, \mathcal{L}_1(tz - \gamma) - \mathcal{L}_1(-\gamma)\Big) \\
&\qquad\qquad + \int_{|\xi|<t^{-1}\varepsilon} A_3\Big(\log|z - \alpha|^2, \log|z - t\beta|^2, \mathcal{L}_1(-\gamma)\Big),
\end{aligned}$$

and the second integral tends to zero because $\mathcal{L}_1(-\gamma)$ is a constant.

For small $t$, the first integral can be represented as a sum of the integral over the disk of radius $\sqrt{t^{-1}\varepsilon}$ and that over the annulus $\sqrt{t^{-1}\varepsilon} < |z| < t^{-1}\varepsilon$. In the second integral, replace $\log|z - \beta|^2$ by $\log|z - \beta|^2 - \log|z - \alpha|^2$. The integral over the disk is small since $\mathcal{L}_1(tz - \gamma)$ and also its derivatives are small; the integral over the annulus is small because $\log|z - \beta|^2 - \log|z - \alpha|^2|$ and its derivatives are small. □

**Lemma 2.3.11** *Let $\alpha$, $\beta$ and $\gamma$ be points on an elliptic curve, not all three coincident. Then the limit of $\mathcal{L}_{1,1}(t\alpha, t\beta, t\gamma; \tau)$ as $t \to 0$ equals $D_2\left(\frac{\gamma-\alpha}{\beta-\alpha}\right)$.*

**Proof** We fix some rather small $\varepsilon$ and represent the integral as the sum of the integral over the disk of radius $\varepsilon$ around 0 and the integral over the complement of this disk inside the elliptic curve. The second integral tends to zero, as $A_3$ is antisymmetric. Inside the disk, $\mathcal{L}_1(\xi - t\alpha)$ equals $\log|\xi - t\alpha|^2 + \varphi(\xi, t\alpha)$, where $\varphi(\xi, t\alpha)$ is a smooth function,and the same is true for $\mathcal{L}_1(\xi - t\beta)$ and for $\mathcal{L}_1(\xi - t\gamma)$. As in the proof of the preceding lemma, only the summand $A_3(\log|\xi - t\alpha|^2, \log|\xi - t\beta|^2, \log|\xi - t\gamma|^2)$ gives a nontrivial contribution in the limit. Therefore

$$\begin{aligned}
&\lim_{t\to 0}\mathcal{L}_{1,1}(t\alpha, t\beta, t\gamma; \tau)\\
&\quad = \frac{1}{4\pi}\lim_{t\to 0}\int_{|\xi|<\varepsilon} A_3\Big(\log|\xi - t\alpha|^2, \log|\xi - t\beta|^2, \log|\xi - t\gamma|^2\Big)\\
&\quad = \frac{1}{4\pi}\lim_{t\to 0}\int_{|\xi|<t^{-1}\varepsilon} A_3\Big(\log|t^{-1}\xi - \alpha|^2, \log|t^{-1}\xi - \beta|^2, \log|t^{-1}\xi - \gamma|^2\Big)\\
&\quad = \frac{1}{4\pi}\int_{\mathbb{C}} A_3\Big(\log|z - \alpha|^2, \log|z - \beta|^2, \log|z - \gamma|^2\Big)\\
&\quad = D_2\Big(\frac{\gamma - \alpha}{\beta - \alpha}\Big). \quad \square
\end{aligned}$$

**Remark 2.3.12** $\mathcal{L}_{1,1}(\alpha, \beta, \gamma; \tau)$ is clearly invariant under "diagonal" translation of the arguments and changing the sign of the arguments, so that the elliptic $(1, 1)$-logarithm is a function on the moduli space $\overline{\mathcal{M}}_{1,3}$ of curves of genus 1 with three marked points.

# 3 From the Kronecker series to the elliptic $(1, 1)$-logarithm

The reduction of the Kronecker series $\mathcal{K}_2$ to the elliptic $(1, 1)$-logarithm for an elliptic curve with complex multiplication splits up into two steps. We first represent $\mathcal{K}_2(\xi; \tau)$ as an integral over the square of the elliptic curve. This representation is valid for any elliptic curve and for any point $\xi$ on it. We then reduce the integral over the square of the elliptic curve to an integral over the elliptic curve itself; this is possible for a curve with complex multiplication, and uses the existence of an extra projection of the square of the curve onto itself.

## 3.1 An integral representation of the Kronecker series

We start from the simplest example which illustrates the main idea of this representation.

**Proposition 3.1.1**

$$\mathcal{K}_2(\alpha;\tau) = \frac{\tau-\overline{\tau}}{2\pi i}\int\limits_{E_\tau\times E_\tau} \mathcal{L}_1(\alpha-\eta_1-\eta_2)\frac{\partial\mathcal{L}_1(\eta_1)}{\partial\eta_1}\frac{\partial\mathcal{L}_1(\eta_2)}{\partial\overline{\eta}_2}\frac{\mathrm{d}\eta_1\wedge\mathrm{d}\overline{\eta}_1}{\tau-\overline{\tau}}\wedge\frac{\mathrm{d}\eta_2\wedge\mathrm{d}\overline{\eta}_2}{\tau-\overline{\tau}}. \tag{9}$$

**Proof** We use the Fourier expansions of $\mathcal{L}_1$ and its derivatives

$$\frac{\tau-\overline{\tau}}{2\pi i}\int\Big(-\frac{\tau-\overline{\tau}}{2\pi i}\sideset{}{_e'}\sum_{w_1\in L}\frac{\chi_{\alpha-\eta_1-\eta_2}(w_1)}{|w_1|^2}\Big)\times$$

$$\Big(\sideset{}{_e'}\sum_{w_2\in L}\frac{\chi_{\eta_1}(w_2)}{w_2}\Big)\Big(-\sideset{}{_e'}\sum_{w_3\in L}\frac{\chi_{\eta_2}(w_3)}{\overline{w}_3}\Big)\times\frac{d\eta_1\wedge d\overline{\eta}_1}{\tau-\overline{\tau}}\wedge\frac{d\eta_2\wedge d\overline{\eta}_2}{\tau-\overline{\tau}}.$$

The integral of the term with $w_1\neq w_2, w_1\neq w_3$ vanishes, as an integral of a nontrivial harmonic over a torus, so we get the integral

$$\frac{\tau-\overline{\tau}}{2\pi i}\int\Big(-\frac{\tau-\overline{\tau}}{2\pi i}\sideset{}{_e'}\sum_{w_1\in L}\frac{\chi_\alpha(w_1)}{|w_1|^2}\Big)\Big(\frac{1}{-w_1}\Big)\Big(-\frac{1}{-\overline{w}_1}\Big)\frac{d\eta_1\wedge d\overline{\eta}_1}{\tau-\overline{\tau}}\wedge\frac{d\eta_2\wedge d\overline{\eta}_2}{\tau-\overline{\tau}}$$

$$=\Big(\frac{\tau-\overline{\tau}}{2\pi i}\Big)^2\sum_{w_1\in L}\frac{\chi_\alpha(w_1)}{|w_1|^4}(-1)^2=\mathcal{K}_2(\xi;\tau). \quad\square$$

At the end of this section we discuss a more symmetric representation of $\mathcal{K}_2$ for any $\tau$ and $\xi$; this result will not be used further. We now state the main result of this section.

**Lemma 3.1.2** *Let $\tau$ be a fixed point of $M=\left(\begin{smallmatrix}a&b\\c&d\end{smallmatrix}\right)\neq 0,1$; we set $m=\det M$ and $n=\det(M-1)$. Write $\eta_3$ for the expression $\eta_1+M_\tau(\eta_2)-\alpha$ and $\eta_4$ for $\eta_1+\eta_2-\beta$. Then*

$$\int\limits_{E_\tau^2} A_3(\mathcal{L}_1(\eta_1),\mathcal{L}_1(\eta_2),\mathcal{L}_1(\eta_3))\Big(\frac{\mathrm{d}\eta_4\wedge\mathrm{d}\overline{\eta}_4}{\tau-\overline{\tau}}\Big)=\frac{\pi i c(\tau-\overline{\tau})}{m}\mathcal{K}_2(-\alpha;\tau);$$

$$\int\limits_{E_\tau^2} A_3(\mathcal{L}_1(\eta_1),\mathcal{L}_1(\eta_2),\mathcal{L}_1(\eta_4))\Big(\frac{\mathrm{d}\eta_3\wedge\mathrm{d}\overline{\eta}_3}{\tau-\overline{\tau}}\Big)=-\pi i c(\tau-\overline{\tau})\mathcal{K}_2(-\beta;\tau);$$

$$\int\limits_{E_\tau^2} A_3(\mathcal{L}_1(\eta_1),\mathcal{L}_1(\eta_3),\mathcal{L}_1(\eta_4))\Big(\frac{\mathrm{d}\eta_2\wedge\mathrm{d}\overline{\eta}_2}{\tau-\overline{\tau}}\Big)=\frac{\pi i c(\tau-\overline{\tau})}{mn}\mathcal{K}_2(M_\tau(\beta)-\alpha;\tau);$$

$$\int\limits_{E_\tau^2} A_3(\mathcal{L}_1(\eta_2),\mathcal{L}_1(\eta_3),\mathcal{L}_1(\eta_4))\Big(\frac{\mathrm{d}\eta_1\wedge\mathrm{d}\overline{\eta}_1}{\tau-\overline{\tau}}\Big)=-\frac{\pi i c(\tau-\overline{\tau})}{n}\mathcal{K}_2(\beta-\alpha;\tau).$$

(10)

**Proof** We only prove the third equation; the others can be proved by the same considerations. The last factor of the integrand has type $(1,1)$ and is closed, so we can replace the expression $A_3(\mathcal{L}_1(\eta_1), \mathcal{L}_1(\eta_3), \mathcal{L}_1(\eta_4))$ by $\frac{1}{2}\mathcal{L}_1(\eta_3)\mathrm{d}\mathcal{L}_1(\eta_1) \wedge \mathrm{d}\mathcal{L}_1(\eta_4)$. Thus we can calculate the integral

$$\frac{1}{2}\int\Big(-\frac{\tau-\overline{\tau}}{2\pi i}\sum_{w_3\in L}{}'_e \frac{\chi_{\eta_1+M_\tau(\eta_2)-\alpha}(w_3)}{|w_3|^2}\Big) \times$$
$$\Bigg(\Big(\sum_{w_1\in L}{}'_e \frac{\chi_{\eta_1}(w_1)\mathrm{d}\eta_1}{w_1}\Big) \wedge \Big(-\sum_{w_4\in L}{}'_e \frac{\chi_{\eta_1+\eta_2-\beta}(w_4)(\mathrm{d}\overline{\eta}_1+\mathrm{d}\overline{\eta}_2)}{\overline{w}_4}\Big)$$
$$+\Big(\sum_{w_1\in L}{}'_e \frac{\chi_{\eta_1}(w_1)\mathrm{d}\overline{\eta}_1}{\overline{w}_1}\Big) \wedge \Big(-\sum_{w_4\in L}{}'_e \frac{\chi_{\eta_1+\eta_2-\beta}(w_4)(\mathrm{d}\eta_1+\mathrm{d}\eta_2)}{w_4}\Big)\Bigg) \wedge \frac{\mathrm{d}\eta_2\wedge\mathrm{d}\overline{\eta}_2}{\tau-\overline{\tau}}.$$

Since $\chi_{M_\tau}(\eta_2)(w) = \chi_{\eta_2}(M(w))$, only the terms with $w_3 + w_1 + w_4 = 0$ and $M(w_3) + w_4 = 0$ give nontrivial contributions in the integral. So we get

$$\frac{1}{2}\frac{(\tau-\overline{\tau})^2}{2\pi i}\sum_{w_3\in L}{}' \frac{\chi_{-\alpha}(w_3)}{|w_3|^2} \times \Bigg(\frac{1}{(M-1)(w_3)} \times \frac{\chi_{-\beta}(-M(w_3))}{-\overline{M(w_3)}}$$
$$-\frac{1}{\overline{(M-1)(w_3)}} \times \frac{\chi_{-\beta}(-M(w_3))}{-M(w_3)}\Bigg) \times \int \frac{\mathrm{d}\eta_1\wedge\mathrm{d}\overline{\eta}_1}{\tau-\overline{\tau}} \wedge \frac{\mathrm{d}\eta_2\wedge\mathrm{d}\overline{\eta}_2}{\tau-\overline{\tau}}$$
$$= -\frac{1}{2}\frac{(\tau-\overline{\tau})^2}{2\pi i}\sum_{w_3\in L}{}' \frac{\chi_{M(\beta)-\alpha}(w_3)}{|w_3|^4} \times$$
$$\Big(\frac{1}{(c\tau+d-1)(c\overline{\tau}+d)} - \frac{1}{(c\overline{\tau}+d-1)(c\tau+d)}\Big)$$
$$= \frac{1}{2}\frac{(\tau-\overline{\tau})^2}{2\pi i}\frac{c(\tau-\overline{\tau})}{nm}\sum_{w_3\in L}{}' \frac{\chi_{M(\beta)-\alpha}(w_3)}{|w_3|^4}$$
$$= \frac{\pi i c(\tau-\overline{\tau})}{mn}\mathcal{K}_2(M_\tau(\beta)-\alpha;\tau).$$

We used above the following simple formula: for any isogeny of $E_\tau$ to itself, $|c\tau+d|^2 = ad - bc = \det M$. □

**Remark 3.1.3** For general values of $\tau$ any isogeny of $E_\tau$ to itself is multiplication by some integer, and the r.-h.s. vanishes. Hence the result is only interesting for an elliptic curve with complex multiplication.

**3.1.4** The function $\frac{2\pi i}{\tau-\overline{\tau}}\mathcal{K}_2$ can be treated as a component of the vector valued function

$$\mathcal{L}_3(\xi;\tau) = \frac{\tau-\overline{\tau}}{2\pi i}\begin{pmatrix} \sum_{w\in L}{}' \frac{\chi_\xi(w)}{w^3\overline{w}} \\ \sum_{w\in L}{}' \frac{\chi_\xi(w)}{w^2\overline{w}^2} \\ \sum_{w\in L}{}' \frac{\chi_\xi(w)}{w\overline{w}^3} \end{pmatrix},$$

(the elliptic trilogarithm), taking values in the symmetric square $S^2(\mathcal{H})$ of the homology group $\mathcal{H}$ of the elliptic curve $E_\tau$ with complex coefficients. This space is isomorphic to a direct summand of the second cohomology group of the square of the elliptic curve. A "natural" basis of this space is

$$f_1 = \frac{\mathrm{d}\overline{\eta}_1 \wedge \mathrm{d}\overline{\eta}_2}{(\tau-\overline{\tau})^2}, \quad f_2 = \frac{\mathrm{d}\eta_1 \wedge \mathrm{d}\overline{\eta}_2 - \mathrm{d}\eta_2 \wedge \mathrm{d}\overline{\eta}_1}{(\tau-\overline{\tau})^2}, \quad f_3 = \frac{\mathrm{d}\eta_1 \wedge \mathrm{d}\eta_2}{(\tau-\overline{\tau})^2}.$$

**Proposition 3.1.5**

$$\mathcal{L}_3(\xi;\tau) = \int_{E_\tau^2} A_3\left(\mathcal{L}_1(\eta_1), \mathcal{L}_1(\eta_2), \mathcal{L}_1(\eta_1+\eta_2-\xi)\right) \wedge \begin{pmatrix} f_1 \\ f_2 \\ f_3 \end{pmatrix}. \tag{11}$$

The proof is a straightforward calculation.

## 3.2 Reduction to the elliptic $(1,1)$-logarithm

**3.2.1** Consider the current $\Phi = \frac{1}{2\pi^2}A_4\big(\mathcal{L}_1(\eta_1), \mathcal{L}_1(\eta_2), \mathcal{L}_1(\eta_3), \mathcal{L}_1(\eta_4)\big)$ on $E_\tau^2$, where, as above, $\eta_3 = \eta_1 + M_\tau(\eta_2) - \alpha$ and $\eta_4 = \eta_1 + \eta_2 - \beta$ for some isogeny $M = \left(\begin{smallmatrix} a & b \\ c & d \end{smallmatrix}\right) \neq 0, 1$. The wedge product of currents is not defined, but for general $\alpha$ and $\beta$ the divisors of the singularities of $\mathcal{L}_1$'s are in general position, and the wedge product is well defined; the formula for the differential of the wedge product also holds.

**3.2.2** The differential $\mathrm{d}\Phi$ of $\Phi$ equals

$$\begin{aligned} \frac{1}{2\pi^2}\times\Big(&-\overline{\partial}\partial\mathcal{L}_1(\eta_1) \wedge A_3\Big(\mathcal{L}_1(\eta_2), \mathcal{L}_1(\eta_3), \mathcal{L}_1(\eta_4)\Big) \\ &+ \overline{\partial}\partial\mathcal{L}_1(\eta_2) \wedge A_3\Big(\mathcal{L}_1(\eta_1), \mathcal{L}_1(\eta_3), \mathcal{L}_1(\eta_4)\Big) \\ &- \overline{\partial}\partial\mathcal{L}_1(\eta_3) \wedge A_3\Big(\mathcal{L}_1(\eta_1), \mathcal{L}_1(\eta_2), \mathcal{L}_1(\eta_4)\Big) \\ &+ \overline{\partial}\partial\mathcal{L}_1\Big(\eta_4) \wedge A_3(\mathcal{L}_1(\eta_1), \mathcal{L}_1(\eta_2), \mathcal{L}_1(\eta_3)\Big)\Big), \end{aligned}$$

because the $(4,0)$-part and the $(0,4)$-part vanish on a surface. We split the $\overline{\partial}\partial\mathcal{L}_1$ into a $\delta$-function part and a smooth part. Integrals with $\delta$-functions are integrals over the elliptic curve and are equal to sums of values of the elliptic $(1,1)$-logarithm. Integrals of smooth parts of the $\partial\overline{\partial}\mathcal{L}_1$ are calculated in the preceding lemma.

**3.2.3** We write $B_1, \dots, B_4$ for the $\delta$-parts of the four components of $\mathrm{d}\Phi$ and $B'_1, \dots, B'_4$ for the smooth ones. We first calculate the integrals using $\delta$-functions:

$$\begin{aligned}\int B_1 &= -\frac{2\pi i}{2\pi^2}\int_{E_\tau} A_3\Big(\mathcal{L}_1(\eta_2), \mathcal{L}_1(\eta_3), \mathcal{L}_1(\eta_4)\Big)_{|\eta_1=0} \\ &= \frac{1}{\pi i}\int_{E_\tau} A_3\Big(\mathcal{L}_1(\eta_2), \mathcal{L}_1(M_\tau\eta_2-\alpha), \mathcal{L}_1(\eta_2-\beta)\Big) \\ &= \frac{1}{\pi i}\int_{E_\tau}\sum_{\alpha':M_\tau(\alpha')=\alpha} A_3\Big(\mathcal{L}_1(\eta_2), \mathcal{L}_1(\eta_2-\alpha'), \mathcal{L}_1(\eta_2-\beta)\Big) \\ &= -4i\sum_{\alpha':M_\tau(\alpha')=\alpha}\mathcal{L}_{1,1}(0,\alpha',\beta).\end{aligned}$$

The same consideration shows that

$$\begin{aligned}B_2 &= 4i\mathcal{L}_{1,1}(0,\alpha,\beta), \\ B_3 &= -4i\sum_{\substack{\alpha':M_\tau(\alpha')=\alpha \\ \gamma':(M_\tau-1)(\gamma')=\alpha-\beta}}\mathcal{L}_{1,1}(\alpha',0,\gamma'), \quad \text{and} \\ B_4 &= 4i\sum_{\gamma':(M_\tau-1)(\gamma')=\alpha-\beta}\mathcal{L}_{1,1}(0,\beta,\gamma').\end{aligned}$$

**3.2.4** Since the smooth part of $\overline{\partial}\partial\mathcal{L}_1(\xi;\tau)$ equals $2\pi i\frac{\mathrm{d}\xi\wedge\mathrm{d}\overline{\xi}}{\tau-\overline{\tau}}$, the integrals of $B'_j$ were already calculated in (10).

**3.2.5** The integral of the differential of a current over a compact variety is zero, so that we get $\sum B_i = -\sum B'_i$, and we have proved the following result:

**Proposition 3.2.6** *Let $\tau$ be a fixed point of $M \neq 0, 1$, with $m = \det M$ and*

$n = \det(M-1)$. *Suppose that* $\alpha \neq 0$ *and* $\beta \neq 0, \alpha, M_\tau(\alpha)$. *Then*

$$\frac{c(\tau-\overline{\tau})}{4i}\Big(\frac{1}{m}\mathcal{K}_2(-\alpha;\tau)+\mathcal{K}_2(-\beta;\tau)+\frac{1}{mn}\mathcal{K}_2(M_\tau(\beta)-\alpha;\tau)+\frac{1}{n}\mathcal{K}_2(\beta-\alpha;\tau)\Big)$$
$$= - \sum_{\alpha' \in M^{-1}(\alpha)} \mathcal{L}_{1,1}(0,\alpha',\beta) + \mathcal{L}_{1,1}(0,\alpha,\beta)$$
$$- \sum_{\substack{\alpha' \in M^{-1}(\alpha)\\ \gamma' \in (M-1)^{-1}(\alpha-\beta)}} \mathcal{L}_{1,1}(\alpha',0,\gamma') + \sum_{\gamma' \in (M-1)^{-1}(\alpha-\beta)} \mathcal{L}_{1,1}(0,\beta,\gamma').$$

As the function $\mathcal{K}_2$ is continuous, we can "degenerate" this formula:

**3.2.1 Theorem.** *Let* $\tau$ *be a fixed point of* $M \neq 0, 1$ *and* $m$, $n$ *as above. Then*

$$\frac{c(\tau-\overline{\tau})}{4i}\left(\frac{m+n+1}{mn}\mathcal{K}_2(-\alpha;\tau)+\mathcal{K}_2(0;\tau)\right)$$
$$= - \sum_{\substack{\alpha' \in M^{-1}(\alpha)\\ \gamma' \in (M-1)^{-1}(\alpha)}} \mathcal{L}_{1,1}(\alpha',0,\gamma'), \quad \textit{for } \alpha \neq 0; \quad (12)$$

$$\frac{c(\tau-\overline{\tau})}{4i}\frac{(m+1)(n+1)}{mn}\mathcal{K}_2(0;\tau)$$
$$= -D_2\left(\frac{ad-bc}{c\tau+d}\right) - \sum_{\substack{\alpha' \in \mathrm{Ker}(M)\backslash 0\\ \gamma' \in \mathrm{Ker}(M-1)\backslash 0}} \mathcal{L}_{1,1}(\alpha',0,\gamma'). \quad (13)$$

**Proof** The first formula is the result of substituting $\beta = 0$; and the second one is the limit of the first as $\alpha \to 0$. This completes the proof of Theorem A. □

# 4 From the elliptic $(1,1)$-logarithm to the dilogarithm

In this section we relate the elliptic $(1,1)$-logarithm to the dilogarithm. Specifically, we express the combination $\mathcal{L}_{1,1}(0,\alpha,\beta;\tau)+\mathcal{L}_{1,1}(0,\alpha,-\beta;\tau)$ as a sum of dilogarithms for any torsion points $\alpha$ and $\beta$ on *any* elliptic curve $E_\tau$. The proof uses the representation of an elliptic curve as a covering of degree 2 of the projective line.

## 4.1 The dilogarithm as a combination of elliptic $(1,1)$-logarithms

We start by expressing the dilogarithm as a combination of elliptic $(1,1)$-logarithms.

**4.1.1** The Weierstrass $\wp$-function maps the elliptic curve as a double cover of $\mathbb{P}^1$. Suppose that $\pm\alpha$ on the elliptic curve are the inverse images of a point $a$ on $\mathbb{P}^1$, that is, $\wp(\pm\alpha) = a$; similarly, suppose that $\wp(\pm\beta) = b$, $\wp(\pm\gamma) = c$. Then by 2.2.4,

$$D_2\left(\frac{c-a}{b-a}\right) = \frac{1}{4\pi}\int_{\mathbb{P}^1} A_3(\log|z-a|^2, \log|z-b|^2, \log|z-c|^2)$$
$$= \frac{1}{2}\frac{1}{4\pi}\int_{E_\tau} A_3(\log|\wp(\xi)-\wp(\alpha)|^2, \log|\wp(\xi)-\wp(\beta)|^2, \log|\wp(\xi)-\wp(\gamma)|^2).$$

The extra factor of $1/2$ reflects the number of branches.

**4.1.2** The "theta decomposition" of 2.3.4 implies that

$$\log|\wp(\xi)-\wp(\alpha)|^2 = \mathcal{L}_1(\xi+\alpha) + \mathcal{L}_1(\xi-\alpha) - 2\mathcal{L}_1(\xi) - 2\mathcal{L}_1(\alpha).$$

We substitute this expression into $A_3$ and integrate. A straightforward computation gives

**Lemma 4.1.3**

$$D_2\left(\frac{\wp(\gamma)-\wp(\alpha)}{\wp(\beta)-\wp(\alpha)}\right) = \mathcal{L}_{1,1}(\alpha,\beta,\gamma) + \mathcal{L}_{1,1}(-\alpha,\beta,\gamma) + \mathcal{L}_{1,1}(\alpha,-\beta,\gamma)$$
$$+ \mathcal{L}_{1,1}(\alpha,\beta,-\gamma) - 2\Big(\mathcal{L}_{1,1}(0,\beta,\gamma) + \mathcal{L}_{1,1}(0,-\beta,\gamma) + \mathcal{L}_{1,1}(\alpha,0,\gamma)$$
$$+ \mathcal{L}_{1,1}(-\alpha,0,\gamma) + \mathcal{L}_{1,1}(\alpha,\beta,0) + \mathcal{L}_{1,1}(-\alpha,\beta,0)\Big). \quad (14)$$

## 4.2 The elliptic $(1,1)$-logarithms as a combination of dilogarithms

Now we combine the expressions of the preceding lemma to cancel almost all terms on the r.-h.s.

**Theorem 4.2.1** *For any two nontrivial torsion points $\alpha \neq \pm\beta$ on an elliptic curve $E_\tau$, say $m\alpha = n\beta = 0$ for some $m, n \in \mathbb{N}$. Then*

$$\sum_{k=1}^{m}\sum_{l=1}^{n} D_2\Big(\frac{\wp(k\alpha+l\beta)-\wp(\alpha)}{\wp(\beta)-\wp(\alpha)}\Big)$$
$$= -2mn(\mathcal{L}_{1,1}(\alpha,\beta,0) + \mathcal{L}_{1,1}(-\alpha,\beta,0)). \quad (15)$$

**Sketch proof** We must check that all except the two last terms on the r.-h.s. of (14) cancel after summation. We show that the first one cancels:

$$\begin{aligned}\mathcal{L}_{1,1}(\alpha,\beta,k\alpha+l\beta) &= \mathcal{L}_{1,1}(\alpha-(\alpha+\beta),\beta-(\alpha+\beta),k\alpha+l\beta-(\alpha+\beta))\\ &= \mathcal{L}_{1,1}(-\beta,-\alpha,(k-1)\alpha+(l-1)\beta)\\ &= \mathcal{L}_{1,1}(\beta,\alpha,(1-k)\alpha+(1-k)\beta)\\ &= -\mathcal{L}_{1,1}(\alpha,\beta,(1-k)\alpha+(1-k)\beta);\end{aligned}$$

hence the first summands with arguments $(k,l)$ and $(1-k,1-l)$ only differ by the sign and the first summands cancel on averaging. The arguments for the other terms are similar.

This completes the proof of Theorem B and hence of the formula for $\mathcal{K}_2(0,\tau)$ in the Main Theorem.

**Remark 4.2.2** We have proved that the *combination*

$$\mathcal{L}_{1,1}(\alpha,\beta,0)+\mathcal{L}_{1,1}(-\alpha,\beta,0)$$

is equal to a sum of dilogarithms for torsion points $\alpha$ and $\beta$. It is even true that the single term $\mathcal{L}_{1,1}(\alpha,\beta,0)$ is equal to a combination of dilogarithms; but we do not need this in this expression for the Main Theorem, and it is rather complicated. We will derive it in Section 6.

# 5 Vanishing of the map $\delta$

In this section we prove that the argument of the dilogarithm in the Main Theorem belongs to the kernel of the map $\delta$ of 1.2.3.

## 5.1 Values of the $\theta$-function at torsion points

**5.1.1** The theta function

$$\widetilde{\theta}(\xi,\tau)=\frac{\theta(\xi,\tau)}{\eta(\tau)}=q^{1/12}(z^{\frac{1}{2}}-z^{-\frac{1}{2}})\prod_{j=1}^{\infty}(1-q^jz)(1-q^jz^{-1})$$

is not elliptic. It is only *quasiperiodic*"

$$\widetilde{\theta}(\xi+1,\tau)=-\widetilde{\theta}(\xi,\tau)\quad\text{and}\quad\widetilde{\theta}(\xi+\tau,\tau)=-z^{-1}q^{-1/2}\times\widetilde{\theta}(\xi,\tau);$$

but for any torsion point $\xi=r\tau+s$ with $r,s\in\frac{1}{N}\mathbb{Z}$ of order $N$, we can redefine the *value* of the theta function at this point by the formula:

$$\widetilde{\theta}[\xi](\tau)=z^{1/2+r}q^{-r^2/2-r}\widetilde{\theta}(\xi,\tau).$$

Translating the argument by a point of the lattice multiplies this value by some root of unity. We write $\equiv$ for equality modulo multiplication by a root of unity.

**Remark 5.1.2** If $\tau$ is imaginary quadratic over $\mathbb{Q}$, the numbers $\widetilde{\theta}[\xi](\tau)$ are *algebraic.*

**5.1.3** We have the theta decomposition

$$\wp(\alpha;\tau) - \wp(\beta;\tau) \equiv \frac{\widetilde{\theta}'(0,\tau)^2 \times \widetilde{\theta}[\alpha-\beta](\tau) \times \widetilde{\theta}[\alpha+\beta](\tau)}{\widetilde{\theta}[\alpha](\tau)^2 \times \widetilde{\theta}[\beta](\tau)^2}.$$

for any two torsion $\alpha$ and $\beta$ points.

**5.1.4** Let $\alpha$ be a torsion point on the elliptic curve $E_{M(\tau)}$. Then

$$\prod_{\beta\in M^{-1}\alpha} \widetilde{\theta}[\beta](\tau) \equiv \widetilde{\theta}[\alpha](M(\tau)) \quad \text{if } \alpha \neq 0; \text{ and}$$

$$\widetilde{\theta}'(0,\tau) \times \prod_{\beta\in \operatorname{Ker} M\backslash 0} \widetilde{\theta}[\beta](\tau) \equiv \frac{ad-bc}{c\tau+d} \times \widetilde{\theta}'(0,M(\tau)).$$

## 5.2 Computations

**5.2.1** We first calculate the value of the map $\delta$ on $\frac{\wp(\gamma)-\wp(\alpha)}{\wp(\beta)-\wp(\alpha)}$

$$\delta\left(\frac{\wp(\gamma)-\wp(\alpha)}{\wp(\beta)-\wp(\alpha)}\right) = \left(\frac{\wp(\gamma)-\wp(\alpha)}{\wp(\beta)-\wp(\alpha)}\right) \wedge \left(\frac{\wp(\gamma)-\wp(\beta)}{\wp(\alpha)-\wp(\beta)}\right)$$
$$\equiv \left(\frac{\widetilde{\theta}[\alpha-\gamma]\times\widetilde{\theta}[\alpha+\gamma]\times\widetilde{\theta}[\beta]^2}{\widetilde{\theta}[\alpha-\beta]\times\widetilde{\theta}[\alpha+\beta]\times\widetilde{\theta}[\gamma]^2}\right) \wedge \left(\frac{\widetilde{\theta}[\beta-\gamma]\times\widetilde{\theta}[\beta+\gamma]\times\widetilde{\theta}[\alpha]^2}{\widetilde{\theta}[\alpha-\beta]\times\widetilde{\theta}[\alpha+\beta]\times\widetilde{\theta}[\gamma]^2}\right).$$

Thus the answer is the same as the result of the following procedure:

1. Define the map $\nu$ by the formula

$$\{\alpha,\beta,\gamma\} \to -\Big(\{\alpha-\beta\}\wedge\{\beta-\gamma\} + \{\beta-\gamma\}\wedge\{\gamma-\alpha\} + \{\gamma-\alpha\}\wedge\{\alpha-\beta\}\Big).$$

2. Apply $\nu$ to the arguments of the function $\mathcal{L}_{1,1}$ on the r.-h.s. of (14).

3. Apply the map $\bigwedge^2\widetilde{\theta}\colon \xi_1\wedge\xi_2 \to \widetilde{\theta}[\xi_1]\wedge\widetilde{\theta}[\xi_2] \in \bigwedge^2\mathbb{C}^*$ to the result of the previous step.

**5.2.2** The map $\bigwedge^2 \widetilde{\theta} \circ \nu$ satisfies the same properties as $\mathcal{L}_{1,1}$ under translations or permutations of arguments. Hence after summation over $\gamma = j\alpha + k\beta$, we get

$$\delta\left(\sum_{k=1}^{m}\sum_{l=1}^{n}\left\{\frac{\wp(k\alpha+l\beta)-\wp(\alpha)}{\wp(\beta)-\wp(\alpha)}\right\}\right)$$
$$\equiv -2mn \wedge^2 \widetilde{\theta} \circ \nu(\{\alpha,\beta,0\}+\{-\alpha,\beta,0\})$$
$$\equiv 2mn(\beta)\Big(\widetilde{\theta}[\alpha-\beta]\wedge\widetilde{\theta}[\beta]+\widetilde{\theta}[\beta]\wedge\widetilde{\theta}[\alpha]+\widetilde{\theta}[\alpha]\wedge\widetilde{\theta}[\alpha-\beta]$$
$$+\widetilde{\theta}[\alpha+\beta]\wedge\widetilde{\theta}[\beta]+\widetilde{\theta}[\beta]\wedge\widetilde{\theta}[\alpha]+\widetilde{\theta}[\alpha]\wedge\widetilde{\theta}[\alpha+\beta]\Big).$$

**5.2.3** Finally, we perform the last summation:

$$\delta\left(\sum_{\substack{\alpha\in\operatorname{Ker}M\backslash 0\\ \beta\in\operatorname{Ker}(M-1)\backslash 0}}\sum_{k=1}^{m}\sum_{l=1}^{n}\left\{\frac{\wp(k\alpha+l\beta)-\wp(\alpha)}{\wp(\beta)-\wp(\alpha)}\right\}\right)$$
$$\equiv 4mn\sum_{\substack{\alpha\in\operatorname{Ker}M\backslash 0\\ \beta\in\operatorname{Ker}(M-1)\backslash 0}}\Big(\widetilde{\theta}[\alpha-\beta]\wedge\widetilde{\theta}[\beta]+(\widetilde{\theta}[\beta]\wedge\widetilde{\theta}[\alpha]+\widetilde{\theta}[\alpha]\wedge\widetilde{\theta}[\alpha-\beta]\Big).$$

now sum the first terms over the $\alpha$, the second terms over the $\alpha$ and $\beta$ and the third terms over the $\beta$. We get

$$4mn\left(\sum_{\beta\in\operatorname{Ker}(M-1)\backslash 0}\Big(\widetilde{\theta}[M\beta]\wedge(\widetilde{\theta}[\beta]-\widetilde{\theta}[\beta]\wedge(\widetilde{\theta}[\beta]\Big)\right.$$
$$+\left(\frac{(a-1)(d-1)-bc}{c\tau+d-1}\right)\wedge\left(\frac{ad-bc}{c\tau+d}\right)$$
$$\left.+\sum_{\alpha\in\operatorname{Ker}M\backslash 0}\Big(\widetilde{\theta}[\alpha]\wedge\widetilde{\theta}[(M-1)\alpha]-\widetilde{\theta}[\alpha]\wedge\widetilde{\theta}[\alpha]\Big)\right)$$
$$=4mn\times\left(\frac{(a-1)(d-1)-bc}{c\tau+d-1}\right)\wedge\left(\frac{ad-bc}{c\tau+d}\right).$$

This expression is the negative of $\delta$ of the first term $4mn\left\{\frac{ad-bc}{c\tau+d}\right\}$ in the formula (14).

# 6 Values of the Kronecker double series at torsion points

In this section we express the value of the Kronecker double series $\mathcal{K}_2(\alpha;\tau)$ at a CM point $\tau$ and a torsion point $\alpha$ on the elliptic curve $E_\tau$ as a sum of dilogarithms. In (12) we proved that this value of the Kronecker double series can be reduced to a combination of $\mathcal{L}_{1,1}(\alpha', 0, \gamma')$ (with $\alpha' \in M^{-1}(\alpha)$ and $\gamma' \in (M-1)^{-1}(\alpha)$), and $\mathcal{K}_2(0,\tau)$. So we will prove that $\mathcal{L}_{1,1}(\alpha', \gamma', 0)$ equals a sum of dilogarithms.

## 6.1 Reduction to "standard" functions

**6.1.1** Let $\alpha$ be a torsion point of exact order $N(\alpha) = 2^{\rho(\alpha)} \times N_0(\alpha)$, with odd $N_0(\alpha)$. Denote by $f_\alpha$ a function with divisor $N(\alpha) \times (\alpha - 0)$. Clearly, $\log|f_\alpha(\eta)|^2 = \text{const} + N\mathcal{L}_1(\eta - \alpha) - N\mathcal{L}_1(\eta)$. This yields the formula

**Lemma 6.1.2** *For any three distinct torsion points $\alpha$, $\beta$ and $\gamma$*

$$\frac{1}{4\pi}\int_{E_\tau} A_3(f_\alpha, f_\beta, f_\gamma) = N(\alpha)N(\beta)N(\gamma) \times \Big(\mathcal{L}_{1,1}(\alpha,\beta,\gamma) - \mathcal{L}_{1,1}(0,\beta,\gamma) - \mathcal{L}_{1,1}(\alpha,0,\gamma) - \mathcal{L}_{1,1}(\alpha,\beta,0)\Big). \quad (16)$$

**Lemma 6.1.3** *For any distinct nontrivial torsion points $\alpha$ and $\beta$ on the elliptic curve $E_\tau$*

$$\sum_{k=1}^{N(\alpha)}\sum_{l=1}^{N(\beta)} \frac{1}{N(\alpha)^2 N(\beta)^2 N(k\alpha + l\beta)} \int_{E_\tau} A_3(f_\alpha, f_\beta, f_{k\alpha+l\beta}) = -\mathcal{L}_{1,1}(\alpha,\beta,0) \quad (17)$$

The proof is parallel to that of (15). We now reduce the integrals on the l.-h.s. of the previous equation to "dilogarithmic" integrals. First, we have a decomposition of $f_\alpha(\eta)$ into a product of "standard" functions $g_\xi = \wp(\eta-\xi) - \wp(\xi)$. Let $\lambda = \lambda(\alpha)$ be a natural number such that $2^\lambda \equiv 1 \mod (N_0)$ and write $\alpha_0$ for the point $2^\rho\alpha$.

**Lemma 6.1.4**

$$f_\alpha(\eta)^{-\frac{2^\lambda - 1}{N_0}} = \text{const}\prod_{i=0}^{\rho-1}\big(\wp(\eta - 2^i\alpha) - \wp(2^i\alpha)\big)^{2^{\rho-i-1}(2^\lambda-1)} \times \prod_{j=0}^{\lambda-1}\big(\wp(\eta - 2^j\alpha_0) - \wp(2^j\alpha_0)\big)^{2^{\lambda-j-1}} \quad (18)$$

**Proof** Compare the divisors of both sides. $\square$

**Remark 6.1.5** If $F$ is a function on an elliptic curve and $\xi$ a torsion point, we can define the operation of *averaging with a factor* 2 by the formula

$$\begin{aligned}\mathrm{Av}_2^{\xi}(F)(\xi) &= \sum_{j=0}^{\infty} 2^{-j} F(2^j \xi) \\ &= \sum_{j=0}^{\rho(\xi)-1} 2^{-j} F(2^j \xi) + \sum_{j=\rho(\xi)}^{\rho(\xi)+\lambda(\xi)-1} \frac{2^{-j}}{1-2^{-\lambda(\xi)}} F(2^j \xi). \qquad (19)\end{aligned}$$

The last equality is nothing more than the formula for the sum of a geometric progression. Hence we can rewrite the statement of the preceding lemma formally as

$$\log(f_\alpha(\eta)) = -\frac{N(\alpha)}{2} \mathrm{Av}_2^{\alpha}(\log(\wp(\eta-\alpha)-\wp(\alpha)))$$

the superscript $\alpha$ denotes the variable over which we average.

## 6.2 From standard functions to dilogarithms

**Lemma 6.2.1** *Let $k \subset K$ be a quadratic field extension, and write $\sigma$ for the involution of $K$ over $k$. Then*

$$f \times \frac{g^\sigma - g}{fg^\sigma - gf^\sigma} + g \times \frac{f^\sigma - f}{gf^\sigma - fg^\sigma} = 1 \quad \textit{and} \quad \frac{g^\sigma - g}{fg^\sigma - gf^\sigma} \in k.$$

*for any $f, g \in K \setminus k$.*

The proof is obvious.

**6.2.2** We apply this lemma to the extension $\mathbb{C}(\mathbb{P}^1) \subset \mathbb{C}(E)$ and "standard" functions. The involution $\sigma$ is given by changing sign of the argument of a function.

**Lemma 6.2.3** *The "standard" functions $g_\alpha(\eta) = \wp(\eta-\alpha)-\wp(\alpha)$ and $g_\beta(\eta) = \wp(\eta-\beta)-\wp(\beta)$ satisfy*

$$\frac{g_\beta^\sigma(\eta) - g_\beta(\eta)}{g_\alpha(\eta) g_\beta^\sigma(\eta) - g_\beta(\eta) g_\alpha^\sigma(\eta)} = K(\alpha,\beta) \times \frac{(\wp(\eta)-\wp(\alpha))^2}{\wp(\eta) - C(\alpha,\beta)},$$

*where*

$$C(\alpha,\beta) = \frac{\wp(\alpha)\wp'(\beta)\left(\wp(\alpha+\beta)-\wp(\alpha)\right) + \wp(\beta)\wp'(\alpha)\left(\wp(\alpha+\beta)-\wp(\beta)\right)}{\wp'(\beta)\left(\wp(\alpha+\beta)-\wp(\alpha)\right) + \wp'(\alpha)\left(\wp(\alpha+\beta)-\wp(\beta)\right)},$$

*and*

$$K(\alpha,\beta) = \frac{-\wp'(\beta)}{(\wp(\alpha)-\wp(\beta)) \times \left(\wp'(\beta)\left(\wp(\alpha+\beta)-\wp(\alpha)\right) + \wp'(\alpha)\left(\wp(\alpha+\beta)-\wp(\beta)\right)\right)}$$

**Proof** We consider separately the numerator and denominator on the l.-h.s. The numerator $g^\sigma_\beta - g_\beta$ is odd, so it vanishes at half-periods. On the other hand it has double poles at $\eta = \pm\beta$. Hence,

$$g^\sigma_\beta - g_\beta = c \times \frac{\wp'(\eta)}{(\wp(\eta) - \wp(\beta))^2}.$$

The constant $c$ can be calculated by considering the leading term at $\eta = \beta$, and is equal to $-\wp'(\beta)$.

The denominator $g_\alpha g^\sigma_\beta - g_\beta g^\sigma_\alpha$ is also odd. It has a zero of order $\geq 2$ at 0, and hence (by oddness) of order $\geq 3$. On the other hand, it has double poles at $\eta = \pm\alpha$ and $\pm\beta$. Therefore,

$$g_\alpha g^\sigma_\beta - g_\beta g^\sigma_\alpha = c' \times \frac{\wp'(\eta) \times (\wp(\eta) - C(\alpha, \beta))}{(\wp(\eta) - \wp(\alpha))^2 \times (\wp(\eta) - \wp(\beta))^2}$$

for some constants $c'$ and $C(\alpha, \beta)$. These constants can be calculated by computing the leading terms at $\alpha$ and $\beta$;

$$c' = -\big(\wp(\alpha) - \wp(\beta)\big)\big(\wp'(\beta)\big(\wp(\alpha+\beta) - \wp(\alpha)\big) - \wp'(\alpha)\big(\wp(\alpha+\beta) - \wp(\beta)\big)\big).$$

□

**6.2.4** We write $F_{\alpha\beta}$ for the function $\frac{g^\sigma_\beta - g_\beta}{g_\alpha g^\sigma_\beta - g_\beta g^\sigma_\alpha}$. Thus the "non-obvious" zeros $\pm\xi_{\alpha\beta}$ of the function $F_{\alpha\beta}$ are defined by the condition $\wp(\pm\xi_{\alpha\beta}) = C(\alpha, \beta)$ (and $\wp'(\pm\xi_{\alpha\beta}) = \pm\sqrt{4C(\alpha,\beta)^3 - c_4(\tau)C(\alpha,\beta) - c_6(\tau)}$, where $c_4$ and $c_6$ are the coefficients of the Weierstrass equation).

**Lemma 6.2.5** *For any three points $\alpha$, $\beta$ and $\gamma$ on an elliptic curve*

$$\begin{aligned}
&g_\alpha \wedge g_\beta \wedge g_\gamma + g^\sigma_\alpha \wedge g^\sigma_\beta \wedge g^\sigma_\gamma \\
&\quad = (g_\alpha F_{\alpha\beta}) \wedge (g_\beta F_{\beta\alpha}) \wedge g_\gamma + (g^\sigma_\alpha F_{\alpha\beta}) \wedge (g^\sigma_\beta F_{\beta\alpha}) \wedge g^\sigma_\gamma \\
&\quad - \Big((g_\alpha F_{\alpha\gamma}) \wedge F_{\beta\alpha} \wedge (g_\gamma F_{\gamma\alpha}) + (g^\sigma_\alpha F_{\alpha\gamma}) \wedge F_{\beta\alpha} \wedge (g^\sigma_\gamma F_{\gamma\alpha}) \\
&- F_{\alpha\beta} \wedge (g_\beta F_{\beta\gamma}) \wedge (g_\gamma F_{\gamma\beta}) + F_{\alpha\beta} \wedge (g^\sigma_\beta F_{\beta\gamma}) \wedge (g^\sigma_\gamma F_{\gamma\beta})\Big) - F_{\alpha\beta} \wedge F_{\beta\alpha} \wedge (g_\gamma g^\sigma_\gamma) \\
&\quad + \Big((g_\alpha g^\sigma_\alpha) \wedge F_{\beta\alpha} \wedge F_{\gamma\alpha} + F_{\alpha\gamma} \wedge F_{\beta\alpha} \wedge (g_\gamma g^\sigma_\gamma) + 2 \times (F_{\alpha\gamma} \wedge F_{\beta\alpha} \wedge F_{\gamma\alpha}) \\
&\quad + F_{\alpha\beta} \wedge (g_\beta g^\sigma_\beta) \wedge F_{\gamma\beta} + F_{\alpha\beta} \wedge F_{\beta\gamma} \wedge (g_\gamma g^\sigma_\gamma) + 2 \times (F_{\alpha\beta} \wedge (F_{\beta\gamma}) \wedge F_{\gamma\beta})\Big)
\end{aligned}$$

The proof is a straightforward computation.

**6.2.6** The first six terms on the r.-h.s. of the previous equation are of the form $\varphi\wedge(1-\varphi)\wedge\psi$, so that by (6) the integral of $A_3$ of any such term equals a sum of dilogarithms. The last seven terms contain only $\sigma$-invariant functions, so the corresponding integrals can be reduced to integrals over $\mathbb{CP}^1$ and are also equal to combinations of dilogarithms. On the other hand, the integral of $A_3$ of the second term on the l.h.s is equal to the corresponding integral of the first one for obvious geometric reasons.

## 6.3 Results

If we combine all previous results, we get the following.

**Theorem 6.3.1** *For any two distinct nontrivial torsion points $\alpha$ and $\beta$ on an elliptic curve, we have*

$$\mathcal{L}_{1,1}(\alpha,\beta,0) = D_2(\Phi(\alpha,\beta)), \tag{20}$$

*where*

$$\Phi(\alpha,\beta) = \frac{1}{\operatorname{ord}(\alpha)\operatorname{ord}(\beta)} \sum_{\gamma\in\langle\alpha,\beta\rangle\setminus 0} \mathrm{Av}_2^{\alpha}\,\mathrm{Av}_2^{\beta}\,\mathrm{Av}_2^{\gamma}\,\Phi_1(\alpha,\beta,\gamma);$$

*here $\langle\alpha,\beta\rangle$ denotes the subgroup generated by $\alpha$ and $\beta$, $\mathrm{Av}_2^{\xi}(F)(\xi)$ is defined by (19) for a torsion point $\xi$, and*

$$\begin{aligned}
\Phi_1(\alpha,\beta,\gamma) &= \{G_{\alpha\beta}(\gamma)\}_2 + 8\Big\{\frac{\wp(\gamma)-\wp(\alpha)}{\wp(\beta)-\wp(\alpha)}\Big\}_2 \\
&+ \mathrm{Alt}_{\alpha,\beta}\Big(2\{G_{\alpha\gamma}(\beta)\}_2 + 2\{G_{\alpha\gamma}(-\beta)\}_2 - \{G_{\alpha\gamma}(\xi_{\beta\alpha})\}_2 - \{G_{\alpha\gamma}(-\xi_{\beta\alpha})\}_2\Big) \\
&+ 4\,\mathrm{Cyc}_{\alpha,\beta,\gamma}\Big(\Big\{\frac{\wp(2\gamma)-\wp(\alpha)}{\wp(\beta)-\wp(\alpha)}\Big\}_2\Big) - 4\,\mathrm{Cyc}_{\alpha,\beta,\gamma}\Big(\Big\{\frac{C(\alpha,\gamma)-\wp(\alpha)}{\wp(\beta)-\wp(\alpha)}\Big\}_2\Big) \\
&- 2\,\mathrm{Alt}_{\alpha,\beta,\gamma}\Big(\Big\{\frac{\wp(2\gamma)-\wp(\alpha)}{C(\beta,\gamma)-\wp(\alpha)}\Big\}_2\Big) + \mathrm{Alt}_{\alpha,\beta}\Big(2\Big\{\frac{C(\alpha,\gamma)-\wp(\alpha)}{C(\alpha,\beta)-\wp(\alpha)}\Big\}_2 \\
&+ 2\Big\{\frac{\wp(\gamma)-C(\alpha,\gamma)}{C(\alpha,\beta)-C(\alpha,\gamma)}\Big\}_2 + \Big\{\frac{\wp(2\gamma)-C(\alpha,\gamma)}{C(\alpha,\beta)-C(\alpha,\gamma)}\Big\}_2 + \Big\{\frac{C(\alpha,\gamma)-\wp(2\alpha)}{C(\alpha,\beta)-\wp(2\alpha)}\Big\}_2\Big),
\end{aligned}$$

*where*

$$G_{\alpha,\beta}(\eta) = \Big(\wp(\eta-\alpha)-\wp(\alpha)\Big) \times K(\alpha,\beta)\frac{(\wp(\eta)-\wp(\alpha))^2}{\wp(\eta)-C(\alpha,\beta)}.$$

*Here $C(\alpha,\beta)$ and $K(\alpha,\beta)$ are as in Lemma 6.2.3., the points $\pm\xi_{\alpha\beta}$ are solutions of the equation $\wp(\pm\xi_{\alpha\beta}) = C(\alpha,\beta)$; $\mathrm{Alt}_{\alpha,\beta}$,$\mathrm{Alt}_{\alpha,\beta,\gamma}$, $\mathrm{Cyc}_{\alpha,\beta,\gamma}$ denote the (anti)symmetrization with respect to $S_2$, $S_3$, $A_3$ (with a factor 1).*

**Theorem 6.3.2** *Let $\tau$ be a fixed point of $M = \begin{pmatrix} a & b \\ c & d \end{pmatrix} \neq 0, 1$; $m = \det M$, $n = \det(M - 1)$. Let $\alpha$ be a torsion point on the curve $E_\tau$. Then*

$$\frac{c(\tau - \overline{\tau})}{i}\Big(\frac{m+n+1}{mn}\mathcal{K}_2(\alpha;\tau) + \mathcal{K}_2(0;\tau)\Big) = D_2\Big(\sum_{\substack{\alpha' \in M^{-1}(\alpha) \\ \gamma' \in (M-1)^{-1}(\alpha)}} \Phi(\alpha', \gamma')\Big), \quad (21)$$

*where $\Phi$ is defined in the preceding theorem. The argument of the dilogarithm belongs to the kernel of the map $\delta\colon \{x\}_2 \to x \wedge (1 - x)$.*

We have proved all statements of this theorem except the last one. It can be proved by the same consideration as in Section 5.

# References


[D1] C. Deninger, Higher regulators and Hecke $L$-series of imaginary quadratic fields. I, *Invent. Math.* **96** (1989) 1–69. II, *Ann. of Math.(2)* **132** (1990) 131–158

[D2] C. Deninger, Higher operations in Deligne cohomology, *Invent. Math.* **120** (1995) 289–315

[G1] A.B. Goncharov, Chow polylogarithms and regulators, *Math. Res. Lett.* **2** (1995) 95–112

[G2] A.B. Goncharov, Deninger's conjecture of $L$-function of elliptic curves at $s = 3$, Algebraic geometry, 4 *J.Math.Sci.* **81** (1996) 2631–2656

[W] Andre Weil, *Elliptic functions according to Eisenstein and Kronecker*, Springer-Verlag, Berlin-New York 1976



Andrey Levin,
L. D. Landau Institute of Theoretical Physics,
Russian Academy of Sciences,
117940 Moscow, Russia
e-mail: andrl@landau.ac.ru and alevin@mpim-bonn.mpg.de

# References

# On Shafarevich–Tate groups and the arithmetic of Fermat curves

William G. McCallum    Pavlos Tzermias

*To Sir Peter Swinnerton-Dyer on his 75th birthday.*

## 1 Introduction

Let $\mathbb{Q}$ denote the field of rational numbers and $\overline{\mathbb{Q}}$ a fixed algebraic closure of $\mathbb{Q}$. For a fixed prime $p$ such that $p \geq 5$, choose a primitive $p$th root of unity $\zeta$ in $\overline{\mathbb{Q}}$ and let $K = \mathbb{Q}(\zeta)$. If $a$, $b$ and $c$ are integers such that $0 < a, b, a+b < p$ and $a + b + c = 0$, let $F_{a,b,c}$ denote a smooth projective model of the affine curve

$$y^p = x^a(1-x)^b \tag{1.1}$$

and let $J_{a,b,c}$ be the Jacobian of $F_{a,b,c}$. Then $J_{a,b,c}$ has complex multiplication induced by the birational automorphism $(x, y) \mapsto (x, \zeta y)$ of $F_{a,b,c}$. Let $\lambda$ denote the endomorphism $\zeta - 1$ of $J_{a,b,c}$. Note that $\lambda^{p-1}$ is, up to a unit in $\mathbb{Z}[\zeta]$, multiplication by $p$ on $J_{a,b,c}$.

We are interested in the Shafarevich–Tate group of $J_{a,b,c}$ over $K$, which we denote simply by Ш. In [McC88], the first author studied the restriction of the Cassels–Tate pairing

$$Ш[\lambda] \times Ш[\lambda] \longrightarrow \mathbb{Q}/\mathbb{Z} \tag{1.2}$$

and showed that Ш$[\lambda]$ is nontrivial in certain cases depending on the reduction type of the minimal regular model of $F_{a,b,c}$ over $\mathbb{Z}_p[\zeta]$. The purpose of this paper is to extend those results by carrying out higher descents, and to derive some consequences for the arithmetic of Fermat curves using the techniques of the second author.

First we recall the main result of [McC88]. The possible reduction types for $F_{a,b,c}$ are as shown in Figure 1 [McC82], with the proper transform of the special fiber of the model (1.1) indicated. The wild type is further divided into split and nonsplit, according to whether the two tangent components are

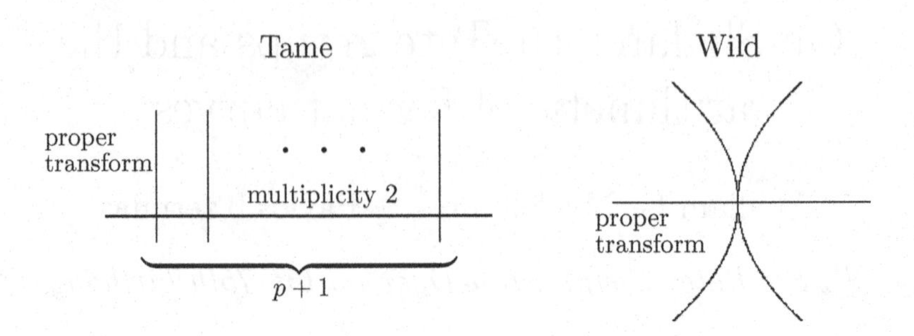


Figure 1: Reduction types of $F_{a,b,c}$

defined over the finite field $\mathbb{F}_p$ or conjugate over a quadratic extension. The reduction type can be computed as follows. For a rational number $x$ of $p$-adic valuation 0, let $q(x) = (x^{p-1} - 1)/p$, viewed as an element of $\mathbb{F}_p$. Then $F_{a,b,c}$ is

$$\begin{cases} \text{tame} & \text{if } -2abcq(a^a b^b c^c) = 0, \\ \text{wild split} & \text{if } -2abcq(a^a b^b c^c) \in \mathbb{F}_p^{\times 2}, \\ \text{wild nonsplit} & \text{if } -2abcq(a^a b^b c^c) \notin \mathbb{F}_p^{\times 2}. \end{cases}$$

Let $M_K$ be the set of finite places of $K$ and let $w$ denote the unique place of $K$ above $p$. Define

$$\begin{aligned} U &= \{x \in K^\times/K^{\times p} : v(x) \equiv 0 \pmod p \text{ for all } v \in M_K\}, \\ V &= K_w^\times/K_w^{\times p}. \end{aligned} \tag{1.3}$$

Let $\pi$ be the uniformizer of $K_w$ defined by

$$\pi^{p-1} = -p \quad \text{and} \quad \frac{\pi}{1-\zeta} \equiv 1 \pmod w. \tag{1.4}$$

If $\kappa\colon \mathrm{Gal}(K/\mathbb{Q}) \to \mathbb{Z}_p^\times$ is the Teichmüller character, let $V(i)$ denote the intersection of the $\kappa^i$th eigenspace of $V$ with the subgroup of $V$ generated by units congruent to 1 modulo $\pi^i$. Thus $V(i)$ is one-dimensional if $2 \le i \le p$.

**Theorem 1.1 ([McC88])** *Suppose that $F_{a,b,c}$ is wild split, $p \equiv 1 \pmod 4$, and the image of $U$ is nontrivial in both $V((p-1)/2)$ and $V((p+3)/2)$. Then*

$$Ш[\lambda]/\lambda Ш[\lambda^2] \simeq \mathbb{Z}/p\mathbb{Z} \oplus \mathbb{Z}/p\mathbb{Z}.$$

The condition on $U$ is satisfied if $p \nmid B_{(p-1)/2}B_{(p+3)/2}$, where $B_k$ is the $k$th Bernoulli number. As noted in [McC88], the technique used to prove Theorem 1.1 applies to the pairing

$$Ш[\lambda^2] \times Ш[\lambda] \longrightarrow \mathbb{Q}/\mathbb{Z} \tag{1.5}$$

and yields information about $Ш[\lambda^2]$.

**Theorem 1.2** *Suppose that either of the following conditions is satisfied:*

*(a)* $F_{a,b,c}$ *is wild split and* $p \equiv 3 \pmod 4$*;*

*(b)* $F_{a,b,c}$ *is wild nonsplit or tame and the image of* $U$ *in either* $V((p+1)/2)$ *or* $V((p+3)/2)$ *is trivial.*

*Then the pairing* (1.5) *is trivial. Thus* $Ш[\lambda^2]/\lambda Ш[\lambda^3] = 0$, *that is,* $Ш[\lambda^3]$ *is a free module over* $\mathbb{Z}[\zeta]/(\lambda^3)$.

As discussed in [McC88], the hypothesis on $U$ in condition (b) of the theorem is quite mild, since for $U$ to be nontrivial in $V(k)$ with $k > 1$ and odd requires that $p$ divides $B_{p-k}$.

**Corollary 1.3** *If one of conditions (a) or (b) of Theorem 1.2 is satisfied, and if* $|Ш[p^\infty]| < p^3$, *then* $Ш[p^\infty] = 0$.

Under the conditions of Theorem 1.2, it is natural to ask which occurs more often: $|Ш[p]| = 0$ or $|Ш[p]| \geq p^3$. To explore this question, we compute

$$Ш[\lambda^3] \times Ш[\lambda] \longrightarrow \mathbb{Q}/\mathbb{Z}. \tag{1.6}$$

**Theorem 1.4** *Suppose that* $p \geq 19$ *is regular,* $p \equiv 3 \pmod 4$, $F_{a,b,c}$ *is tame or wild nonsplit and*

$$q(a^a b^b c^c)^3 + abcB_{p-3} \not\equiv 0 \pmod p. \tag{1.7}$$

*Then the pairing* (1.6) *is nontrivial. Thus* $Ш[\lambda^3] \neq 0$ *(and hence, by Corollary 1.3,* $|Ш[p^\infty]| \geq p^3$*).*

For example, the curve $y^{19} = x^2(1-x)$ satisfies the conditions of the theorem. Modest numerical experiments suggest that about half the curves satisfy the conditions. More precisely, there are about $p/6$ isomorphism classes of curves $F_{a,b,c}$ for a given prime $p$, and heuristically about half of them are tame or wild nonsplit. The incongruence (1.7) is usually satisfied for these curves; for example, it is satisfied for all such curves if $p < 100$ (and $p \equiv 3 \pmod 4$).

The next result shows that, in certain cases, one can combine Theorems 1.2 and 1.4 to describe the exact structure of $Ш[p^\infty]$:

**Theorem 1.5** *Suppose that* $p$, $a$, $b$ *and* $c$ *are chosen to satisfy the hypotheses of both Theorems 1.2 and 1.4. If, in addition, the free* $\mathbb{Z}[\zeta]/(\lambda^3)$*-module* $Ш[\lambda^3]$ *has rank* 2, *then*

$$Ш[p^\infty] = Ш[\lambda^3] \simeq (\mathbb{Z}[\zeta]/(\lambda^3))^2.$$

In Section 6 we establish the following application of the above results:

**Theorem 1.6** *Let $p = 19$, $a = 7$, $b = 1$. Then*

1. $Ш[p^\infty] \simeq (\mathbb{Z}[\zeta]/(\lambda^3))^2$.

2. *The Mordell–Weil rank of $J_{7,1,-8}$ over $\mathbb{Q}$ equals 1.*

3. *The only quadratic points (i.e. algebraic points whose field of definition is a quadratic extension of $\mathbb{Q}$) on the Hurwitz–Klein curve $F_{7,1,-8}$ and also on the Fermat curve $X^{19} + Y^{19} + Z^{19} = 0$ are those described by Gross and Rohrlich in [GR78].*

We also note that, by combining Theorem 1.4 with Faddeev's bounds in [Fad61], one gets that the Mordell–Weil rank (over $\mathbb{Q}$) of any tame or wild nonsplit quotient of the Fermat curve $F_{19}$ or $F_{23}$ is at most 2.

Lim [Lim95] has also stated a result attempting to improve on [McC88] in certain cases. However, in Section 6, we show that the hypotheses of Propositions A and B of [Lim95] are never simultaneously satisfied.

# 2 Formulas for the pairings

We recall the situation and notation of [McC88]. For $\phi \in \mathcal{O}_K$ and $F$ a field containing $K$, we write $\delta = \delta_{\phi,F}$ for the coboundary map $J(F) \to H^1(F, J[\phi])$. The $\phi$-Selmer group $S_\phi \subset H^1(K, J[\phi])$ is defined to be the subgroup whose specialization to each completion $K_v$ of $K$ lies in the image of $\delta_{\phi,K_v}$. It sits in an exact sequence

$$0 \to J(K)/\phi J(K) \to S_\phi \to Ш[\phi] \to 0.$$

For $\phi, \psi \in \mathrm{End}(J)$, we have a pairing

$$S_\phi \times S_{\hat{\psi}} \to \mathbb{Q}/\mathbb{Z}, \tag{2.1}$$

described in [McC88], which is a lift of the restriction of the Cassels pairing to $Ш[\phi] \times Ш[\hat{\psi}]$. An expression for the pairing (2.1) is given in [McC88], under a certain splitting hypothesis.

We use[McC88] to derive formulas for the pairings (1.5) and (1.6). The formula for (1.5) is a straightforward consequence of Theorem 2.6 in [McC88]; the formula for (1.6) takes more work. The point is that $J[\lambda^3] \subset J(K)$ (Greenberg [Gre81]), so that it is possible to express the pairings (1.2) and (1.5) as purely local pairings at $w$, as explained in [McC88]. However, by [Gre81] and Kurihara [Kur92], the $\lambda^4$-torsion on $J_{a,b,c}$ is not in general defined over $K$, introducing an essentially global aspect to the calculation of (1.6).

For technical reasons, it is convenient to replace $\lambda$ with an endomorphism (which we also denote by $\lambda$) that is congruent modulo $\lambda^5$ to the uniformizer $\pi$ defined by (1.4), since then

$$\lambda^\delta \equiv \kappa(\delta)\lambda \pmod{\lambda^5}, \quad \delta \in \mathrm{Gal}(K/\mathbb{Q}).$$

In particular, we have $\hat{\lambda} \equiv -\lambda$ modulo $\lambda^5$, and we will often replace $\hat{\lambda}$ with $-\lambda$ without mention in what follows, in cases where we are dealing with a module killed by $\lambda^5$. Furthermore, it suffices to prove Theorems 1.2 and 1.4 with this new choice of $\lambda$. Since $\lambda/\hat{\lambda}$ is a unit, $Ш[\lambda] = Ш[\hat{\lambda}]$, and we can proceed by computing the pairing $\langle\ ,\ \rangle_k$ mentioned in (2.1) with $\phi = \lambda^k$ and $\psi = \hat{\lambda}$.

The local formula for the Cassels–Tate pairing is expressed in terms of certain local descent maps as follows. Given a $p$-torsion point $Q$ in $J(\overline{K})$ we denote by $D_Q$ a divisor defined over $K(Q)$ representing $Q$ and by $f_Q$ a function on $F_{a,b,c}$ whose divisor is $pD_Q$. Evaluating $f_Q$ on divisors induces a map $\iota_Q\colon J(F) \to F^\times/F^{\times p}$ for any field $F$ containing $K(Q)$.

By [Gre81],

$$K(J[\lambda^3]) = K \quad \text{and} \quad K(J[\lambda^4]) = L = K(\eta_{p-3}^{1/p}), \tag{2.2}$$

where $\eta_{p-3}$ is a generator for the $\kappa^{p-3}$-eigenspace of the cyclotomic units in $K$. Let $\tilde{\Delta} \subset \mathrm{Gal}(L/\mathbb{Q})$ be a subgroup projecting isomorphically to $\mathrm{Gal}(K/\mathbb{Q})$. For $i = 1, 2, 3, 4$, we choose points $P_i$ of order $\lambda^i$ on $J$ and a generator $\sigma$ for $G = \mathrm{Gal}(L/K)$ such that

1. $P_1$ is the point represented by the divisor $(0,0) - \infty$;
2. $\lambda P_i = P_{i-1}$ for $i = 2, 3, 4$;
3. $P_i$ is an eigenvector for the action of $\tilde{\Delta}$ with character $\kappa^{1-i}$;
4. $\sigma P_4 = P_4 + P_1$.

For $i \leq 4$, let $e_{\lambda^i}(P, Q)$ be the $\lambda^i$ Weil pairing on $J[\lambda^i]$. We have an isomorphism $J[\lambda^i] \simeq \mu_p^i$ defined over $K(P_i)$ (and thus over $K$ for $i \leq 3$), given by

$$Q \mapsto (e_{\lambda^i}(Q, P_1), \dots, e_{\lambda^i}(Q, P_i)). \tag{2.3}$$

With this identification, by [McC88, Lemma 2.2], we have

$$\delta_i = \delta_{\lambda^i, K(P_i)} = \iota_{P_1} \times \cdots \times \iota_{P_i}.$$

Since $J$ has good reduction outside $p$ and $\lambda$ has degree $p$, we can regard $S_{\lambda^i}$ as a subgroup of $H^1(K(p)/K, J[\lambda^i])$, where $K(p)/K$ is the maximal extension of $K$ unramified outside $p$. As explained in Section 7 of [McC88], we can also regard $S_{\lambda^i}$ as a subgroup of $U^i$ for $i \leq 3$, where $U$ is as defined in (1.3). For $a, b \in K_w^\times$, denote by $(a, b)$ the Hilbert symbol.

**Proposition 2.1** *Let $a \in S_{\lambda^2}$, $b \in S_{\hat{\lambda}}$, $a_w = \delta(x_w)$, $x_w \in J(K_w)$. Then*

$$\zeta^{p\langle a,b\rangle_2} = (\iota_{P_3}(x_w), b_w).$$

**Proof** This follows from [McC88, Theorem 2.6], with $\phi = \lambda^2$ and $\psi = \lambda$. □

For a number field $F$ we denote by $\mathcal{O}'_F$ the ring of $p$-integers in $F$. Suppose $F \subset K(p)$ and let $C$ be the ideal class group of $\mathcal{O}'_F$. Since the group $\mathcal{O}'^{\times}_{K(p)}$ is $p$-divisible, we have an exact sequence

$$0 \to \mu_p \to \mathcal{O}'^{\times}_{K(p)} \xrightarrow{p} \mathcal{O}'^{\times}_{K(p)} \to 0,$$

which induces a long exact sequence of Galois cohomology

$$\begin{aligned} \cdots \to H^{i-1}(K(p)/F, \mathcal{O}'^{\times}_{K(p)}) \xrightarrow{p} H^{i-1}(K(p)/F, \mathcal{O}'^{\times}_{K(p)}) \to \\ H^i(K(p)/F, \mu_p) \to H^i(K(p)/F, \mathcal{O}'^{\times}_{K(p)}) \xrightarrow{p} H^i(K(p)/F, \mathcal{O}'^{\times}_{K(p)}) \to \cdots \end{aligned}$$

If $i = 1$ then, since $H^1(K(p)/F, \mathcal{O}'^{\times}_{K(p)})$ is isomorphic to $C$, we obtain the exact sequence

$$0 \to \mathcal{O}'^{\times}_F / \mathcal{O}'^{\times p}_F \to H^1(K(p)/F, \mu_p) \to C[p] \to 0. \tag{2.4}$$

Also, by [NSW00, VIII.3], it follows that $H^2(K(p)/F, \mathcal{O}'^{\times}_{K(p)})[p^\infty]$ can be identified with the subgroup $\mathrm{Br}(K(p)|F)[p^\infty]$ of $\mathrm{Br}(F)[p^\infty]$. Setting $i = 2$ in the above long exact cohomology sequence gives another exact sequence

$$0 \to C/pC \to H^2(K(p)/F, \mu_p) \to \mathrm{Br}(K(p)/F)[p] \to 0. \tag{2.5}$$

**Lemma 2.2** *Every element of $H^1(K(p)/K, J[\lambda^k])$ lifts to $H^1(K, J[\lambda^{k+1}])$. Moreover, if $p$ is regular, it lifts to $H^1(K(p)/K, J[\lambda^{k+1}])$.*

**Proof** Let $a \in H^1(K(p)/K, J[\lambda^k])$, and let $\delta a \in H^2(K(p)/K, J[\lambda])$ be the coboundary of $a$ for the sequence

$$0 \to J[\lambda] \to J[\lambda^{k+1}] \to J[\lambda^k] \to 0. \tag{2.6}$$

Then the inflation of $\delta a$ in $H^2(K, J[\lambda]) \simeq H^2(K, \mu_p) = \mathrm{Br}(K)[p]$ has zero invariant at every place not dividing $p$. Thus it is zero by the Brauer–Hasse–Noether theorem (since there is only one place of $K$ dividing $p$). For the second statement, we argue in the same way, using (2.5). □

We recall the definition of $\langle\ ,\ \rangle_3$. Let $a \in S_{\lambda^3}$ and $b \in S_{\hat{\lambda}}$. Lift $a$ to an element $a_1$ of $H^1(K, J[\lambda^4])$ (which is possible by Lemma 2.2). For each place

$v$ of $K$, lift $a_v$ to an element $a_{v,1}$ that is in the image of $\delta$. Then $a_{1,v} - a_{v,1}$ is the image of an element $c_v \in H^1(K_v, J[\lambda])$, and

$$\langle a, b\rangle = \sum_v c_v \cup b_v$$

where the cup product is with respect to the local pairing

$$H^1(K_v, J[\lambda]) \times H^1(K_v, J[\hat{\lambda}]) \to \mathbb{Q}/\mathbb{Z}.$$

If $p$ is regular, $L/K$ is totally ramified at $w$, and there is a unique extension of $w$ to $L$, that we also denote by $w$. Let $N' = \sum_{i=1}^{p-1} i\sigma^i$.

**Proposition 2.3** *Let $a \in S_{\lambda^3}$, $b \in S_\lambda$, $a_w = \delta(x_w)$, $x_w \in J(K_w)$. Suppose that $\lambda_*^2 a$, regarded as an element of $\mathcal{O}_K^\times/\mathcal{O}_K^{\times p}$, can be written as $N_{L/K}\epsilon$ for some $\epsilon \in \mathcal{O}_L^\times$. Then there exists a $\lambda^4$-torsion point $P_4$, and an element $c_w \in K_w^\times$ such that*

$$\zeta^{p\langle a,b\rangle_3} = (c_w, b_w),$$

*and the image of $c_w$ in $L_w^\times/L_w^{\times p}$ satisfies*

$$c_w = \iota_{P_4}(x_w)^{-1}\eta N'\epsilon,$$

*where $\eta \in H^1(K(p)/L, \mu_p)^G$. In addition, if $a$ and $b$ are eigenvectors for the action of $\Delta$, we may assume that $c_w$ is also.*

**Proof** Consider the sequence

$$0 \to J[\lambda] \to J[\lambda^4] \to J[\lambda^3] \to 0 \tag{2.7}$$

and the commutative diagram with exact rows

$$\begin{array}{ccccc}
0 \to H^1(K(p)/L, J[\lambda])^G & \longrightarrow & H^1(K(p)/L, J[\lambda^4])^G & \xrightarrow{\lambda_*} & H^1(K(p)/L, J[\lambda^3])^G \\
\uparrow{\scriptstyle \mathrm{res}_{L/K}} & & \uparrow{\scriptstyle \mathrm{res}_{L/K}} & & \uparrow{\scriptstyle \mathrm{res}_{L/K}} \\
H^1(K(p)/K, J[\lambda]) & \longrightarrow & H^1(K(p)/K, J[\lambda^4]) & \xrightarrow{\lambda_*} & H^1(K(p)/K, J[\lambda^3]).
\end{array}$$

The top row is exact because (2.7) splits over $L$, and hence the sequence

$$0 \to H^1(K(p)/L, J[\lambda]) \to H^1(K(p)/L, J[\lambda^4]) \to H^1(K(p)/L, J[\lambda^3])$$

is exact. By Lemma 2.2, $a$ lifts to an element $a_1 \in H^1(K(p)/K, J[\lambda^4])$. Let $a_1' \in H^1(K(p)/L, J[\lambda^4])^G$ be any lift of $\mathrm{res}_{L/K}\, a$ ($\mathrm{res}_{L/K}\, a_1$ itself is one such). Then

$$\mathrm{res}_{L/K}\, a_1 = a_1'\eta \quad \text{and} \quad \eta \in H^1(K(p)/L, \mu_p)^G. \tag{2.8}$$

We now construct a candidate for $a_1'$. Under the identification (2.3) between $J[\lambda^i]$ and $\mu_p^i$, the map $\lambda^{i-1} : J[\lambda^i] \to J[\lambda]$ corresponds to projection on the first component. Hence under the identification $H^1(K, J[\lambda^3]) = (K^\times/K^{\times p})^3$, $a$ corresponds to an element $(x_1, x_2, x_3) \in (K^\times/K^{\times p})^3$, and $\lambda_*^2 a = x_1$. Moreover, in the identification

$$H^1(L, J[\lambda^4]) \simeq (L^\times/L^{\times p})^4,$$

the action of $\sigma$ on $H^1(L, J[\lambda^4])$ is intertwined with

$$(t_1, t_2, t_3, t_4) \mapsto (t_1^\sigma, t_2^\sigma, t_3^\sigma, t_4^\sigma t_1^\sigma), \quad t_i \in L^\times/L^{\times p}.$$

Thus $(t_i)$ is fixed by $G$ if

$$t_i^\sigma = t_i, \quad i = 1, 2, 3, \quad \text{and} \quad t_4^{\sigma - 1} = t_1^{-1}.$$

By hypothesis, $x_1 = N_{L/K}\epsilon$, $\epsilon \in \mathcal{O}_L^\times$. Then

$$a_1' = (x_1, x_2, x_3, N'\epsilon) \tag{2.9}$$

is an equivariant lift of $(x_1, x_2, x_3)$.

Now let $a_{w,1}$ be the local lift of $a$ given by $a_{w,1} = \delta_4(x_w)$. Then

$$\mathrm{res}_{L_w/K_w}\, a_{w,1} = (x_1, x_2, x_3, \iota_{P_4}(x_w)). \tag{2.10}$$

Thus, from equations (2.8), (2.9), and (2.10), we get

$$\mathrm{res}_{L_w/K_w}(c_w) = \mathrm{res}_{L_w/K_w}(a_{1,w} - a_{w,1}) = \iota_{P_4}(x_w)^{-1}\eta N'\epsilon.$$

The last statement of the proposition is clear, since at each stage in the calculation we can choose eigenvectors, and the maps $\lambda$ and $\iota_{P_k}$ are also eigenvectors for the action of $\Delta$, by the choices we have made of $\lambda$ and $P_k$. □

# 3 The local approximation

Let $P_i$ be as in the previous section, $i = 1, 2, 3, 4$, let $D_i$ be a divisor on $F_{a,b,c}$ representing $P_i$, and let $f_i$ be a function whose divisor is $pD_i$. Take $D_i$ and $f_i$ to be defined over $K = \mathbb{Q}(\zeta)$ if $i = 1, 2, 3$ and over $L = K(\eta_{p-3}^{1/p})$ if $i = 4$. The maps $\iota_{P_i}$ in Propositions 2.1 and 2.3 are computed by evaluating $f_i$ on certain divisors. We use the approximation technique in [McC88] to find expansions for $f_i$ on $p$-adic discs in $F_{a,b,c}$. Given a function $f$ whose divisor is divisible by $p$ we approximate $f$ on an affinoid $Y$ in $F_{a,b,c}$ using the fact that

$$\frac{df}{f} \equiv \omega \quad (\mathrm{mod}_Y\ p). \tag{3.1}$$

for some holomorphic differential $\omega$ on $F_{a,b,c}$ ([McC88], Theorem 5.2). For general facts about rigid analysis, we refer the reader to [BGR84].

We recall the notion of congruence used in (3.1). If $Y$ is a one-dimensional affinoid defined over an extension $F$ of $\mathbb{Q}_p$ with uniformizer $\pi_F$, we let $A(Y)$ be the ring of rigid analytic functions on $Y$, $M(Y)$ the quotient field of $A(Y)$, and $D(Y)$ the module of Kähler differentials of $M(Y)$. We define sub-$\mathcal{O}_F$-modules

$$\begin{aligned} A^0(Y) &= \{f \in A(Y) : |f(x)| \leq 1 \text{ for all } x \in Y(\mathbb{C}_p)\} \\ M^0(Y) &= \{f/g : f \in A^0(Y), g \in A^0(Y) \setminus \pi_F A^0(Y)\} \\ D^0(Y) &= \{\sum_i f_i\, dg_i : f_i, g_i \in M^0(Y)\}. \end{aligned}$$

If $f, g \in A(Y)$, $c \in F$, we say that $f \equiv g \pmod{_Y\, c}$ if $(f - g) \in cA^0(Y)$, and similarly we define the notion of congruence on $Y$ in $M(Y)$ and $D(Y)$. In order to deduce from (3.1) information about power series expansions of $f$ on closed discs in $Y$, we need the following lemmas.

**Lemma 3.1** *Suppose that $Y$ is a one-dimensional affinoid over a finite extension $F$ of $\mathbb{Q}_p$, $Y$ has good reduction, and $Z$ is an affinoid contained in $Y$, isomorphic to a closed disc. If $\omega \in D^0(Y)$ is a differential with at worst simple poles on $Y$ that is regular on $Z$, then $\omega \in D^0(Z)$.*

**Proof** Since $Z$ is isomorphic to a closed disc, it is contained in a residue class $U$ of $Y$ (or is equal to $Y$, in which case there is nothing to prove). It is clear from the definitions that $D^0(Y)|_U = D^0(U)$, hence $\omega \in D^0(U)$. Furthermore, since $Y$ has good reduction, $U$ is isomorphic to an open disc. Choose a parameter $t$ for $U$ such that $Z$ is the disc $|t| \leq |c| < 1$ for some $c \in F$, and write

$$\omega = g\, dt + \sum_{i=1}^{n} \frac{a_i}{t - b_i}\, dt, \quad g \in \mathcal{O}_F[[t]], \quad a_i, b_i \in \mathcal{O}_F, \quad |c| < |b_i| < 1.$$

Expanding the polar terms in powers of $t/b_i$ and setting $t = cs$, we get $\omega = f\, ds$ for some $f \in \mathcal{O}_F[[s]]$. Since $s$ is a parameter on $Z$, this proves the lemma. □

**Lemma 3.2** *Suppose that $f$ is a function whose divisor is divisible by $p$. Let $Y$ be an affinoid with good reduction contained in $F_{a,b,c}$ and let $Z$ be a $p$-adic disc contained in $Y$ such that there is a function on $F_{a,b,c}$ restricting to a parameter on $Z$. If $\omega$ satisfies the congruence (3.1)* $\pmod{_Y\, p}$*, then it satisfies the same congruence* $\pmod{_Z\, p}$.

**Proof** With notation as in [McC88], we have

$$\frac{df}{f} = \omega + p\eta, \quad \eta \in D^0(Y).$$

Let $g$ be a function on $F_{a,b,c}$ such that $f/g^p$ is regular on $Z$ (we can construct $g$ using a parameter on $Z$ as in the hypotheses). Since a suitable scalar multiple of $g$ is in $M^0(Y)$, $d\log g \in D^0(Y)$. Thus $\eta - d\log g \in D^0(Y)$ and is also regular on $Z$, and hence is in $D^0(Z)$ by Lemma 3.1. Thus

$$\frac{df}{f} \equiv \frac{df}{f} - p\frac{dg}{g} = \omega + p\Big(\eta - \frac{dg}{g}\Big) \equiv \omega \quad (\mathrm{mod}_Z\ p). \quad \square$$

We apply these considerations to the affinoid $Y$ introduced in [McC88] and defined as follows. Choose $\pi_K = \pi$ as the uniformizer for $K_w$. Let $s$ and $t$ be the functions on $F_{a,b,c}$ defined by

$$x = -\frac{a}{c}(1 + \pi^{(p-1)/2}s) \tag{3.2}$$

$$y = (-1)^c a^a b^b c^c (1 + \pi t). \tag{3.3}$$

Let $Y$ be the affinoid defined over $L_w$ by the inequalities

$$|t| \leq |\pi_L^{-1}|, \quad |s| \leq 1.$$

A basis of holomorphic differentials on $F_{a,b,c}$ is

$$\omega_k = E_k \frac{x^{\left[\frac{ka}{p}\right]}(1-x)^{\left[\frac{kb}{p}\right]}}{y^k}\, dx, \quad k \in H_{a,b,c},$$

for some constants $E_k$ and where $H_{a,b,c}$ is a certain subset of $\{1, 2, \ldots, p-1\}$ of cardinality $(p-1)/2$ (we can identify $H_{a,b,c}$ with the CM-type of $F_{a,b,c}$). We can and do choose the constants $E_k$ so that $\omega_k$ has expansion

$$\omega_k \equiv ds \quad (\mathrm{mod}_Y\ \pi_L), \tag{3.4}$$

(note that this normalization is different from that of [McC88]). Now $P_1$ is the $\lambda$-torsion point represented by the divisor $(0,0) - \infty$, and we choose $f_1 = x$. In [McC88] it was shown that

$$\frac{df_1}{f_1} \equiv \pi^{(p-1)/2} \sum_{k \in H_{a,b,c}} b_k \omega_k \quad (\mathrm{mod}_Y\ p) \tag{3.5}$$

for some $p$-adic integers $b_k$ satisfying

$$\sum_{k \in H_{a,b,c}} b_k k^i \equiv \begin{cases} F & i = 0 \\ 0 & 1 \leq i \leq (p-3)/2 \end{cases} \quad (\mathrm{mod}\ \pi_K), \quad F \in \mathbb{Z}/p\mathbb{Z}^\times. \tag{3.6}$$

Note that although it was assumed that $F_{a,b,c}$ is wild split in Section 5 of [McC88], there is nothing in the definition of $Y$ or the calculation showing (3.5) and (3.6) that uses this assumption. It is only at the end of that section that the assumption comes in.

**Lemma 3.3** *If*

$$\sum_{k \in H_{a,b,c}} u_k \omega_k \equiv \sum_{k \in H_{a,b,c}} v_k \omega_k \ (\mathrm{mod}_Y\ \pi^{n+(p-3)/2})$$

*then*

$$u_k \equiv v_k \pmod{\pi^n}, \quad k \in H_{a,b,c}.$$

**Proof** See pages 658–659 of [McC88]. □

**Proposition 3.4** *We have*

$$\frac{df_3}{f_3} \equiv \sum_{k \in H_{a,b,c}} c_k \omega_k \ (\mathrm{mod}_Y\ p), \quad c_k \equiv 0 \pmod{\pi^{(p-5)/2}}$$

*and*

$$\frac{df_4}{f_4} \equiv \sum_{k \in H_{a,b,c}} d_k \omega_k \ (\mathrm{mod}_Y\ p), \quad d_k \equiv -\pi^{(p-7)/2} \frac{b_k}{k^3} \pmod{\pi^{(p-5)/2}}, \qquad (3.7)$$

*where the $b_k$ are as in equation* (3.5).

**Proof** We have

$$\lambda_*^2 \frac{df_3}{f_3} \equiv \frac{df_1}{f_1} \ (\mathrm{mod}_Y\ p).$$

Since $\zeta_* \omega_k = \zeta^{-k} \omega_k$, we have $\lambda_* \omega_k = \lambda^\sigma \omega_k$, for some $\sigma \in \mathrm{Gal}(K_w/\mathbb{Q}_p)$. Hence it follows from Lemma 3.3 and (3.5) that

$$\lambda^{2\sigma} c_k \equiv \pi^{(p-1)/2} b_k \pmod{\pi^{(p+1)/2}}.$$

Thus $c_k \equiv 0 \pmod{\pi^{(p-5)/2}}$, as claimed. Similarly, we have $\lambda_*^3(df_4/f_4) \equiv (df_1/f_1)$, so $\lambda^{3\sigma} d_k \equiv \pi^{(p-1)/2} b_k \pmod{\pi^{(p+1)/2}}$. Furthermore, since $\zeta^\sigma = \zeta^{-k}$, it follows from our choice of $\lambda$ that $\lambda^\sigma/\pi \equiv -k \pmod{\pi}$ for $1 \leq k \leq p-1$, so we get equation (3.7). □

**Lemma 3.5**

$$-\sum_{k \in H_{a,b,c}} \frac{b_k}{k^3} \equiv F(q(a^a b^b c^c)^3 + abcB_{p-3}) \pmod{\pi},$$

*where $F$ is as in* (3.6).

**Proof** Let $n = (p-1)/2$. Define

$$\Gamma_k(x_1, \ldots, x_n) = \det \begin{bmatrix} 1 & 1 & \ldots & 1 \\ x_1 & x_2 & \ldots & x_n \\ x_1^2 & x_2^2 & \ldots & x_n^2 \\ \vdots & \vdots & \ddots & \vdots \\ x_1^{n-2} & x_2^{n-2} & \ldots & x_n^{n-2} \\ x_1^{n-1+k} & x_2^{n-1+k} & \ldots & x_n^{n-1+k} \end{bmatrix}.$$

Then an elementary linear algebra calculation using (3.6) gives

$$\sum_{k \in H_{a,b,c}} \frac{b_k}{k^3} \equiv F\Gamma_3(H_{a,b,c}^{-1})/\Gamma_0(H_{a,b,c}^{-1}) \pmod{\pi}.$$

Let $S_i(x_1, \ldots, x_n)$ be the $i$th symmetric function. Then

$$\Gamma_3 = \Gamma_0(S_1^3 - 2S_1S_2 + S_3).$$

This can be proved by the usual method: the determinant vanishes if $x_i = x_j$ for $i \neq j$, or if there is a polynomial of degree $n+2$ vanishing on the $x_i$, and with no term of degree $n-1$, $n$, or $n+1$. Thus, if the roots of the polynomial are $x_1, \ldots, x_n, \alpha, \beta$, then

$$\begin{aligned} \alpha + \beta + S_1 &= 0, \\ S_2 + (\alpha+\beta)S_1 + \alpha\beta &= 0, \\ \alpha\beta S_1 + \alpha S_2 + \beta S_2 + S_3 &= 0. \end{aligned}$$

Eliminating $\alpha$ and $\beta$ gives the condition $S_1^3 - 2S_1S_2 + S_3 = 0$. Now, we have

$$S_1(H_{a,b,c}^{-1}) \equiv -q(a^a b^b c^c), \tag{3.8}$$

$$S_2(H_{a,b,c}^{-1}) \equiv 0, \tag{3.9}$$

$$S_3(H_{a,b,c}^{-1}) \equiv -\frac{B_{p-3}}{3}(a^3 + b^3 + c^3) \equiv -abcB_{p-3}. \tag{3.10}$$

It is explained in [McC88], Lemma 5.24, how equation (3.8) follows from [Van20, 17]; equation (3.9) follows from parity considerations; and equation (3.10) follows from [Van20, 16], in exactly the same way as (3.8) follows from [Van20, 17]. □

We now define $p$-adic discs in $Y$, to which we apply Lemma 3.2. Let $X$ be the sub-affinoid of $Y$ defined by $|t| \leq 1$ in the wild case and by $|s| \leq |\pi_K|$ in the tame case. Let $E_w$ be the quadratic unramified extension of $K_w$. If $F_{a,b,c}$ is wild, $X$ is isomorphic to a union of two closed discs, which are defined over $K_w$ in the split case and over $E_w$ in the nonsplit case. Furthermore, $T = t$ is a parameter on each disc. If $F_{a,b,c}$ is tame, then $X$ is isomorphic to a union of $p$ closed discs defined over $K_w$, and $T = s/\pi_K$ is a parameter on each disk. For proofs of these facts we refer the reader to[McC88] (where $T = s'$ in the tame case). We denote by $Z$ any of the discs that are components of $X$, with parameter $T$. We can write

$$f_i|_X = C_i u_i(T) v_i(T^p) g_i(T)^p, \quad i = 1, 2, 3, 4,$$

where $u_i$ and $v_i$ are unit power series with constant term 1 and integer coefficients, $u_i$ has no terms in $T^p$, and $g_i$ is a monic polynomial with integer coefficients. Furthermore, these conditions uniquely determine the $u_i$, $v_i$ and $g_i$. Then

$$\frac{df_i}{f_i} \equiv \frac{du_i}{u_i} \pmod{_Z\, p}. \tag{3.11}$$

For a $p$-adic field $H$ we denote by $U_H[[T]]$ the power series in $\mathcal{O}_H[[T]]$ which are congruent to 1 modulo the maximal ideal in $\mathcal{O}_H[[T]]$.

**Theorem 3.6** *Let $Z$ be any of the discs that are components of $X$ and let $T$ be a parameter on $Z$. Then*

$$u_i = 1 + \pi^{(p+3)/2-i} D_i T + O(\pi^{(p+5)/2-i} T), \tag{3.12}$$

*where $|D_i| \leq 1$, $i = 1, 2, 3, 4$. Moreover, $|D_1| = 1$, and under the hypotheses of Theorem 1.4, $|D_4| = 1$ and*

$$\frac{D_4}{D_1} \equiv q(a^a b^b c^c)^3 + abcB_{p-3}.$$

*Finally, $u_i$ for $i = 1, 2, 3$ are defined over $E_w$, and*

$$u_4 \in 1 + \pi_K^{(p-5)/2} D_4 T U_{E_w}[[T]] + \pi_K^{(p+1)/2} \pi_L^{-3} E D_1 T U_{F_w}[[T]],$$

*where $E \in \mathbb{Z}/p\mathbb{Z}^\times$ is independent of the triple $(a, b, c)$ and $F_w$ is the quadratic unramified extension of $L_w$.*

**Proof** In both wild and tame cases we have $\omega_k \equiv \pi D\, dT \pmod{_Z\, \pi^2}$ for all $k \in H_{a,b,c}$, with $D \in \mathbb{Z}/p\mathbb{Z}^\times$ independent of $k$. This follows from our normalization (3.4), since in the tame case we have

$$s = \pi T, \tag{3.13}$$

and in the wild case it follows from [McC88, (5.6)], where it is shown that the expansion of $s$ in terms of $t$ on either of the discs in $X$ is

$$s^2 = \frac{-q(a^a b^b c^c) 2b}{ac} + \pi \frac{2b}{ac}(t^p - t) + O(\pi^2). \tag{3.14}$$

The statements about $D_1$ follow from (3.2), (3.13) (in the tame case) and (3.14) (in the wild case), since $f_1 = x$. The statement about $D_2$ was proved in [McC88, Theorem 5.13]. Although this theorem is stated only for the wild split case, the consequence (3.12) is easily seen to hold also in the other cases (the part of Theorem 5.13 specific to the wild split case translates into the statement $|D_2| = 1$ in the current notation, and we do not need it here). The statements about $D_3$ and $D_4$ follow from Proposition 3.4, Lemma 3.5, and (3.11). The statement about the ratio $D_4/D_1$ follows from (3.5), the case $i = 0$ of (3.6), (3.7), and Lemma 3.5, taking note of the normalization (3.4) and the fact that $ds \equiv \text{unit} \times \pi dT \pmod{_Y \pi^2}$. The statements about the fields of definition follow from the fact that $f_i$ is defined over $K$ for $i = 1, 2, 3$ and $f_4$ is defined over $L$, and that the discs $Z$ are always defined over $E_w$. The final statement follows from considerations of ramification theory. Locally, we have $\eta_{p-3} = 1 + a\pi^{p-3}$ modulo $p$th powers, so the (upper) conductor of $L_w/K_w$ is 3. Now, it follows from the properties of the $P_i$ that $u_4^{\sigma-1} \equiv u_1$ modulo $\mathcal{O}_{F_w}[[T]]^{\times p} U_{F_w}[[T^p]]$, and, since $u_1 \in 1 + \pi^{(p+1)/2} T D_1 U_{F_w}[[T]]$, this implies the final statement with $E$ such that $(\sigma - 1)\pi_L^{-3} \equiv E^{-1} \pmod{\pi_L}$. □

# 4 Computation of the Cassels pairing

Recall the local descent maps

$$\delta_i = \iota_{P_1} \times \cdots \times \iota_{P_i} \colon J(K_w) \to (K_w^\times / K_w^{\times p})^i$$

described in Section 2. We start by noting a couple of properties that follow from the choice of $P_i$ made in Section 2. First, we have

$$\iota_{P_i} \circ \lambda = \iota_{P_{i-1}}, \quad i = 2, 3, 4. \tag{4.1}$$

Second, for $i = 1, 2, 3$ we have, from eigenspace considerations,

$$\iota_{P_i}(J(K_w)(k)) \subset V(k - i + 1). \tag{4.2}$$

Let $A \subset J(K_w)$ be the subgroup generated by divisors supported on the discs $|T| \leq |\pi_K|$ in $X$. Let

$$V[i,j] = \bigoplus_{i \leq k \leq j} V(k).$$

Note that $V(i) = 0$ for $i > p$.

**Proposition 4.1** *We have*

$$\iota_{P_i}(A) \subset V\big[(p+5)/2 - i, p\big], \quad i = 1, 2, 3. \tag{4.3}$$

*If $F_{a,b,c}$ is wild split, we have*

$$\iota_{P_i}(J(K_w)) \subset V\big[(p+1)/2 - i, p\big], \quad i = 1,2,3. \tag{4.4}$$

*If $F_{a,b,c}$ is wild nonsplit or tame, we have*

$$\iota_{P_i}(J(K_w)) \subset V\big[(p+3)/2 - i, p\big], \quad i = 1,2,3. \tag{4.5}$$

*Furthermore, in the case $i = 1$, the inclusions in (4.3) and (4.5) are equalities.*

**Proof** The inclusions in (4.3) follow immediately from Theorem 3.6, as does the claim that the inclusion is equality in the case $i = 1$. Now Faddeev [Fad61] proved that

$$\operatorname{im} \iota_{P_1} = \begin{cases} V((p-1)/2) \oplus V[(p+3)/2, p] & \text{if } F_{a,b,c} \text{ is wild split,} \\ V[(p+1)/2, p] & \text{otherwise.} \end{cases}$$

This implies the statements (4.4) and (4.5) in the case $i = 1$, and also that the image of $A$ in $J(K_w)/\lambda J(K_w)$ has codimension 1, and the eigenvalue of the quotient is $\kappa^{(p-1)/2}$ in the wild split case and $\kappa^{(p+1)/2}$ in the other cases. The remaining statements now follow from (4.1), (4.2), and (4.3). □

**Proposition 4.2** *If $F_{a,b,c}$ is wild nonsplit or tame, then*

$$\delta_2(J(K_w)) = V\big[(p+1)/2, p\big]^2.$$

**Proof** From (4.1) with $i = 2$ we have

$$\operatorname{im} \delta_2 \cap (K^\times/K^{\times p}) \times 1 = \operatorname{im} \delta_1 \times 1 = V\big[(p+1)/2, p\big] \times 1 \tag{4.6}$$

Furthermore, given $u \in V[(p+3)/2, p]$, we can find $a \in A$ such that $\iota_{P_1}(a) = u$, and $\iota_{P_2}(a) \in V[(p+1)/2, p] \subset \operatorname{im} \iota_{P_1}$. Thus, modifying $a$ by $\lambda J(K_w)$, we can choose it so that $\iota_{P_2}(a) = 1$. Hence

$$\operatorname{im} \delta_2 \supset 1 \times V\big[(p+3)/2, p\big]. \tag{4.7}$$

Now it follows from local duality that $\operatorname{im} \delta_2$ must be maximal isotropic with respect to the cup product pairing on $(K^\times/K^{\times p})^2 = H^1(K, J[\lambda^2])$ induced by the Weil pairing on $J[\lambda^2]$. Since $\lambda^2 = \bar{\lambda}^2$, the Weil pairing is skew symmetric. Thus the pairing on $(K^\times/K^{\times p})^2$ is a nonzero multiple of $((a_1, b_1), (a_2, b_2)) \mapsto (a_1, b_2)_w (b_1, a_2)_w^{-1}$, where $(\ ,\ )_w$ denotes the Hilbert symbol at $w$. The only maximal isotropic subgroup satisfying (4.6) and (4.7) is the one given in the statement of the proposition. □

Define a subspace $V_{\text{global}} \subset V$ by

$$V_{\text{global}} = \bigoplus_{\substack{2 \le i \le p-3 \\ i \text{ even}}} V(i).$$

**Proposition 4.3** *Assume $p \geq 11$ and $F_{a,b,c}$ is wild nonsplit or tame. There exists a point $x \in A$ such that $\iota_{P_1}(x)$ generates $V((p+5)/2)$ and $\iota_{P_i}(x) \in V_{\text{global}}$ for $i = 2, 3$.*

**Proof** It follows from Proposition 4.1 that $A$, regarded as a $\mathbb{Z}_p[\zeta]$-submodule of $J(K_w)$, has codimension at most 1. Hence $\delta_3(A)$ has codimension $\leq 3$ as a $\mathbb{F}_p$-vector space in $\delta_3(J(K_w))$. By Proposition 4.1 we can choose $x \in A$ such that $\iota_{P_1}(x)$ generates $V((p+5)/2)$. This condition leaves freedom to modify $x$ by anything in $\lambda J(K_w)$, which would change $\delta_3(x)$ by anything in $\operatorname{im} \delta_2$. Thus, modifying $x$ as needed, we can ensure that $\iota_{P_i}(x) \in V_{\text{global}}$, $i = 2, 3$. The number of degrees of freedom in performing this modification is equal to the dimension of $\operatorname{im} \delta_2 \cap V_{\text{global}}$, which is at least 4 if $p \geq 11$, by Proposition 4.2. Thus we can ensure that $x$ remains in $A$ when making the modification. □

**Computation of the Cassels pairing for Theorem 1.2** We now use Proposition 2.1 to show that the pairing $\langle \, , \, \rangle_2$ is trivial under the hypotheses of Theorem 1.2. In the next section we explain how this implies the theorem.

Denote by $\ell_i \colon S_{\lambda^i} \to J(K_w)/\lambda^i J(K_w)$ the localization map. We claim that, under the hypotheses of Theorem 1.2, $\iota_{P_1}(\ell_1(S_\lambda)) \subset V[(p+3)/2, p]$ or $\iota_{P_1}(\ell_1(S_\lambda)) \subset V((p+1)/2) \oplus V[(p+5)/2, p]$. Now, $V(i)$ pairs nontrivially with $V(j)$ under the Hilbert pairing if and only if $i + j \equiv p \pmod{p-1}$. Thus, it follows from our claim and from (4.2) that $\iota_{P_3}(\ell_3(J(K_w))$ pairs trivially with $\iota_{P_1}(\ell_1(J(K_w)))$.

To see the claim, note that if hypothesis (a) of Theorem 1.2 is satisfied, namely that $F_{a,b,c}$ is wild split and $p \equiv 3 \pmod 4$, then, by [Fad61], $\iota_{P_1}(\ell_1(S_\lambda)) \subset V((p-1)/2) \oplus V[(p+3)/2, p]$. Furthermore, we can eliminate $V((p-1)/2)$ as a possibility, because $\ell_1$ factors through $H^1(K(p)/K, \mu_p) \to H^1(K_w, \mu_p)$. Since $(p-1)/2$ is odd, it follows from (2.4) that $\ell_1$ can have nontrivial image in $V((p-1)/2)$ only if $C((p-1)/2)$ is nontrivial, which would imply $p \mid B_{(p+1)/2}$. This never happens if $p \equiv 3 \pmod 4$.

If hypothesis (b) of Theorem 1.2 is satisfied, namely that $F_{a,b,c}$ is wild nonsplit or tame and the image of $U$ in either $V((p+1)/2)$ or $V((p+3)/2)$ is trivial, then the claim follows immediately from (4.5).

The proof of Theorem 1.4 uses the following lemma.

**Lemma 4.4** *Suppose $p \geq 5$ is regular, and let $L$ be as in (2.2). Then*

1. *the map $H^1(K(p)/L, \mu_p) \to H^1(L_w, \mu_p)$ is injective*

2. *the norm map $N_{L/K} \colon \mathcal{O}_L^\times \to \mathcal{O}_K^\times$ is surjective*

3. *$H^1(K(p)/L, \mu_p)^G(i) = 0$ if $i$ is odd and $i \neq 1$, or if $i = p-1$.*

**Proof** Let $H_K$ (resp. $H_L$) be the Hilbert class field of $K$ (resp. $L$). Since $L/K$ is unramified outside $p$, and there is only prime of $L$ above $p$, it follows that $\mathrm{Gal}(H_L/L)/(\sigma - 1) \simeq \mathrm{Gal}(H_K/K)$. Therefore $p$ does not divide the order of the class group $C_L \simeq \mathrm{Gal}(H_L/L)$. The injectivity statement follows, since anything in the kernel would generate an unramified Kummer extension of $L$ of degree $p$. Furthermore, every unit of $K$ is a local norm everywhere except possibly at the prime above $p$, and therefore is a local norm there also by the product formula. Thus it is a global norm. The surjectivity of the norm map follows by a standard argument using $\mathrm{Gal}(L/K)$ cohomology of the sequences

$$1 \to \mathcal{O}_L^\times \to L^\times \to P_L \to 1 \quad \text{and} \quad 1 \to P_L \to I_L \to C_L \to 1,$$

where $I_L$ and $P_L$ are the groups of ideals and principal ideals respectively. Finally, by (2.4),

$$H^1(K(p)/K, \mu_p) = \mathcal{O}_K'^\times/\mathcal{O}_K'^{\times p} \quad \text{and} \quad H^1(K(p)/L, \mu_p) = \mathcal{O}_L'^\times/\mathcal{O}_L'^{\times p}.$$

Moreover, the cokernel of $\mathcal{O}_K^\times/\mathcal{O}_K^{\times p}$ in $(\mathcal{O}_L^\times/\mathcal{O}_L^{\times p})^G$ is $H^2(L/K, \mu_p) \simeq \mathbb{Z}/p\mathbb{Z}$, with $\mathrm{Gal}(K/\mathbb{Q})$ acting via $\kappa^{p-3}$, since it acts on $G$ via $\kappa^3$. Since $p - 3$ is even and $(\mathcal{O}_K^\times/\mathcal{O}_K^{\times p})(i) = 0$ if $i$ is odd and $i \neq 1$, or if $i = p-1$, the third statement of the lemma follows. □

**Proof of Theorem 1.4** We exhibit $a \in S_{\lambda^3}$ and $b \in S_\lambda$ which pair non-trivially under the Cassels pairing.

Since $p$ is regular, the exact sequence [McC88, 7.3] identifies $U$ with $\mathcal{O}_K^\times/\mathcal{O}_K^{\times p}$. Thus $S_{\lambda^i} \subset (\mathcal{O}_K^\times/\mathcal{O}_K^{\times p})^i$ for $i \leq 3$. The Selmer group is the subgroup obtained by imposing the local conditions at $w$. Since $(p+1)/2$ is even, we can choose an element $b \in \mathcal{O}_K^\times/\mathcal{O}_K^{\times p}$ which generates $V((p+1)/2)$, and $b$ satisfies the local condition by Proposition 4.1, so $b \in S_\lambda$.

As for $a$, by Proposition 4.3 there exists $a_w = (a_{w,1}, a_{w,2}, a_{w,3}) = \delta_3(x)$, $x \in A$, such that $a_{w,1}$ generates $V((p+5)/2)$ and $a_{w,2}, a_{w,3} \in V_{\mathrm{global}}$. Using a suitable projector, we may further assume that $x$ is an eigenvector for the action of $\Delta$. Choose eigenvectors $a_i \in \mathcal{O}_K^\times/\mathcal{O}_K^{\times p}$ specializing to $a_{w,i}$ for $i = 1, 2, 3$ and define $a \in S_{\lambda^3}$ by $a = (a_1, a_2, a_3)$.

Now, by Lemma 4.4, $\lambda_*^2 a = a_1 \in V((p+5)/2)$ is the norm of a global unit $\epsilon$ in $\mathcal{O}_L^\times$, and by Proposition 2.3, the Cassels pairing of $a$ and $b$ is the Hilbert pairing $(c_w, b_w)$, where $c_w \in K^\times/K^{\times p}$ is an eigenvector and

$$c_w = \iota_{P_4}(x)^{-1}\eta N'\epsilon \quad \text{in } L_w^\times/L_w^{\times p}, \tag{4.8}$$

where $\eta \in H^1(K(p)/L, \mu_p)^G$. We can identify the precise eigenspace in which $c_w$ lies as follows. Since $a_{w,1} = \iota_{P_1}(x)$, and since $P_1$ is fixed by $\Delta$, $x$ has eigenvalue $\kappa^{(p+5)/2}$. Then, since $\lambda$ has eigenvalue $\kappa$ (modulo $\lambda^5$), it follows that $\delta_3(x)$, and hence $c_w$, have eigenvalue $\kappa^{(p+5)/2-3} = \kappa^{(p-1)/2}$. Thus we may assume without loss of generality that $c_w$, $\eta$, $N'\epsilon$, and $\iota_{P_4}(x)$ are eigenvectors for a lift $\tilde{\Delta}$ of $\Delta = \mathrm{Gal}(K_w/\mathbb{Q}_p)$ to $\mathrm{Gal}(L_w/\mathbb{Q}_p)$, with eigenvalue $\kappa^{(p-1)/2}$. Since $\eta \in H^1(K(p)/L, \mu_p)^G$, its projection onto an eigenspace $(L_w^\times/L_w^{\times p})(i)$ with $i > 1$ odd is trivial, by Lemma 4.4. This applies in particular to $i = (p-1)/2$, so that the image of $\eta$ in $L_w^\times/L_w^{\times p}$ is trivial.

Under the Hilbert pairing the $\kappa^{(p-1)/2}$ and $\kappa^{(p+1)/2}$ eigenspaces of $K_w^\times/K_w^{\times p}$ pair nontrivially. Thus, to prove that the pairing $(c_w, b_w)$ is nontrivial, it suffices to show that $c_w$ is not a $p$th power, and for that it suffices to show that its image in $L_w^\times/L_w^{\times p}$ is nontrivial.

Since $x \in A$, we may choose a divisor $D$ supported on $|T| \leq |\pi|$ such that $a_{w,i} = f_i(D)$, $1 \leq i \leq 3$. Since $D$ is supported on $|T| \leq |\pi|$, we have

$$u = f_4(D) \equiv u_4(D) \pmod{L_w^{\times p}(1 + \pi^p \mathcal{O}_{L_w})}.$$

From the Galois properties of the $P_i$, we have

$$u \in (L_w^\times/L_w^{\times p})((p-1)/2)), \quad (\sigma - 1)u = v, \tag{4.9}$$

where $v$ is the image in $L_w^\times/L_w^{\times p}$ of a generator of $V((p+5)/2)$. Since $p \geq 19$, $(p+5)/2$ is less than $p-3$, and thus $v \neq 0$. Thus the subspace of $L_w^\times/L_w^{\times p}$ satisfying the conditions (4.9) is two-dimensional, with generators $u_1$ and $u_2$, where $u_1$ is the image of a generator of $V((p-1)/2)$ with expansion $u_1 = 1 + \pi_K^{(p-1)/2} + O(\pi_K^{(p+1)/2})$, and $u_2 \in (L_w^\times/L_w^{\times p})((p-1)/2)$ has expansion $u_2 = 1 + \pi_K^{(p+5)/2}\pi_L^{-3} + O(\pi_K^{(p+5)/2})$. Thus $u = u_1^\alpha u_2^\beta$ for some $\alpha, \beta \in \mathbb{Z}/p\mathbb{Z}$. Expanding the binomial series, we get

$$u = 1 + \alpha\pi_K^{(p-1)/2} + \beta\pi_K^{(p+5)/2}\pi_L^{-3} + O(\pi_K^{p-1}). \tag{4.10}$$

We can now use Theorem 3.6 to evaluate $u_4$ at $D$. Comparing appropriate coefficients (note that $D$ is supported on $|T| \leq |\pi|$), we see that

$$\frac{\alpha}{\beta} = \frac{D_4}{ED_1} = \frac{1}{E}\gamma(a,b,c) = \frac{1}{E}(q(a^a b^b c^c)^3 + abcB_{p-3}). \tag{4.11}$$

Now, we may replace $(a,b,c)$ by any $(a',b',c') \equiv (ta, tb, tc) \pmod{p}$, for $t \in \mathbb{F}_p^\times$. It is easily seen, using the property $q(xy) \equiv q(x)+q(y)$, that $\gamma(ta, tb, tc) \equiv t^3\gamma(a,b,c)$. Thus, from (4.11), we see that by varying $t$ appropriately we may ensure that $u$, and hence $c_w$, varies in $L_w^\times/L_w^{\times p}$, and, in particular, takes on nonzero values. Hence there exists a choice of $t$ such that the pairing is nontrivial for the curve $F_{a',b',c'}$. However, this curve is isomorphic to $F_{a,b,c}$, and hence the pairing must be nontrivial in that case as well. $\square$

# 5 Shafarevich–Tate groups

The proofs of Theorem 1.2 and Theorem 1.5 follow from the computations of the Cassels–Tate pairing by means of the following proposition.

**Proposition 5.1** *For all positive integers $m$ and $n$, the restriction of the Cassels–Tate pairing induces a perfect pairing*

$$\Big(Ш[\lambda^m]/(\lambda^n Ш[\lambda^{n+m}])\Big) \times \Big(Ш[\lambda^n]/(\lambda^m Ш[\lambda^{n+m}])\Big) \longrightarrow \mathbb{Q}/\mathbb{Z}.$$

Let $Ш_{\mathrm{div}}$ denote the maximal divisible subgroup of $Ш$, i.e. $x \in Ш_{\mathrm{div}}$ if and only if for every nonzero integer $n$ there exists $y \in Ш$ such that $x = ny$. Let $Ш_{\mathrm{red}}$ denote the quotient group $Ш/Ш_{\mathrm{div}}$. Note that:

**Lemma 5.2** *$Ш_{\mathrm{div}}$ is a divisible group in the usual sense that multiplication by any nonzero $n \in \mathbb{Z}$ is surjective on it.*

**Proof** The argument is standard: since $Ш[m]$ is finite for all nonzero $m \in \mathbb{Z}$, the groups $NШ[Nm]$, $N > 0$, stabilize for sufficiently large $N$. Thus for every $m$ there is an integer $N(m)$ such that if an element of $Ш[m]$ is divisible by $N(m)$ it is infinitely divisible. Now if $x \in Ш_{\mathrm{div}}[m]$ and $n > 0$, choose $y \in Ш[N(nm)nm]$ such that $N(nm)ny = x$. Then $y' = N(nm)y$ is in $Ш_{\mathrm{div}}[nm]$ and $ny' = x$. □

Note that since $\zeta$ is an automorphism of $Ш$ it preserves $Ш_{\mathrm{div}}$, and hence so does $\mathbb{Z}[\zeta]$. Furthermore, since $\lambda^{p-1}$ is a unit times $p$ in $\mathbb{Z}[\zeta]$, $Ш_{\mathrm{div}}$ is divisible by $\lambda^n$ for any positive $n$.

**Lemma 5.3** *The exact sequence*

$$0 \longrightarrow Ш_{\mathrm{div}} \longrightarrow Ш \longrightarrow Ш_{\mathrm{red}} \longrightarrow 0$$

*induces by restriction an exact sequence*

$$0 \longrightarrow Ш_{\mathrm{div}}[\lambda^n] \longrightarrow Ш[\lambda^n] \longrightarrow Ш_{\mathrm{red}}[\lambda^n] \longrightarrow 0$$

*for any positive integer $n$.*

**Proof** Only the surjectivity is in question. Let $x \in Ш_{\mathrm{red}}[\lambda^n]$. Lift $x$ to $y \in Ш$. Then $\lambda^n y = z \in Ш_{\mathrm{div}}$. By Lemma 5.2, we can find $w \in Ш_{\mathrm{div}}$ such that $\lambda^n w = z = \lambda^n y$. But then $y - w \in Ш[\lambda^n]$ and $y - w$ reduces to $x$ in $Ш_{\mathrm{red}}$. □

It is well known that $Ш_{\mathrm{red}}[p^\infty]$ is a finite group and that the Cassels–Tate pairing induces a perfect pairing

$$[\cdot\,\cdot]\colon Ш_{\mathrm{red}}[p^\infty] \times Ш_{\mathrm{red}}[p^\infty] \longrightarrow \mathbb{Q}/\mathbb{Z}.$$

We now have the following lemma:

**Lemma 5.4** *The annihilator of* $Ш_{\rm red}[\lambda^m]$ *with respect to the latter pairing equals* $\lambda^m Ш_{\rm red}[p^\infty]$*, for all positive integers* $m$*.*

**Proof** It is clear from the definition of the pairing given in [McC88], for example, and from the functoriality properties of the Weil pairing, that $[\zeta a, a'] = [a, \zeta^{-1} a']$. Hence, if $\hat{\lambda} = \zeta^{-1} - 1$, then $\hat{\lambda}^m Ш_{\rm red}[p^\infty]$ annihilates $Ш_{\rm red}[\lambda^m]$. Since $\hat{\lambda}/\lambda$ is a unit in $\mathbb{Z}[\zeta]$, we have $\hat{\lambda}^m Ш_{\rm red}[p^\infty] = \lambda^m Ш_{\rm red}[p^\infty]$. So the kernel $H$ on the right factor of the restricted pairing

$$Ш_{\rm red}[\lambda^m] \times Ш_{\rm red}[p^\infty] \longrightarrow \mathbb{Q}/\mathbb{Z}$$

contains $\lambda^m Ш_{\rm red}[p^\infty]$. Note that the kernel on the left factor of the latter pairing is trivial. Therefore, the cardinalities of $Ш_{\rm red}[\lambda^m]$ and $Ш_{\rm red}[p^\infty]/H$ are equal. But

$$|Ш_{\rm red}[p^\infty]| = |Ш_{\rm red}[\lambda^m]|\ |\lambda^m Ш_{\rm red}[p^\infty]|,$$

hence $H = \lambda^m Ш_{\rm red}[p^\infty]$. □

**Lemma 5.5** *For all positive integers* $m$ *and* $n$*, the restriction of the Cassels–Tate pairing induces a perfect pairing*

$$\left(Ш_{\rm red}[\lambda^m]/(\lambda^n Ш_{\rm red}[\lambda^{n+m}])\right) \times \left(Ш_{\rm red}[\lambda^n]/(\lambda^m Ш_{\rm red}[\lambda^{n+m}])\right) \longrightarrow \mathbb{Q}/\mathbb{Z}.$$

**Proof** By Lemma 5.4, the annihilator of $Ш_{\rm red}[\lambda^m]$ in $Ш_{\rm red}[\lambda^n]$ equals

$$\lambda^m Ш_{\rm red}[p^\infty] \cap Ш_{\rm red}[\lambda^n] = \lambda^m Ш_{\rm red}[\lambda^{n+m}],$$

and the assertion follows. □

**Proof of Proposition 5.1** By Lemma 5.5, it suffices to show that for all $m$ and $n$ the groups $Ш[\lambda^m]/(\lambda^n Ш[\lambda^{n+m}])$ and $Ш_{\rm red}[\lambda^m]/(\lambda^n Ш_{\rm red}[\lambda^{n+m}])$ are isomorphic. By Lemma 5.3, we have a commutative diagram

$$\begin{array}{ccccccccc} 0 & \to & Ш_{\rm div}[\lambda^{n+m}] & \longrightarrow & Ш[\lambda^{n+m}] & \longrightarrow & Ш_{\rm red}[\lambda^{n+m}] & \to & 0 \\ & & \downarrow \alpha = \lambda^n & & \downarrow \beta = \lambda^n & & \downarrow \gamma = \lambda^n & & \\ 0 & \to & Ш_{\rm div}[\lambda^m] & \longrightarrow & Ш[\lambda^m] & \longrightarrow & Ш_{\rm red}[\lambda^m] & \to & 0 \end{array}$$

where the horizontal sequences are exact. By the Snake Lemma, we get an exact sequence

$$0 \to \mathrm{Ker}(\alpha) \to \mathrm{Ker}(\beta) \to \mathrm{Ker}(\gamma) \to$$
$$\to \mathrm{Coker}(\alpha) \to \mathrm{Coker}(\beta) \to \mathrm{Coker}(\gamma) \to 0.$$

By Lemma 5.2, we have $\mathrm{Coker}(\alpha) = 0$, hence $\mathrm{Coker}(\gamma)$ is isomorphic to $\mathrm{Coker}(\beta)$, and this completes the proof. □

**Proof of Theorem 1.2** By the structure theorem for torsion modules over Dedekind domains we have a $\mathbb{Z}[\zeta]$-module decomposition

$$Ш[\lambda^3] \simeq (\mathbb{Z}[\zeta]/(\lambda))^{t_1} \oplus (\mathbb{Z}[\zeta]/(\lambda^2))^{t_2} \oplus (\mathbb{Z}[\zeta]/(\lambda^3))^{t_3},$$

where $t_1$, $t_2$ and $t_3$ are nonnegative integers. The computations in the previous section show that the pairing (obtained by restricting the Cassels–Tate pairing)

$$Ш[\lambda^2] \times Ш[\lambda] \to \mathbb{Q}/\mathbb{Z}$$

is trivial. By Proposition 5.1 (for $m = 2$ and $n = 1$), we get that the groups $Ш[\lambda^2]/(\lambda Ш[\lambda^3])$ and $Ш[\lambda]/(\lambda^2 Ш[\lambda^3])$ are both trivial. But then

$$(\mathbb{Z}[\zeta]/(\lambda))^{t_1} \oplus (\mathbb{Z}[\zeta]/(\lambda))^{t_2} \oplus (\mathbb{Z}[\zeta]/(\lambda))^{t_3} \simeq Ш[\lambda] = \lambda^2 Ш[\lambda^3] \simeq (\mathbb{Z}[\zeta]/\lambda)^{t_3}$$

so $t_1 = t_2 = 0$, which proves the claim. □

**Proof of Theorem 1.5** Let

$$Ш[\lambda^4] \simeq (\mathbb{Z}[\zeta]/(\lambda))^a \oplus (\mathbb{Z}[\zeta]/(\lambda^2))^b \oplus (\mathbb{Z}[\zeta]/(\lambda^3))^c \oplus (\mathbb{Z}[\zeta]/(\lambda^4))^d.$$

If we show that $d = 0$, then $\lambda^3$ annihilates $Ш[\lambda^4]$, therefore $Ш[\lambda^4] = Ш[\lambda^3]$. By induction, this implies $Ш[p^\infty] = Ш[\lambda^\infty] = Ш[\lambda^3]$. So assume $d \geq 1$. Since the Cassels–Tate pairing on $Ш[\lambda^3] \times Ш[\lambda]$ is nontrivial, Proposition 5.1 implies that $Ш[\lambda^3]/(\lambda Ш[\lambda^4])$ has dimension $\geq 2$ over $\mathbb{F}_p$. Now

$$\lambda Ш[\lambda^4] \simeq (\mathbb{Z}[\zeta]/(\lambda))^b \oplus (\mathbb{Z}[\zeta]/(\lambda^2))^c \oplus (\mathbb{Z}[\zeta]/(\lambda^3))^d.$$

Counting $\mathbb{F}_p$-dimensions, we get $6-(b+2c+3d) \geq 2$, therefore $b+2c+3d \leq 4$. This implies $d = 1$ and $c = 0$. Therefore,

$$Ш[\lambda^4] \simeq (\mathbb{Z}[\zeta]/(\lambda))^a \oplus (\mathbb{Z}[\zeta]/(\lambda^2))^b \oplus (\mathbb{Z}[\zeta]/(\lambda^4)).$$

This implies that

$$(\mathbb{Z}[\zeta]/(\lambda))^2 = \lambda^2(\mathbb{Z}[\zeta]/(\lambda^3))^2 \simeq \lambda^2 Ш[\lambda^3] \subseteq \lambda^2 Ш[\lambda^4] \simeq \mathbb{Z}[\zeta]/(\lambda^2),$$

a contradiction. □

# 6 Tame reduction

Although it is not strictly necessary for Theorem 1.6, we take the opportunity to prove a general lemma on tame reduction, since it clears up some confusion in the literature. In [Lim95], an attempt was made to improve the result of

[McC88] on the existence of nontrivial elements in Ш$[\lambda]$ in the wild split case, under the additional hypothesis that the Jacobian of the Fermat curve in question is nonsimple. However, as Lemma 6.1 shows, nonsimple Jacobian and wild split reduction over $\mathbb{Z}_p[\zeta]$ are incompatible properties, so the Mordell–Weil rank estimates given in the last section of [Lim95] are incorrect. As far as we can tell, the problem lies in the use of the function $q(x)$ which computes the reduction type (see the introduction). Here as well as in [McC88], $q$ is evaluated on triples $(a,b,c)$ of integers such that $0 < a, b, a+b < p$ and $a+b+c=0$. In [Lim95] however, $q$ is evaluated on triples $(a,b,c)$ such that $0 < a, b, a+b < p$ and $a+b+c=p$. While it does not make any difference which of the two types of triples one chooses to define the curve $F_{a,b,c}$, it does make a difference which type of triple one uses to evaluate $q$ and hence the reduction type. We have the following lemma:

**Lemma 6.1** *Let $(a,b,c)$ be such that $J_{a,b,c}$ is nonsimple. Then $F_{a,b,c}$ has tame reduction over $\mathbb{Z}_p[\zeta]$.*

**Proof** By [KR78], $J_{a,b,c}$ is nonsimple if and only if $p \equiv 1 \pmod 3$ and $F_{a,b,c}$ is isomorphic to $F_{1,r,-(r+1)}$, where $r^2+r+1=0$ in $\mathbb{F}_p$. By definition of $q(x)$, it therefore suffices to show that $(r+1)^{(r+1)(p-1)} - r^{r(p-1)} \equiv 0 \pmod{p^2}$. Since 6 divides $p-1$, it suffices to show that

$$(r+1)^{6(r+1)} - r^{6r} \equiv 0 \pmod{p^2}.$$

Let $k$ be an integer such that $r^2+r+1 = pk$. Then $(r+1)^2 = pk + r$. Therefore,

$$(r+1)^6 = (pk+r)^3 \equiv r^3 + 3r^2pk \pmod{p^2}.$$

Hence $(r+1)^{6(r+1)} \equiv (r^3+3r^2pk)^{r+1} \equiv (r^{3(r+1)} + 3r^2pk(r+1)r^{3r}) \pmod{p^2}$. Now note that $r^{3r}r^2(r+1) \equiv -r \pmod p$ since $r$ is a cube root of unity modulo $p$, so that $3r^2pk(r+1)r^{3r} \equiv -3rpk \pmod{p^2}$. Hence, $(r+1)^{6(r+1)} \equiv (r^{3r+3} - 3rpk) \pmod{p^2}$. Therefore,

$$(r+1)^{6(r+1)} - r^{6r} \equiv (r^{3r}(r^3 - r^{3r}) - 3rpk) \pmod{p^2}.$$

Since $r^3 = pk(r-1)+1$, we get $r^{3r} \equiv (rpk(r-1)+1) \pmod{p^2}$, so $r^3 - r^{3r} \equiv -pk(r-1)^2 \pmod{p^2}$. Hence,

$$(r+1)^{6(r+1)} - r^{6r} \equiv -pk(r^{3r}(r-1)^2 + 3r) \pmod{p^2}.$$

Since $r^{3r}(r-1)^2 + 3r \equiv 0 \pmod p$, this proves the proposition. □

**Remark** A less computational proof of Lemma 6.1 was suggested to us by Dino Lorenzini. The argument goes as follows: To show that the reduction is tame, it suffices, by work of McCallum, to show that the degree of the minimum extension $M/K_w^{unr}$ such that $J_{a,b,c}$ has good reduction over $M$ is prime to $p$. It is known that this minimum degree is at most $2g + 1$. Now suppose $J_{a,b,c}$ is isogenous to the product of two abelian varieties of smaller dimension. Then $M$ is the compositum of the corresponding minimum extensions for the factors. Each of the latter extensions has degree strictly less than $p$, so the degree of their compositum is prime to $p$.

**Proof of Theorem 1.6** By Lemma 6.1, the reduction is tame in this case. By Theorem 1.4 and Proposition 5.1, the $\mathbb{F}_p$-dimension of $Ш[\lambda]/(\lambda^3 Ш[\lambda^4])$ is $\geq 2$. In particular, the $\mathbb{F}_p$-dimension of $Ш[\lambda]$ is $\geq 2$. Since $p$ is regular, the results of Faddeev ([Fad61]) show that the Selmer group $S_\lambda$ is 3-dimensional over $\mathbb{F}_p$. On the other hand, Gross and Rohrlich ([GR78]) have shown that the Mordell–Weil rank of $J_{7,1,-8}$ over $\mathbb{Q}$ is nonzero. Therefore, the rank equals 1 and $Ш[\lambda]$ is 2-dimensional over $\mathbb{F}_p$. By Theorem 1.2 it follows that $Ш[\lambda^3]$ has rank 2 over $\mathbb{Z}[\zeta]/(\lambda^3)$. Theorem 1.5 then implies that $Ш[p^\infty] \simeq (\mathbb{Z}[\zeta]/(\lambda^3))^2$. The statement about quadratic points on $F_{7,1,-8}$ and on the Fermat curve $X^{19} + Y^{19} + Z^{19} = 0$ follows immediately from Corollary 2.2 and Theorem 1.3 of [Tze02]. □

# References

[BGR84] Siegfried Bosch, Ulrich Güntzer and Reinhold Remmert, *Non-Archimedean analysis*, A systematic approach to rigid analytic geometry, Springer, 1984

[Fad61] D. K. Faddeev, *Invariants of divisor classes for the curves $x^k(1-x) = y^l$ in an $l$-adic cyclotomic field*, Trudy Mat. Inst. Steklov. **64** (1961) 284–293

[GR78] Benedict H. Gross and David E. Rohrlich, *Some results on the Mordell–Weil group of the Jacobian of the Fermat curve*, Invent. Math. **44** (1978) 201–224

[Gre81] Ralph Greenberg, *On the Jacobian variety of some algebraic curves*, Compositio Math. **42** (1981) 345–359

[KR78] Neal Koblitz and David Rohrlich, *Simple factors in the Jacobian of a Fermat curve*, Canad. J. Math. **30** (1978) 1183–1205

[Kur92] Masato Kurihara, *Some remarks on conjectures about cyclotomic fields and k-groups of $\mathbb{Z}$*, Compositio Math. **81** (1992) 223–226

[Lim95] Chong-Hai Lim, *The Shafarevich–Tate group and the Jacobian of a cyclic quotient of a Fermat curve*, Arch. Math. (Basel) **64** (1995) 17–21

[McC82] William G. McCallum, *The degenerate fiber of the Fermat curve*, Number theory related to Fermat's last theorem (Cambridge, Mass., 1981) (N. Koblitz, ed.), Progress in Mathematics, Birkhaüser, 1982, pp. 57–70

[McC88] William G. McCallum, *On the Shafarevich–Tate group of the Jacobian of a quotient of the Fermat curve*, Invent. Math. **93** (1988) 637–666

[NSW00] Jürgen Neukirch, Alexander Schmidt and Kay Wingberg, *Cohomology of Number Fields*, Springer-Verlag, Berlin, 2000

[Tze02] Pavlos Tzermias, *Low-degree points on Hurwitz–Klein curves*, Recommended for publication in Trans. Amer. Math. Soc. 2002

[Van20] Hugh S. Vandiver, *A property of cyclotomic integers and its relation to Fermat's last theorem*, Ann. of Math. **21** (1919–1920) 73–80

William G. McCallum,
Department of Mathematics,
P.O. Box 210089, 617 N. Santa Rita,
The University of Arizona,
Tucson, AZ 85721-0089, USA
e-mail: wmc@math.arizona.edu

Pavlos Tzermias,
Department of Mathematics,
University of Tennessee,
Knoxville, TN 37996-1300, USA
e-mail: tzermias@math.utk.edu

# Cascades of projections from log del Pezzo surfaces

Miles Reid     Kaori Suzuki

*To Peter Swinnerton-Dyer, in admiration*

**Abstract**

One of the best-loved tales in algebraic geometry is the saga of the blowup of $\mathbb{P}^2$ in $d \le 8$ general points and its anticanonical embedding. If a del Pezzo surface $F$ with log terminal singularities has a large anticanonical system $|-K_F|$, it can likewise be blown up many times to produce cascades of del Pezzo surfaces; as in the ancient fable, a blowup can be viewed as a projection from a bigger weighted projective space to a smaller one, leading in nice cases to weighted hypersurfaces or other low codimension Gorenstein constructions. The simplest examples already give several beautiful cascades, that we exploit as test cases for practice in the study of various kinds of projections and unprojections. We believe that these calculations will eventually have more serious applications to Fano 3-folds of Fano index $\ge 2$, involving 1001 lovely and exotic adventures.

# 1 The story of $\overline{\mathbb{F}}_3$

Once upon a time, there was a surface $F = \overline{\mathbb{F}}_3$, known to all as the cone over the twisted cubic, or as $\mathbb{P}(1,1,3) = \operatorname{Proj} k[u_1, u_2, v]$, where $\operatorname{wt} u_1, u_2 = 1$, $\operatorname{wt} v = 3$. The anticanonical class of $F$ is $-K_F = \mathcal{O}_{\overline{\mathbb{F}}_3}(5)$, so that its anticanonical ring $R(F, -K_F)$ is the fifth Veronese embedding or truncation $k[u_1, u_2, v]^{(5)}$. We see that this ring is generated by

$$\begin{aligned} x_1, \dots, x_9 &= S^5(u_1,u_2), S^2(u_1,u_2)v && \text{in degree 1,} \\ y_1, y_2 &= u_1v^3, u_2v^3 && \text{in degree 2,} \\ z &= v^5 && \text{in degree 3,} \end{aligned}$$

where, as usual, we write $S^d(u_1,u_2) = \{u_1^d, u_1^{d-1}u_2, \dots, u_2^d\}$ for the set of monomials of degree $d$ in $u_1, u_2$.

Note that the two generators $y_1, y_2$ in degree 2 are essential as orbifold coordinates or *orbinates* at the singular point. This point is simple and well known, but we spell it out, as it is essential for the enjoyment of our narrative: at $P = P_v = (0,0,1) \in \mathbb{P}(1,1,3)$, only $v \neq 0$. We take a cube root $\xi = \sqrt[3]{v}$, thus introducing a $\mathbb{Z}/3$ Galois extension of the homogeneous coordinate ring. The homogeneous ratios $u_1/\xi, u_2/\xi$ are coordinates on a copy of $\mathbb{C}^2$, which is a $\mathbb{Z}/3$ cover of an affine neighbourhood of $P$; hence $P$ is a quotient singularity of type $\frac{1}{3}(1,1)$. In our truncated subring $R(F, -K_F)$, only $z \neq 0$ at $P$, and the same orbinates are provided by the homogeneous ratios $y_1/z^{2/3}, y_2/z^{2/3}$. In the projective embedding given by $R(F, -K_F)$, since the orbinates are naturally forms of degree 2, we think of $P$ as a quotient singularity of type $\frac{1}{3}(2,2)$.

There are many ways of seeing that the Hilbert function of $R(F, -K_F)$ is given by

$$P_n = h^0(F, -nK_F) = 1 + \frac{25}{3}\binom{n+1}{2} - \begin{cases} \frac{1}{3} & \text{if } n \equiv 1 \bmod 3 \\ 0 & \text{otherwise} \end{cases}$$

for all $n \geq 0$, and thus the Hilbert series is

$$P_F(t) := \sum P_n t^n = \frac{1 + 7t + 9t^2 + 7t^3 + t^4}{(1-t)^2(1-t^3)}.$$

You can do this as an exercise in orbifold RR ([YPG], Chapter III); or another way is to multiply the Hilbert series $1/(1-s)^2(1-s^3)$ of $k[u_1, u_2, v]$ through by $(1-s^5)^2(1-s^{15})$, truncate it to the polynomial consisting only of terms of degree divisible by 5, and substitute $s^5 = t$.

Now let $S = S^{(d)} \to F$ be the blowup of $F$ in $d$ general points $P_i$, for $d \leq 8$. Write $E_i$ for the $-1$-curves over $P_i$. Since $K_S = K_F + \sum E_i$, the anticanonical ring $R(S, -K_S)$ consists of elements of $R(F, -K_F)$ of degree $n$ passing $n$ times through $P_i$. Thus each point imposes one condition in degree 1, 3 in degree 2, etc. Therefore the Hilbert series of $S$ is

$$P_S(t) = P_F(t) - d \times \frac{t}{(1-t)^3} = \frac{1 + (7-d)t + (9-d)t^2 + (7-d)t^3 + t^4}{(1-t)^2(1-t^3)}.$$

In particular $S^{(d)}$ has anticanonical degree $\frac{25-3d}{3} = (8-d) + \frac{1}{3}$. The first cases are listed in Table 1.1; the first three models suggested by the Hilbert function work without trouble. For $S^{(6)}$, the Hilbert function requires 3 generators in degree 1, 2 in degree 2, and 1 in degree 3, and the corresponding Hilbert numerator is

$$(1-t)^3(1-t^2)^2(1-t^3)P_S(t) = 1 - 2t^3 - 3t^4 + 3t^5 + 2t^3 - t^8.$$

| | | | |
|---|---|---|---|
| $d=8$ | $1/3$ | $P_S(t) = \frac{1+t^5}{(1-t)(1-t^2)(1-t^3)}$ | $S_{10} \subset \mathbb{P}(1,2,3,5)$ |
| $d=7$ | $4/3$ | $P_S(t) = \frac{1+2t^2+t^4}{(1-t)^2(1-t^3)}$ | $S_{4,4} \subset \mathbb{P}(1,1,2,2,3)$ |
| $d=6$ | $7/3$ | $P_S(t) = \frac{1+2t^2-2t^3-t^5}{(1-t)^3(1-t^3)}$ | $S_{\mathrm{Pf}} \subset \mathbb{P}(1^3,2^2,3)$ |
| $d=5$ | $10/3$ | $P_S(t) = \frac{1+t^2-4t^3+t^4+t^6}{(1-t)^4(1-t^3)}$ | codim 4 |
| $d=4$ | $13/3$ | $P_S(t) = \frac{1-t^2-4t^3+4t^4+t^5-t^7}{(1-t)^5(1-t^3)}$ | codim 5 |

Table 1.1: The cascade above $S_{10} \subset \mathbb{P}(1,2,3,5)$

This indicates that $S^{(6)} \subset \mathbb{P}(1^3,2^2,3)$ should be defined (in coordinates $x_1,x_2,x_3,y_1,y_2,z$) by the Pfaffians of a $5\times 5$ skew matrix

$$A(S^{(6)}) = \begin{pmatrix} x_1 & x_2 & b_{14} & b_{15} \\ & x_3 & b_{24} & b_{25} \\ & & b_{34} & b_{35} \\ & & & z \end{pmatrix} \quad \text{of degrees} \quad \begin{pmatrix} 1 & 1 & 2 & 2 \\ & 1 & 2 & 2 \\ & & 2 & 2 \\ & & & 3 \end{pmatrix}. \tag{1.1}$$

We see that this works: thus the 3 Pfaffians involving $z$ give $x_i z = \cdots$, so that at the point $P_z = (0,\dots,0,1)$ the three $x_i$ are eliminated as implicit functions, and $P_z$ is a $\frac{1}{3}(2,2)$ singularity with orbinates $y_1, y_2$.

**Remark 1.1** For $S^{(5)}$ and $S^{(4)}$, innocently putting in only the generators required by the Hilbert series suggests the similar codimension 3 Pfaffian models of Table 1.2. However, experience says that they cannot possibly

| | | | |
|---|---|---|---|
| $d=5$ | $\frac{1-4t^3-t^4+t^4+4t^5-t^8}{(1-t)^4(1-t^2)(1-t^3)}$ | $S^{(5)} \subset \mathbb{P}(1^4,2,3)$ | $\begin{pmatrix} 1 & 1 & 1 & 1 \\ & 2 & 2 & 2 \\ & & 2 & 2 \\ & & & 2 \end{pmatrix}$ |
| $d=4$ | $\frac{1-t^2-4t^3+4t^4+t^5-t^7}{(1-t)^5(1-t^3)}$ | $S^{(4)} \subset \mathbb{P}(1^5,3)$ | $\begin{pmatrix} 1 & 1 & 1 & 2 \\ & 1 & 1 & 2 \\ & & 1 & 2 \\ & & & 2 \end{pmatrix}$ |

Table 1.2: Candidate Pfaffian models that don't work

work: each of these is a *mirage* of a type encountered many times in the course of previous adventures. For one thing, there is nowhere for a variable of degree 3 to appear in the matrix, so that its Pfaffians define a weighted projective cone with vertex $(0,\dots,0,1)$ over a base $C \subset \mathbb{P}(1^4,2)$ (respectively, $C \subset \mathbb{P}^4$) that is a projectively Gorenstein curve $C$ with $K_C = \mathcal{O}(2)$; the cone point is not log terminal. For another, the anticanonical ring needs two generators of degree 2 to provide orbinates at the singularity of type $\frac{1}{3}(2,2)$. The conclusion is that we have not yet put in enough generators for the graded ring (or, in other contexts, that the variety we seek does not exist). Mirages of this type appear all over the study of graded rings, as discussed in 3.3.

As we see below, $S^{(d)}$ is an explicit construction from $\overline{\mathbb{F}}_3$, and has projections down to $S_{10} \subset \mathbb{P}(1,2,3,5)$ or $S_{4,4} \subset \mathbb{P}(1,1,2,2,3)$, so that we can find out anything we want to know about the rings $R(S, -K_S)$ by working in birational terms, either from above by projecting from $\overline{\mathbb{F}}_3$, or from below by unprojecting from one of the low codimension cases. We first relate without proof what happens. Listen and attend!

Consider $S = S^{(5)}$ first. First, $R(S, -K_S)$ has two generators $y_1, y_2$ and one relation in degree 2; the Hilbert series on its own cannot detect this, because the relation masks the second generator. Once you know about the additional generator, the anticanonical model of $S^{(5)}$ is a codimension 4 construction $S^{(5)} \subset \mathbb{P}(1^4, 2^2, 3)$, with Hilbert numerator

$$(1-t)^4(1-t^2)^2(1-t^3)P_S(t) = 1 - t^2 - 4t^3 + 8t^5 - 4t^7 - t^8 + t^{10};$$

however, there is still more masking going on: although the Hilbert series only demands one relation in degree 2 and 4 in degree 3, there are in fact also 4 relations and 4 syzygies in degree 4, and the ring has the $9 \times 16$ minimal resolution

$$\begin{aligned} \mathcal{O}_S \leftarrow \mathcal{O}_{\mathbb{P}} \leftarrow \mathcal{O}_{\mathbb{P}}(-2) \oplus 4\mathcal{O}(-3) \oplus 4\mathcal{O}(-4) \\ \leftarrow 4\mathcal{O}_{\mathbb{P}}(-4) \oplus 8\mathcal{O}(-5) \oplus 4\mathcal{O}(-6) \leftarrow \cdots \text{(sym.)} \end{aligned} \tag{1.2}$$

The syzygy matrixes in this complex have $4 \times 4$ blocks of zeros (of degree 0). We represent this by writing out the Hilbert numerator as the expression

$$1 \quad - t^2 - 4t^3 - 4t^4 \quad + 4t^4 + 8t^5 + 4t^6 \quad - 4t^6 - 4t^7 - t^8 \quad + t^{10},$$

where the spacing is significant. Likewise, $S^{(4)}$ is the codimension 5 construction $S^{(4)} \subset \mathbb{P}(1^5, 2^2, 3)$, with $14 \times 35$ resolution represented by

$$\begin{aligned} 1 \quad - 3t^2 - 6t^3 - 5t^4 \quad + 2t^3 + 12t^4 + 15t^5 + 6t^6 \\ - 6t^5 - 15t^6 - 12t^7 - 2t^8 \quad + 5t^7 + 6t^8 + 3t^9 \quad - t^{11}. \end{aligned} \tag{1.3}$$

These assertions can be justified either by viewing $S^{(d)}$ as projected from

$F = \overline{\mathbb{F}}_3$, or as unprojected from $S^{(d+1)}$. For convenience, we do $S^{(5)}$ from below, and $S^{(4)}$ from above (but we could do either case by the other method, with slightly longer computations).

Projecting from a general $P \in S^{(5)}$ blows $P$ up to a $-1$-curve $l = \mathbb{P}^1$ contained in the Pfaffian model of $S^{(6)} \subset \mathbb{P}(1^3, 2^2, 3)$. Inversely, $S^{(5)}$ is obtained as the Kustin–Miller unprojection of $l \subset S^{(6)}$ (see Papadakis and Reid [PR]): the ring of $S^{(5)}$ is generated over that of $S^{(6)}$ by adjoining 1 unprojection variable $x = x_4$ of degree $k_S - k_l = -1 - (-2) = 1$, with unprojection equations $x \cdot g_i = \cdots$, for the generators $g_i$ of $I_l$. Now $l$ is clearly a complete intersection of 4 hypersurfaces of degrees $1, 2, 2, 3$ (it is $x_3 = y_1 = y_2 = z = 0$ up to a coordinate change). The ring of $S^{(5)}$ thus has equations the old equations of $S^{(6)}$ of degrees $3, 3, 4, 4, 4$ (the Pfaffians (1.1) defining $A(S^{(6)})$), together with 4 unprojection equations of degrees $2, 3, 3, 4$. The numerical shape of the resolution (1.2) comes from this and Gorenstein symmetry. The same result can be obtained by applying the Kustin–Miller construction directly: the projective resolution of the ring of $S^{(6)}$ is the Buchsbaum–Eisenbud complex $L_\bullet$ of the matrix $A(S^{(6)})$, and that of $l$ is the Koszul complex $M_\bullet$ of the regular sequence defining $l$. Then $R(S^{(5)})$ arises from a homomorphism $L_\bullet \to M_\bullet$ extending the map $\mathcal{O}_{S^{(6)}} \twoheadrightarrow \mathcal{O}_l$. For details, see Papadakis [P2].

We justify $S^{(4)}$ in the other direction, by projecting down from $F$. We can choose coordinates to put a general set of 4 points in the form

$$\{P_1, \ldots, P_4\} \subset F = \mathbb{P}(1,1,3) \quad \text{given by} \quad f_4(u_1, u_2) = v = 0.$$

The anticanonical ring of the 4-point blowup $S^{(4)}$ is then generated by

$$\begin{aligned} x_1, \ldots, x_5 &= \{u_1 f, u_2 f, S^2(u_1, u_2) v\} && \text{in degree 1,} \\ y_1, y_2 &= u_1 v^3, u_2 v^3 && \text{in degree 2,} \\ z &= v^5 && \text{in degree 3.} \end{aligned}$$

The ideal of relations between these can be studied by explicit elimination (we used computer algebra, but it is not at all essential); one finds that it is generated by

$$\operatorname{rank} \begin{pmatrix} * & x_1 & x_2 & y_0 \\ x_1 & x_3 & x_4 & y_1 \\ x_2 & x_4 & x_5 & y_2 \\ y_0 & y_1 & y_2 & z \end{pmatrix} \le 1, \quad \text{where} \quad y_0 = q(x_3, x_4, x_5). \tag{1.4}$$

Taking $y_0$ as a variable gives the second Veronese embedding of the one point blowup of the 3-fold wps $\mathbb{P}(\frac{1}{2}, \frac{1}{2}, \frac{1}{2}, \frac{3}{2})$. Thus $S^{(4)}$ is a hypersurface of weighted degree 2 in this curious weighted quasihomogenous variety. The second Veronese embedding of the one point blowup of $\mathbb{P}^3$ is a well known codimension 5 del Pezzo variety appearing in other myths, and its equations have a $14 \times 35$ resolution. We check that this agrees with (1.3).

**Exercise 1.2** Chronicle the fate of $\overline{\mathbb{F}}_5$ and its $d$-point blowup $S^{(d)} \to \overline{\mathbb{F}}_5$ for $d \le 9$. [Hint: the Hilbert series is

$$\begin{aligned} P(t) &= \frac{1+9t+9t^2+11t^3+9t^4+9t^5+t^6}{(1-t)^2(1-t^5)} - d \times \frac{t}{(1-t)^3} \\ &= \frac{1+(9-d)t+(9-d)t^2+(11-d)t^3+(9-d)t^4+(9-d)t^5+t^6}{(1-t)^2(1-t^5)}. \end{aligned}$$

The singularity polarised by $-K = A$ is of type $\frac{1}{5}(3,3)$, so that $S^{(d)}$ is in $\mathbb{P}(1^{11-d},3,3,5)$. Thus $d=9$ gives $S_{6,6} \subset \mathbb{P}(1,1,3,3,5)$ and $d=8$ gives a nice Pfaffian in $\mathbb{P}(1,1,1,3,3,5)$, with Hilbert numerator

$$1-2t^4-3t^6+3t^7+2t^9-t^{13},$$

etc.]

These surfaces have a singularity of type $\frac{1}{5}(3,3)$; we were disappointed at first to observe that none of these is a hyperplane section $S \in |A|$ for a Mori Fano 3-fold $X$ of Fano index 2. For then $X$ would have a quotient singularity of type $\frac{1}{5}(1,3,3)$, which is unfortunately not terminal. For further disappointment, see 3.2.2.

# 2 The ingenious history of $\frac{1}{5}(2,4)$

Let $T$ be a del Pezzo surface polarised by $-K_T = \mathcal{O}_T(A)$ with a quotient point $P \in T$ of type $\frac{1}{5}(2,4)$ as its only singularity. (Up to isomorphism, $P$ is the quotient singularity $\frac{1}{5}(1,2)$, but to give sections of $-K_T$ weight 1, and make $\mathcal{O}_T(A) = -K_T$ the preferred generator of the local class group, we twist $\mu_5$ by an automorphism so that $d\xi \wedge d\eta$ is in the $\varepsilon \mapsto \varepsilon$ character space, and thus $\operatorname{wt}\xi = 2, \operatorname{wt}\eta = 4 \bmod 5$.) By an exercise in the style of [YPG], Chapter III, we see that

$$P_n(T) = 1 + \binom{n+1}{2} A^2 - \begin{cases} 0 & n \equiv 0 \mod 5 \\ 2/5 & n \equiv 1 \mod 5 \\ 1/5 & n \equiv 2 \mod 5 \\ 2/5 & n \equiv 3 \mod 5 \\ 0 & n \equiv 4 \mod 5 \end{cases}$$

Trying $n = 1$ gives $A^2 \equiv 2/5 \bmod \mathbb{Z}$. Putting these values in a Hilbert series as usual and setting $A^2 = k + \frac{2}{5}$ gives

$$\begin{aligned}
P(t) &= \frac{1}{1-t} + \frac{t}{(1-t)^3}A^2 - \frac{1}{5}\cdot\frac{2t+t^2+2t^3}{1-t^5} \\
&= \frac{1}{1-t} + \frac{t}{(1-t)^3}k \\
&\quad + \frac{1}{5}\cdot\frac{2t(1+t+t^2+t^3+t^4)-(1-t)^2(2t+t^2+2t^3)}{(1-t)^2(1-t^5)} \\
&= \frac{1-t+t^2+t^4-t^5+t^6}{(1-t)^2(1-t^5)} + \frac{t}{(1-t)^3}k.
\end{aligned}$$

The case $k = 0$ gives

$$\begin{aligned}
\frac{1-t+t^2+t^4-t^5+t^6}{(1-t)^2(1-t^5)} &= \frac{1+t^3+t^4+t^7}{(1-t)(1-t^2)(1-t^5)} \\
&= \frac{1-t^6-t^8+t^{14}}{(1-t)(1-t^2)(1-t^3)(1-t^4)(1-t^5)},
\end{aligned}$$

that is, $T_{6,8} \subset \mathbb{P}(1,2,3,4,5)$.

This surface turns out to be the bottom of a cascade of six projections, whose head is the surface $T = T_6 \subset \mathbb{P}(1,1,3,5)$ with $-K_T = A = \mathcal{O}(4)$. We guessed this as follows: by the standard dimension count for del Pezzo surfaces, we expect $T_{6,8}$ to contain a finite number of $-1$-curves not passing through the singularity. Contracting $k$ disjoint $-1$-curves gives a surface with $K_T^2 = A^2 = k + \frac{2}{5}$ and the above Hilbert series. For $k = 6$, we see that $A^2 = 6 + \frac{2}{5} = \frac{32}{5}$ is divisible by $4^2$, and we guess that $A = 4B$, leading to a surface with the Hilbert series of $T = T_6 \subset \mathbb{P}(1,1,3,5)$. Hindsight is the only justification for this guesswork.

One sees that the minimal resolution $\widetilde{T} \to T$ is the scroll $\mathbb{F}_3$ blown up in two points on a fibre, and that $T$ is obtained from this by contracting the chain of $\mathbb{P}^1$s with self-intersection $(-3,-2)$ coming from the negative section and the birational transform of the fibre (see Figure 2.1).

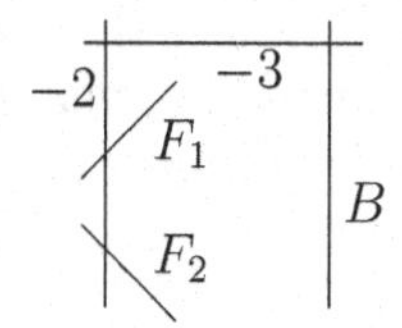


Figure 2.1: Resolution of $T = T_6 \subset \mathbb{P}(1,1,3,5)$

We start by calculating the anticanonical ring of the head of the cascade, $T = T_6 \subset \mathbb{P}(1,1,3,5)$. Take coordinates $u_1, u_2, v, w$ in $\mathbb{P}(1,1,3,5)$, and take the defining equation of $T$ to be

$$u_2w = f_6(u_1, v) = av^2 + bvu_1^3 + cu_1^6 = l_1(v, u_1^3)l_2(v, u_1^3); \tag{2.1}$$

we could normalise the right-hand side to $(v - u_1^3)(v + u_1^3)$. We use this relation to eliminate any monomial divisible by $u_2w$. Write $B$ for the divisor class corresponding to $\mathcal{O}_{\mathbb{P}}(1)$ or its restriction to $T$. Since $-K_T = 4B$, the anticanonical embedding of $T$ is the 4th Veronese embedding of $T \subset \mathbb{P}(1,1,3,5)$; one checks that the anticanonical ring is generated by

$$\begin{array}{rcll} x_1, \dots, x_7 &=& S^4(u_1, u_2), (u_1, u_2)v & \text{in degree } 1, \\ y_1, y_2 &=& u_1^3 w, vw & \text{in degree } 2, \\ z &=& u_1^2 w^2 & \text{in degree } 3, \\ t &=& u_1 w^3 & \text{in degree } 4, \\ u &=& w^4 & \text{in degree } 5, \end{array} \tag{2.2}$$

and that its relations are given by the $2 \times 2$ minors of

$$\begin{pmatrix} x_1 & x_2 & x_3 & x_4 & x_6 & y_1 & z & t \\ x_2 & x_3 & x_4 & x_5 & x_7 & A & B & C \\ y_1 & A & B & C & y_2 & z & t & u \end{pmatrix}, \tag{2.3}$$

with

$$A = ax_6^2 + bx_1x_6 + cx_1^2,$$
$$B = ax_6x_7 + bx_2x_6 + cx_1x_2,$$
$$C = ax_7^2 + bx_3x_6 + cx_1x_3.$$

**Theorem 2.1** *For $d \le 6$, write $\sigma\colon T^{(d)} \dashrightarrow T$ for the blowup of $T$ in $d$ general points $P_1, \dots, P_d$. (We elucidate what "general" means in (2.6) below.) Write $E_i$ for the $-1$-curves over $P_i$ and $A^{(d)} = \sigma^*A - \sum E_i$ for the anticanonical class of $T^{(d)}$. Then $T^{(d)}$ is a log del Pezzo surface with only singularity of type $\frac{1}{5}(2,4)$ and $(-K_S)^2 = 6 - d + \frac{2}{5}$.*

*For $d \le 5$, the anticanonical ring of $T^{(d)}$ needs $12-d$ generators of degrees $1^{7-d}, 2^2, 3, 4, 5$, and gives an embedding $T^{(d)} \subset \mathbb{P}(1^{7-d}, 2^2, 3, 4, 5)$ that takes the $E_i$ to disjoint projectively normal lines*

$$E_i \cong \mathbb{P}^1 \subset T^{(d)} \subset \mathbb{P}(1^{7-d}, 2^2, 3, 4, 5).$$

*The anticanonical ring of $T^{(6)}$ needs 5 generators of degrees $1, 2, 3, 4, 5$, and embeds $T^{(6)}$ as the complete intersection $T_{6,8} \subset \mathbb{P}(1,2,3,4,5)$, taking the $E_i$ to disjoint $-1$-curves in $T_{6,8}$ (of course, the $E_i \subset \mathbb{P}(1,2,3,4,5)$ cannot be projectively normal).*

Each inclusion $R(T^{(d)}, A^{(d)}) \subset R(T^{(d-1)}, A^{(d-1)})$ for $d \le 5$ is a Kustin–Miller unprojection in the sense of [PR]. That is, it introduces precisely one new generator of degree 1 with pole along $E_d$, subject only to linear relations. For $d = 6$, see Remark 2.5.

**Proof** As in the analogous recitations for nonsingular del Pezzo surfaces, the proof consists for the most part of restricting to the general curve $C \in |A^{(d)}|$. The restriction $R(T^{(d)}, A^{(d)}) \to R(C, A^{(d)})$ is a surjective ring homomorphism, and is the quotient by the principal ideal $(x_C)$, where $x_C$ is the equation of $C$. Thus the hyperplane section principle applies, and we only have to prove the appropriate generation results for $R(C, A^{(d)})$. In the antique tale, $C$ is a nonsingular elliptic curve, and we win because we know everything about linear systems on it. In our case $C \in |-K_{T^{(d)}}|$ is an elephant, so is again a projectively Gorenstein curve with $K_C = 0$, but it is an orbifold nodal rational curve in a sense we are about to study. Our proof will then boil down to a monomial calculation.

The general curve $C \in |A|$ on $T$ is irreducible and has an ordinary node at $P$, and the two orbinates of $P \in T$ restrict to respective local analytic coordinates on the two branches of the node. In other words, $P \in C$ is locally analytically equivalent to the quotient $\bigl((\xi\eta = 0) \subset \mathbb{C}^2\bigr)/(\frac{1}{5}(2,4))$, where $\xi, \eta$ are as in Remark 2.4. To make formal sense of this, we need to work with the affine cone over $T \supset C$ along the $u$-axis, and the $\mathbb{C}^*$ action on them. The cone over $T$ is nonsingular along the $u$-axis outside the origin, with transverse coordinates $y_2, t$ (see (2.8)) – the $\frac{1}{5}(2,4)$ singularity arises from the $\mathbb{Z}/5$ isotropy. The coefficient of $x_6$ in the equation $(x_C)$ of $C$ is nonzero in general, corresponding to $u_1 v$ in (2.2). Therefore, along the $u$-axis, the cone over $C$ is given locally by $y_2 t =$ higher order terms.

We choose a general curve $C \in |A|$ and $d \le 6$ general points $P_1, \dots, P_d$ contained in $C$. These points are also independent general points of $T$, because $|A|$ is a 6-dimensional linear system on $T$. This choice ensures the existence of an irreducible curve $C \in |A - \sum P_i|$ with the local behaviour at $P$ just described. The birational transform of $C$ on $T^{(d)}$ is an isomorphic curve $C \in |A^{(d)}|$ that we continue to denote by $C$. It is irreducible, therefore nef, and big since $(A^{(d)})^2 > 0$.

The normalisation $n\colon \widetilde{C} \to C \subset T^{(d)}$ is a conventional orbifold curve: it is a rational curve with two marked point $P_1, P_2$, the inverse image of the node of $C$. In calculations, we take $\widetilde{C} = \mathbb{P}^1$, and $P_1 = 0$ and $P_2 = \infty$. It is polarised by $\widetilde{A} = n^*(A^{(d)}) = \frac{3}{5}P_1 + \frac{4}{5}P_2 + (5-d)Q$, where $Q$ is some other point. This is just a notational device to handle the sheaf of graded algebras

$$\mathcal{A} = \bigoplus \mathcal{A}_i \quad \text{with} \quad \mathcal{A}_i = \mathcal{O}_{\mathbb{P}^1}\left(\left[\frac{3i}{5}\right] P_1 + \left[\frac{4i}{5}\right] P_2\right) \otimes \mathcal{O}_{\mathbb{P}^1}((5-d)i).$$

We calculate $R(\widetilde{C}, \widetilde{A})$ in monomial terms (the answer has a nice toric description, see Exercise 2.2).

For $d = 5$, the calculations is as follows: $R(\widetilde{C}, \widetilde{A}) = R(\mathbb{P}^1, \frac{3}{5}P_1 + \frac{4}{5}P_2)$ is generated by

$$\begin{aligned} x &\text{ in degree 1 with } \operatorname{div} x = \tfrac{3}{5}P_1 + \tfrac{4}{5}P_2, \\ y_1, y_2 &\text{ in degree 2 with } \operatorname{div}(y_1, y_2) = (2P_1, 2P_2) + \tfrac{1}{5}P_1 + \tfrac{3}{5}P_2, \\ z &\text{ in degree 3 with } \operatorname{div} z = 3P_1 + \tfrac{4}{5}P_1 + \tfrac{2}{5}P_2, \\ t &\text{ in degree 4 with } \operatorname{div} t = 5P_1 + \tfrac{2}{5}P_1 + \tfrac{1}{5}P_2, \\ u_1, u_2 &\text{ in degree 5 with } \operatorname{div}(u_1, u_2) = (7P_1, 7P_2). \end{aligned} \tag{2.4}$$

Here, in each degree, $|iD|$ is the fractional part $\{\frac{3i}{5}\}P_1 + \{\frac{4i}{5}\}P_2$ plus a linear system $|\mathcal{O}_{\mathbb{P}^1}(k_i)|$, based by elements corresponding to the monomials $S^{k_i}(t_1, t_2)$, of which the middle ones are old, and some of the extreme ones are new generators. Thus in degree 2, $k_2 = 2$, and the monomials $y_1, x^2, y_2$ correspond to $t_1^2, t_1t_2, t_2^2$.

**Exercise 2.2** The generators of $R(\widetilde{C}, \widetilde{A})$ and the relations between them are simply grasped by noting that $u_1, t, z, y_1, x, y_2, u_2$ in (2.4) satisfy

$$u_1 z = t^2, \quad t y_1 = z^2, \quad z x = y_1^2, \quad y_1 y_2 = x^4, \quad x u_2 = y_2^3;$$

this is the Jung–Hirzebruch presentation of the invariant ring of $\mathbb{Z}/(35)$ acting on $\mathbb{C}^2$ by $\frac{1}{35}(1, 12)$, where $[2, 2, 2, 4, 3] = \frac{35}{35-12}$. The case $d = 6$ gives $[2, 2, 4] = \frac{10}{7}$. Generalising this result to the general orbifold curve $(\mathbb{P}^1, \alpha_1 P_1 + \alpha_2 P_2)$ is a little gem of a problem.

The extension of graded rings $R(C, A^{(d)}) \subset R(\widetilde{C}, \widetilde{A})$ is a normalisation, separating two transverse sheets along the $u$-axis. The affine cone over the nonnormal curve $C$ is obtained by glueing the $u_1$ and $u_2$-axes together (different choices of glueing differ by a factor in $\mathbb{C}^*$, and lead to isomorphic rings). The functions compatible with this glueing are those that take the same value on $u_1$ and $u_2$-axes. Thus $R(C, A^{(d)}) \subset R(\widetilde{C}, \widetilde{A})$ is the subring generated as above, but with only one generator $u = u_1 - u_2$ in degree 5 instead of two. This proves the statement on generators of $R(S^{(d)}, A^{(d)})$ for $d = 5$. The cases $d \leq 4$ are similar.

In case $d = 6$, the orbifold divisor on $\widetilde{C} = \mathbb{P}^1$ is

$$\widetilde{A} = n^*(A^{(d)}) = \frac{3}{5}P_1 + \frac{4}{5}P_2 - Q.$$

An identical calculation shows that $R(\widetilde{C}, \widetilde{A})$ is generated by

$$\begin{aligned} y &\text{ in degree 2 with } \operatorname{div} y = \tfrac{1}{5}P_1 + \tfrac{3}{5}P_2, \\ z &\text{ in degree 3 with } \operatorname{div} z = \tfrac{4}{5}P_1 + \tfrac{2}{5}P_2, \\ t &\text{ in degree 4 with } \operatorname{div} t = P_1 + \tfrac{2}{5}P_1 + \tfrac{1}{5}P_2, \\ u_1, u_2 &\text{ in degree 5 with } \operatorname{div}(u_1, u_2) = (2P_1, 2P_2). \end{aligned} \tag{2.5}$$

As before, the nonnormal subring $R(C, A^{(d)})$ is generated by $y, z, t$ and $u = u_1 - u_2$, and one sees that the relations are

$$yt = z^2, \quad zu = t^2 - y^4.$$

That is, $C$ is the complete intersection $C_{6,8} \subset \mathbb{P}(2,3,4,5)$, as required.

This proves the assertion of Theorem 2.1 on the generation of the rings $R(T^{(d)}, A^{(d)})$. This proof uses that $A^{(d)}$ is nef and big, but not that it is ample.

We now prove that $A^{(d)}$ is ample. It is enough to show that the anticanonical morphism of $T^{(d)}$ does not contract any curve $\Gamma$ of $T$, or equivalently, that $T^{(d)}$ does not contain any curve with $A^{(d)}\Gamma = 0$. Now because the generators of $R(T^{(d)}, A^{(d)})$ include elements $y_2, t$ in (2.4) or $y, t$ in (2.5) that give the orbinates at $P \in A$, the anticanonical morphism of $T^{(d)}$ is an isomorphism near $P$, and so $\Gamma$ cannot pass through $P$. On the other hand, a curve with $A^{(d)}\Gamma = 0$ is necessarily a component of a divisor in the mobile linear system $|A^{(d)}|$ if $d \le 5$, or $|2A^{(d)}|$ if $d = 6$.

One sees that $T = T_6 \subset \mathbb{P}(1,1,3,5)$ has a free pencil $|B|$ defined by $(u_1 : u_2)$, with a reducible fibre $u_2 = 0$ that splits into two components $F_i : (u_2 = l_i = 0)$, where, as in (2.1), the equation of $T$ is $u_2 w = l_1(v, u_1^3) l_2(v, u_1^3)$ (compare Figure 2.1). Every effective Weil divisor is linearly equivalent to a positive linear combination of $F_1, F_2$. These satisfy $F_1^2 = F_2^2 = -\frac{2}{5}$ and $F_1F_2 = \frac{3}{5}$, so that $iF_1 + jF_2$ is nef if only if $\frac{2}{3}j < i < \frac{3}{2}j$. Moreover, $iF_1 + jF_2$ can only move away from $P$ if it is Cartier there, which happens if and only if $5 \mid (i+j)$. Next $iF_1 + jF_2$ a component of $|A| = |4F_1 + 4F_2|$ (resp. $|2A|$) implies $i, j \le 4$ (resp. $i, j \le 8$).

Thus for $d \le 5$ we just have to handle $\Gamma \in |2F_1 + 3F_2|$ and $|3F_1 + 2F_2|$. Since $-K_T\Gamma = 4$ and $(\Gamma)^2 = 2$, RR gives $h^0(T, \Gamma) = 4$, and for general points, no 4 of $P_1, \dots, P_d$ are contained in $\Gamma$. This completes the proof if $d \le 5$. For $d = 6$ we also need to consider

$$4F_1 + 6F_2, \quad 5F_1 + 5F_2 \quad \text{and} \quad 7F_1 + 8F_2.$$

The proper transform of a curve $\Gamma \subset T$ will give $A^{(d)}\Gamma = 0$ if $\Gamma \in |A|$ passes through the $P_i$ with multiplicity $a_i$, where

$$\sum a_i = -K_T\Gamma, \quad \sum a_i = (\Gamma)^2.$$

In the 3 cases above, the only solutions are

$$\begin{aligned}
\Gamma = 4F_1 + 6F_2 :&\quad -K_T\Gamma = 8, \quad (\Gamma)^2 = 8, \quad \text{none};\\
\Gamma = 5F_1 + 5F_2 :&\quad -K_T\Gamma = 8, \quad (\Gamma)^2 = 10, \quad (1,1,1,1,2,2);\\
\Gamma = 7F_1 + 8F_2 :&\quad -K_T\Gamma = 12, \quad (\Gamma)^2 = 24, \quad (2,2,2,2,2,2).
\end{aligned}$$

The conclusion is that $A^{(d)}$ is ample if and only

$$\begin{array}{ll}(0) & \text{the } P_i \text{ are distinct and contained in an irreducible curve } C \in |A|;\\ (1) & \text{no 4 } P_i \text{ are contained in any } \Gamma \in |2F_1+3F_2| \text{ or } |3F_1+2F_2|;\\ (2) & |5F_1+5F_2-P_1-P_2-P_3-P_4-2P_5-2P_6| = \emptyset;\\ (3) & |7F_1+8F_2-2\sum P_i| = \emptyset \text{ and } |8F_1+7F_2-2\sum P_i| = \emptyset.\end{array} \tag{2.6}$$

Here conditions (2–3) are only required if $d=6$.

These are open conditions on $P_1,\dots,P_d$, and they should fail in codimension 1. It remains to check that they are satisfied for general $P_1,\dots,P_6$. Write $C$ for the unique curve of $|A|$ through $P_1,\dots,P_6$. Then any divisor $\Gamma$ on $T$ in Case (2) contains $C$: indeed,

$$\left(5F_1+5F_2-P_1-P_2-P_3-P_4-2P_5-2P_6\right)_{|C}$$

has degree 0, but is not linear equivalent to 0 on $C$ for general $P_1,\dots,P_6$ (recall that $C$ is a nodal cubic, so that its nonsingular points correspond to different points of the algebraic group $\operatorname{Pic} C = \mathbb{C}^*$). Thus $\Gamma = C+B$, where $|B| = |F_1+F_2|$ is the pencil of $T$. Clearly, the element of $|B|$ through $P_5$ does not in general pass through $P_6$. The argument in Case (3) is similar: a divisor $\Gamma$ in Case (3) must be of the form $C+D$, where $D \in |3F_1+4F_2-\sum P_i|$. But

$$h^0(T,3F_1+4F_2) < h^0(T,4F_1+4F_2) = h^0(T,-K_T) = 7,$$

(see (2.2)) so that $|3F_1+4F_2|$ does not contain a curve through 6 general points of $T$. QED

**Exercise 2.3** State and prove the analog of Theorem 2.1 for the cascade of Section 1. In other words, prove that the $d$ point blowup of $\overline{\mathbb{F}}_3$ for $d \le 8$ has the properties asserted (without proof!) throughout Section 1.

**Remark 2.4** The monomials in (2.2) map to some of the local generators of the sheaf of algebras $\bigoplus_{i=0}^4 \mathcal{O}_{T,P}(i)$ at the $\frac{1}{5}(2,4)$ singularity. Indeed, write $\xi,\eta$ for $\varepsilon^2$ and $\varepsilon^4$ eigencoordinates on $\mathbb{C}^2$; then $\mathcal{O}_T$ is the sheaf of invariant functions, locally generated by $\xi^5,\xi^3\eta,\xi\eta^2,\eta^5$, whereas the eigensheaves $\mathcal{O}_{T,P}(i)$ are modules over $\mathcal{O}_{T,P}$, and are locally generated by

$$\begin{aligned}\mathcal{O}_{T,P}(1) &\ni \xi^3,\xi\eta,\eta^4\\ \mathcal{O}_{T,P}(2) &\ni \xi,\eta^3\\ \mathcal{O}_{T,P}(3) &\ni \xi^4,\xi^2\eta,\eta^2\\ \mathcal{O}_{T,P}(4) &\ni \xi^2,\eta\\ \mathcal{O}_{T,P}(5) &\ni 1.\end{aligned} \tag{2.7}$$

Then the homogeneous to inhomogeneous correspondence at $P$ (setting $\sqrt[5]{w} = 1$) has the effect

$$u_1 \mapsto \eta \quad \text{and} \quad v \mapsto \xi,$$

so that the generators of $R(T, -K_T)$ map to local generators of $\mathcal{O}_{T,P}(i)$ by

$$\begin{array}{ll} \text{deg 1:} & x_1 = u_1^4 \mapsto \eta^4, \quad x_6 = u_1 v \mapsto \xi\eta, \quad \emptyset \mapsto \xi^3; \\ \text{deg 2:} & y_1 = u_1^3 w \mapsto \eta^3, \quad y_2 = vw \mapsto \xi; \\ \text{deg 3:} & z = u_1^2 w^2 \mapsto \eta^2, \quad x_6 y_2 = u_1 v^2 w \mapsto \xi^2\eta, \quad \emptyset \mapsto \xi^4; \\ \text{deg 4:} & t = u_1 w^3 \mapsto \eta, \quad y_2^2 = v^2 w^2 \mapsto \xi^2; \\ \text{deg 5:} & u = w^4 \mapsto 1. \end{array} \tag{2.8}$$

The remaining generators in (2.7) are hit by monomials in these generators: for example, $\xi^3 \in \mathcal{O}_T(1)$ is first hit by $y_2^3$ in degree 6. Thus $\mathcal{O}_T(i)$ is not always generated by its $H^0$, and not just because the $H^0(\mathcal{O}_X(i))$ are too small. However, by ampleness, $R(T, -K_T)$ maps surjectively to local generators of $\bigoplus_{i=0}^4 \mathcal{O}_T(i)$, so that, for example, the orbinates $\xi$ and $\eta$ must be hit by some generators of $R(T, -K_T)$.

**Remark 2.5 (Detailed calculations of Type II projection)** We hope eventually to use the two cascades of surfaces treated in Sections 1 and 2 as exercises in understanding Type II unprojection as in [Ki], Section 9, and in particular, solve the unfinished calculation in loc. cit., 9.12. The unprojection from $S_{10} \subset \mathbb{P}(1,2,3,5)$ to $S_{4,4} \subset \mathbb{P}(1,1,2,2,3)$ is covered by the equations of [Ki], 9.8. The only little surprise here is that, instead of increasing the codimension by 2, one of the entries in the $5 \times 5$ Pfaffian matrix is a unit, and one of the equations masks the variable of degree 5 as a combination of other variables.

On the other hand, the unprojection from $S_{6,8} \subset \mathbb{P}(1,2,3,4,5)$ leads to a codimension 4 ring, and the calculation is similar to the one unfinished in [Ki], 9.12. The image $\Gamma$ of $\mathbb{P}^1 \hookrightarrow \mathbb{P}(1,2,3,4,5)$ cannot be projectively normal; indeed, if $v_1, v_2$ are coordinates on $\mathbb{P}^1$ and $x, y, z, t, u$ coordinates on $\mathbb{P}(1,2,3,4,5)$, the two rings have monomials

| | $\mathbb{P}(1,2,3,4,5)$ | $\mathbb{P}^1$ |
|---|---|---|
| in degree 1 | $x$ | $v_1, v_2$ |
| in degree 2 | $x^2, y$ | $v_1^2, v_1v_2, v_2^2$ |
| in degree 3 | $x^3, xy, z$ | $v_1^3, v_1^2v_2, v_1v_2^2, v_2^3$ |
| in degree 4 | $x^4, x^2y, y^2, xz, t$ | $S^4(v_1, v_2)$ |

and the restriction map from $\mathbb{P}(1,2,3,4,5)$ to $\Gamma$ clearly misses at least one monomial in each degree $1, 2, 3$. Choose $S_{6,8}$ containing $\Gamma$. Unprojecting it adds one linear generator, and one generator in each degree 2, 3 and 4 corresponding to these missing monomials (this will be explained better in [qG]). The old variables of degree 3 and 4 are masked by equations, and this gives rise to a codimension 4 surface $S' \subset \mathbb{P}(1,1,2,2,3,4,5)$. We still do not know how to complete this calculation directly.

**Remark 2.6** In the projection from $\overline{\mathbb{F}}_3$ of Section 1, we always assumed that the blown up points were in general position. In the classic epic of del Pezzo surfaces, there are lots of interesting degenerations, most simply if 3 points in $\mathbb{P}^2$ become collinear. The simplest way that blowups of $\overline{\mathbb{F}}_3$ degenerate is that two points come to lie on a fibre $l$ of the ruling of $\mathbb{F}_3$. If we project from two points on $l$, the birational transform of the fibre $l$ becomes a $-2$-curve, and contracting it together with the negative section of $\mathbb{F}_3$ gives a $\frac{1}{5}(1,2)$ singularity. Thus all the surfaces in Section 2 are degenerate projections of those in Section 1. For example, $T_6 \subset \mathbb{P}(1,1,3,5)$ is a projection of $\overline{\mathbb{F}}_3$ from 2 points in a fibre (see Figure 2.1). This gives a top down elimination argument as on page 231 that might allows us to complete the tricky Type 2 unprojection calculation just discussed.

This type of contraction between surfaces with log terminal singularities corresponds to the bad links of [CPR], 5.5. We do not make this too precise. The fact that we blow up a point, then unexpectedly contract the line $l$ with negative discrepancy is analogous to Sarkisov links involving an antiflip. The regular kind of blowup of a nonsingular point in a del Pezzo cascade decreases $K_S^2$ by 1, and the Hilbert function $P_S(t)$ by $t/(1-t)^3 = t+3t^2+6t^3+10t^4+\cdots$; whereas the special blowup (of a point contained in a curve of degree 2/3 that is a component of a split fibre of the conic bundle structure) considered here only decreases $K_S^2$ by 14/15, and $P_S(t)$ by

$$\frac{t(1+2t+3t^2+2t^3+3t^4+2t^5+t^6)}{(1-t)(1-t^3)(1-t^5)} = t+3t^2+6t^3+9t^4+14t^5+20t^6+26t^7+\cdots$$

# 3 Final remarks

## 3.1 Why weighted projective varieties?

Nonsingular surfaces over a field $k$ that are rational or ruled over $\overline{k}$ (that is, have $\kappa = -\infty$) are prominent objects of study in birational geometry and in Diophantine geometry. By a theorem of Castelnuovo (a distinguished precursor of Mori theory!), such a surface can be blown down (over $k$) to a minimal surface, which is a del Pezzo surface of rank 1, or a conic bundle over a curve with relative rank 1. In justifying the pre-eminent position of the cubic surfaces among del Pezzo surfaces, Peter Swinnerton-Dyer observes that del Pezzo surfaces of degree $\geq 4$ are in most respects too simple to be interesting, whereas del Pezzo surfaces of degree 2 and 1 tend to be much too difficult. Whereas the cubic surface is associated with the root systems $E_6$, those of degree 2 and 1, the weighted hypersurfaces $S_4 \subset \mathbb{P}(1^3, 2)$ and $S_6 \subset \mathbb{P}(1^2, 2, 3)$, are associated with $E_7$ and $E_8$, and are much more complicated from essentially every point of view (Galois theory, biregular and birational geometry, Diophantine arithmetic, etc.). In Peter's words:

> "if your research adviser gives you a problem involving del Pezzo surfaces of degree 2 and 1, it means he really *hates* you."

In view of this, working with del Pezzo surfaces with cyclic singularities may seem perverse, since it leads to even more exotic weighted projective constructions. For example our model case is $S_{10} \subset \mathbb{P}(1, 2, 3, 5)$, the 8 point blowup of $\overline{\mathbb{F}}_3$. It makes sense to write down the equation of $S_{10}$ over any field, and to ask for its solutions: does this lead to any interesting problems of birational geometry or Diophantine arithmetic? The Galois group of the configuration of eight $-1$-curves is clearly the symmetric groups $S_8$. In contrast to the minimal cubic surface, this is birational over $\overline{k}$ in an obvious way to the conic bundle $\mathbb{F}_3 \to \mathbb{P}^1$, with the marked section, and a set of 8 points defined over $k$. This suggests that our surfaces are actually simpler objects and do not involve especially difficult or interesting Diophantine issues. On a more positive note, log del Pezzo surfaces come in large infinite families, among which we can surely always find some really complicated case for the graduate student who deserves that special attention.

## 3.2 Log del Pezzo surfaces and Fano 3-folds of index 2

### 3.2.1 The fabulous half-elephant

Our main motivation was of course to use log del Pezzo surfaces to study Fano 3-folds of Fano index $f = 2$. The *Fano index* of a Fano 3-fold $X$ in the Mori category is the maximum natural number $f$ such that $-K_X = fA$ with $A$ a Weil divisor of $X$. Our model is the general strategy of Altınok, Brown and Reid [ABR], that uses K3 surfaces as technical background and motivation in the study of Fano 3-folds of index 1. If $X$ is a Fano with $-K_X = 2A$ twice an ample Weil divisor, a sufficiently good surface $S \in |A|$ is a del Pezzo surface (if it exists, see below); an element of $|-K_X|$ is called an elephant, so $S \in |A|$ is a *half-elephant*. In the two cascades of Sections 1 and 2, all the del Pezzo surfaces up to codimension 3 extend in an unobstructed way to Fano 3-folds. Thus for example, we have Fano 3-folds of index 2

$$X_{10} \subset \mathbb{P}(1^2, 2, 3, 5), \quad X_{4,4} \subset \mathbb{P}(1^3, 2^2, 3), \quad \text{and} \quad X_{\mathrm{Pf}} \subset \mathbb{P}(1^4, 2^2, 3)$$

extending the del Pezzo surfaces of Table 1.1. What happens in cases of codimension 4 is a computation based on the same projection cascade that we have not had time to finish; the basic question is to find all Pfaffian 3-folds $X_{\mathrm{Pf}} \subset \mathbb{P}(1^4, 2^2, 3)$ containing a linearly embedded $\mathbb{P}^2 \hookrightarrow \mathbb{P}(1^4, 2^2, 3)$. It seems likely that the single unprojection type for del Pezzo surfaces from codimension 3 to 4 splits into Tom and Jerry cases for Fano 3-folds that are essentially different (compare [Ki], Example 6.4 and 6.8 and [P1]–[P2]).

On the other hand, the codimension 5 surface $S^{(4)} \subset \mathbb{P}(1^5, 2^2, 3)$ of (1.4) probably does not have any extension in degree 1 to a Fano 3-fold of index 2: we conjecture this because it seems hard to incorporate a new variable $x_6$ of degree 1 into the equations (1.4) in a nontrivial way to give a 3-fold having only terminal singularities.

### 3.2.2 A good half-elephant is an extremely rare beast

In contradiction to our initial hopes, most Fano 3-folds $X$ of index 2 do not have a half-elephant, and most log del Pezzo surface $S$ do not extend to a Fano 3-fold of index 2. An obvious necessary global condition is $P_1(X) \geq 1$, but there are also severe local restrictions on the basket of quotient singularities: each quotient singularity $\frac{1}{r}(1, a, r-a)$ in the basket of $X$ must have $2a \cong \pm 1 \bmod r$ (so that when we rewrite the singularity as $\frac{1}{r}(2, 2a, r-2a)$, the equation of $S$ in degree 1 can be one of the orbinates). In slightly different terms, as we saw in 1.2, a del Pezzo surface $S$ with a singularity of type $\frac{1}{r}(a, b)$, polarised by $-K_S = A$, so that $a + b \cong 1 \bmod r$, can only extend to a Fano 3-fold of index 2 if $a + 1$ or $b + 1 \cong 0 \bmod r$ (compare Example 1.2), so that $\frac{1}{r}(1, a, b)$ is terminal.

These conditions restricts the several thousand baskets for index 2 Fanos to just a handful having a possible log del Pezzo surface as half-elephant. Table 3.1 is a preliminary list of a few $f = 2$ Fano 3-folds without any projections from smooth points (not complete, but possibly fairly typical). Apart from Nos. 1 and 2 that we already know from Sections 1–2, the only cases in this list having a good half-elephant are No. 12, $X_{10,12} \subset \mathbb{P}(1, 2, 3, 5, 6, 7)$ and No. 14, $X_{8,10} \subset \mathbb{P}(1, 2, 3, 4, 5, 5)$.

### 3.2.3 Fano 3-folds of index 2 and projections

Quite independently of del Pezzo surfaces, Fano 3-folds of index 2 usually have projections based on blowing up a nonsingular point, so often belong to projection cascades. Suppose that $X$ is a Fano 3-fold in the Mori category (that is, with at worst terminal singularities) and $-K_X = 2A$ with $A$ a Weil divisor. Consider the blowup $\sigma\colon X' \to X$ at a nonsingular point $P \in X$ with exceptional surface $E \cong \mathbb{P}^2$. Then by the adjunction formula for a blowup, $-K_{X'} = 2A'$, where $A' = \sigma^* A - E$. If $A^3 > 1$ and $P \in X$ is

| | | |
|---|---|---|
| 1. | $X_{10} \subset \mathbb{P}(1,1,2,3,5)$ | $\frac{1}{3}(2,2,1)$ |
| 2. | $X_{6,8} \subset \mathbb{P}(1,1,2,3,4,5)$ | $\frac{1}{5}(1,2,4)$ |
| 3. | $X_{10,14} \subset \mathbb{P}(1,2,2,5,7,9)$ | $\frac{1}{9}(2,2,7)$ |
| 4. | $X_{12,14} \subset \mathbb{P}(1,2,3,4,7,11)$ | $\frac{1}{11}(2,4,7)$ |
| 5. | $X_{8,10} \subset \mathbb{P}(1,2,2,3,5,7)$ | $\frac{1}{3}(2,2,1), \frac{1}{7}(2,2,5)$ |
| 6. | $X_{22} \subset \mathbb{P}(1,2,3,7,11)$ | $\frac{1}{3}(2,2,1), \frac{1}{7}(2,3,4)$ |
| 7. | $X_{10,12} \subset \mathbb{P}(1,2,3,4,5,9)$ | $\frac{1}{3}(2,2,1), \frac{1}{9}(2,4,5)$ |
| 8. | $X_{6,10} \subset \mathbb{P}(1,2,2,3,5,5)$ | $2 \times \frac{1}{5}(2,2,3)$ |
| 9. | $X_{8,12} \subset \mathbb{P}(1,2,3,4,5,7)$ | $\frac{1}{5}(1,3,4), \frac{1}{7}(2,3,4)$ |
| 10. | $X_{26} \subset \mathbb{P}(1,2,5,7,13)$ | $\frac{1}{5}(2,2,3), \frac{1}{7}(1,2,6)$ |
| 11. | $X_{6,8} \subset \mathbb{P}(1,2,2,3,3,5)$ | $\frac{1}{3}(2,2,1), \frac{1}{5}(2,2,3)$ |
| 12. | $X_{10,12} \subset \mathbb{P}(1,2,3,5,6,7)$ | $2 \times \frac{1}{3}(2,2,1), \frac{1}{7}(1,2,6)$ |
| 13. | $X_{14,18} \subset \mathbb{P}(2,2,3,7,9,11)$ | $2 \times \frac{1}{3}(2,2,1), \frac{1}{11}(2,2,9)$ |
| 14. | $X_{8,10} \subset \mathbb{P}(1,2,3,4,5,5)$ | $\frac{1}{3}(2,2,1), 2 \times \frac{1}{5}(1,2,4)$ |
| 15. | $X_{12,14} \subset \mathbb{P}(2,2,3,5,7,9)$ | $\frac{1}{3}(2,2,1), \frac{1}{5}(2,2,3), \frac{1}{9}(2,2,7)$ |
| 16. | $X_{10,14} \subset \mathbb{P}(2,2,3,5,7,7)$ | $\frac{1}{3}(2,2,1), 2 \times \frac{1}{7}(2,2,5)$ |
| 17. | $X_{10,12} \subset \mathbb{P}(2,2,3,5,5,7)$ | $2 \times \frac{1}{5}(2,2,3), \frac{1}{7}(2,2,5)$ |
| 18. | $X_{10,12} \subset \mathbb{P}(2,3,3,4,5,7)$ | $4 \times \frac{1}{3}(2,2,1), \frac{1}{7}(2,3,4)$ |
| 19. | $X_{6,6} \subset \mathbb{P}(1,1,2,2,3,5)$ | $\frac{1}{5}(2,2,3)$ |

Table 3.1: Some index 2 Fano 3-folds

general then $A'$ is nef and big, and defines a birational contraction $X' \to \overline{X}$, where $\overline{X}$ is again a (singular) Fano 3-fold of index 2 containing a copy of $E \cong \mathbb{P}^2$ with $\overline{A}_{|E} \cong \mathcal{O}_{\mathbb{P}^2}(1)$; in general, $\overline{X}$ will have finitely many nodes on $E$, corresponding to the lines on $X$ through $P$. The inclusion $R(\overline{X}, \overline{A}) \subset R(X, A)$ is the quasi-Gorenstein unprojection of $E$ (in the sense of [PR] and [qG]). This means that Fano 3-folds of index 2 could in principle be constructed by starting from a variety such as one of Table 3.1, force it to contain an embedded plane $E \cong \mathbb{P}^2$ of degree 1, which can then be contracted to a nonsingular point by an unprojection. This calculation has a number of entertaining features, not the least the question of how to describe embeddings (say) $\mathbb{P}^2 \hookrightarrow \mathbb{P}(1,2,2,5,6,9)$ and codimension 2 complete intersections $X_{10,14}$ containing the image.

The nonsingular case is well known: for example, a Fano 3-fold $X \subset \mathbb{P}^7$ of index 2 and degree 6 has a projection $X \dashrightarrow \overline{X}$, that coincides with the linear projection from a point, whose image is a linear section of the Grassmannian Grass(2, 5) containing a linearly embedded plane $\mathbb{P}^2 \subset \overline{X} \hookrightarrow \text{Grass}(2,5)$. There are two different ways of embedding a plane $\mathbb{P}^2 \hookrightarrow \text{Grass}(2,5)$ related to Schubert conditions, and these give rise to the two families of unprojection called Tom and Jerry, corresponding to the linear section of the Segre embedding of the hyperplane section of $\mathbb{P}^2 \times \mathbb{P}^2$, and $\mathbb{P}^1 \times \mathbb{P}^1 \times \mathbb{P}^2$. See [P1]–[P2] for details.

### 3.2.4 Alternative birational treatments

Whereas Table 3.1 (or a suitable completion), together with unprojection of planes to nonsingular points, could thus provide a basis for a detailed classification of Fano 3-folds of index 2 (or at least for their numerical invariants), it is possible that many of these varieties could be studied more easily by birational methods: in this paper we have mainly concentrated on projections from nonsingular points, but each projection can presumably be completed to a Sarkisov link (Corti [Co]), giving rise to a birational description.

There are alternative birational methods, for example, based on projections from quotient singularities; these may take us outside the Mori category, as with the "Takeuchi program" used by Takagi in his study of Fano 3-folds with singular index 2 (see [T]). Most of the del Pezzo surfaces and Fano 3-folds we treat here in fact have projections of Type I. For example, $X_{6,8} \subset \mathbb{P}(1,1,2,3,4,5)_{x_1,x_2,y,z,t,u}$ has equations

$$ux_1 = A_6(x_2, y, z, t) \quad \text{and} \quad uz = B_8(x_2, y, z, t),$$

so that eliminating $u$ gives a birational map from $X_{6,8}$ to the hypersurface

$$X_9 : (Bx - Az) \subset \mathbb{P}(1,1,2,3,4).$$

Algebraically this is a Type I projection, in fact of the simplest $Bx - Ay$ type (see [Ki], Section 2). However, from the point of view of the Sarkisov program, it is quite different: introducing the weighted ratio $x_2 : y : t$ makes the $(1,2,4)$ blowup at $P$, not the Kawamata blowup – it is the blowup $X_1 \to X$ with exceptional surface $E$ of discrepancy $2/5$, so that $-K_{X_1} = 2(A - 1/5E)$. This preserves the index 2 condition, but introduces a line of $A_1$ singularities along the $y, t$ axis on $X_9$, taking us out of the Mori category. Compare also Example 3.1.

### 3.2.5 How many Fano 3-folds of index $\geq 3$ are there?

Fano 3-folds of index $f \geq 3$ do not form projection cascades – a blowup $X' \to X$ changes the index. Another way of seeing this is to note that for $f \geq 3$, orbifold RR applied to $\chi(-A) = 0$ gives a formula for $A^3$ in terms of the basket of singularities $\mathcal{B} = \{\frac{1}{r}(1, a, r-a)\}$, in much the same what that $\frac{Ac_2}{12}$ is determined by the classic orbifold RR formula for $\chi(\mathcal{O}_X)$:

$$\frac{(-K_X)c_2}{24} = 1 - \sum_{\mathcal{B}} \frac{r^2-1}{12r},$$

(see [YPG], Corollary 10.3).

The numerical invariants of a Fano 3-fold are the data going into the orbifold RR formula, giving the Hilbert series; compare [ABR], Section 4. It consists of $A^3$, $\frac{Ac_2}{12}$ and the basket of singularities $\mathcal{B}$; for $f \geq 3$, the first two rational numbers are determined by $\mathcal{B}$.

Suzuki's Univ. of Tokyo thesis [Su], [Su1] (based in part on Magma programming by Gavin Brown [GRD]) contains lists of the possible numerical invariants of Fano 3-folds of index $f \geq 2$. She proves in particular that $f \leq 19$, with $f = 19$ if and only if $X$ has the same Hilbert series as weighted projective space $\mathbb{P}(3, 4, 5, 7)$ (we conjecture of course that then $X \cong \mathbb{P}(3, 4, 5, 7)$.) For $f = 3, \ldots, 19$, the number of possible numerical types is bounded as follows:

| $f$ | 3 | 4 | 5 | 6 | 7 | 8 | 9 | 10 | 11 | 12 | 13 | 14 | 15 | 16 | 17 | 18 | 19 |
|---|---|---|---|---|---|---|---|---|---|---|---|---|---|---|---|---|---|
| $n_f$ | 12 | 9 | 7 | 1 | 5 | 3 | 2 | 0 | 3 | 0 | 1 | 0 | 0 | 0 | 1 | 0 | 1 |
| $N_f$ | 20 | 24 | 14 | 5 | 11 | 5 | 2 | 1 | | | | | | | | | |

Here $n_f$ is a lower bound, and $N_f$ a rough upper bound: $n_f$ refers to the number of established cases in codimension $\leq 2$, that is, weighted projective spaces, hypersurfaces or codimension 2 complete intersections. $N_f$ is the number of candidate baskets, that includes cases in codimension 4 and 5 that we expect to be able to justify with more work, together with many less reputable candidates.[1] For $f \geq 9$ the number $n_f$ is correct, except for an annoying (and thoroughly disreputable) candidate with $f = 10$.

Rather remarkably, there are no codimension 3 Pfaffians except for the case $S^{(6)}$ of Section 1 (see (1.1)) with $f = 2$; so far we are unable to determine which candidate cases in codimension $\geq 4$ really occur (which accounts for the uncertainties in the list). By analogy with Mukai's results for nonsingular Fanos, one may speculate that Fano 3-folds in higher codimension should often be quasilinear sections of certain "key varieties", such as the weighted Grassmannians treated in Corti and Reid [CR], and there may be some convincing reason why there are few codimension $\geq 3$ cases.

### 3.2.6 How many interesting cascades are there?

For present purposes, for a cascade to be of interest, at least one of the graded rings at the bottom must be explicitly computable; for us to get some

[1] There are currently some problems with the upper bound $N_f$; the rigorous bound is much larger than given here. For details, see Suzuki's thesis [Su1].

benefit, it should realistically have codimension $\leq 3$. Also, we must be able to identify the surface at the top of the cascade, for example, because it has higher Fano index, so is a simpler object in a Veronese embedding. The cascades of Sections 1–2 illustrate how these conditions work in ideal settings. These conditions are restrictive, and probably only allow a small number of numerical cases. Thus, whereas each of $\overline{\mathbb{F}}_k$ for $k = 7, 9, \dots$ is the head of a tall cascade, involving $k + 4$ blowups, a moment's thought along the lines of Exercise 1.2 shows that essentially none of the surfaces in it has anticanonical ring of small codimension. They do not extend to Fano 3-folds of index 2 for the reason given in Exercise 1.2 and 3.2.2.

As another example, consider the Fano 3-fold $X_{10,12} \subset \mathbb{P}(1, 2, 3, 5, 6, 7)$ of Table 3.1, No. 12 and its half-elephant $S_{10,12} \subset \mathbb{P}(2, 3, 5, 6, 7)$. This is a surface with quotient singularities $2 \times \frac{1}{3}(2, 2)$ and $\frac{1}{7}(2, 6)$ and $K^2 = \frac{2}{21}$. Its minimal resolution $\widetilde{S} \to S$ is a surface with $K^2_{\widetilde{S}} = -1$, so is a scroll $\mathbb{F}_n$ blown up 9 times, containing two disjoint $-3$-curves and a disjoint $-3, -2, -2$ chain of curves arising from the $\frac{1}{7}(2, 6)$ singularity. $\widetilde{S}$ can be constructed by blowing up $\mathbb{F}_0 = \mathbb{P}^1 \times \mathbb{P}^1$ 9 times, with 3 of the centres on each of 2 sections, and 3 other centres infinitely near points along a nonsingular arc. It seems likely that if these blowups are chosen generically, this surface contains no $-1$-curves not passing through the singularities. Thus there seem to be more complicated cases in which there is no cascade at all. Now, in what way is $S_{10,12} \subset \mathbb{P}(2, 3, 5, 6, 7)$ so different from $T_{6,8} \subset \mathbb{P}(1, 2, 3, 4, 5)$ of Section 2?

## 3.3 Mirages

Mirages have been a common phenomenon in the study of weighted projective varieties since Fletcher's thesis. The question is to construct a graded ring and a plausible candidate for a variety in weighted projective space having a given Hilbert series. It happens frequently that we can find a graded ring, but it does not correspond to a good variety, for example, because one of the variables cannot appear in any relations for reasons of degree, so that the candidate variety is a weighted cone. See p. 229 and Example 3.1 below for typical cases.

A *mirage* is an unexpected component of a Hilbert scheme, that does not consist of the varieties that we want, but of some degenerate cases, e.g., cones, varieties with index bigger than specified, or varieties condemned to have some extra singularities. The Hilbert scheme of a family of Fano 3-folds may have other components, e.g., consisting of varieties with the same numerical data, but different divisor class group. For example, the second Veronese embedding of our index 2 Fanos $X_{10} \subset \mathbb{P}(1, 1, 2, 3, 5)$ gives an extra component of the family of Fano 3-folds of index 1 with $(-K)^3 = 2 + \frac{2}{3}$.

More generally, it is an interesting open problem to understand what these mirages really are, and to find formal criteria to deal with them systematically in computer generated lists. One clue is to consider how global sections of $\mathcal{O}_X(i)$ correspond to local sections of the sheaf of algebras $\bigoplus \mathcal{O}_{X,P}(i)$ as indicated in Remark 2.4.

**Example 3.1** We work out one final legend that illustrates several points. Looking for a Fano 3-fold $X$ of Fano index $f = 2$ with a $\frac{1}{11}(2,3,8)$ terminal quotient singularity $P \in X$ by our Hilbert series methods gives (we omit a couple of lines of Magma)

$$P_X(t) = \frac{(1-t^6)(1-t^9)(1-t^{10})}{\prod(1-t^{a_i}) : i \in [1,2,2,3,3,5,11]}.$$

That is, the Hilbert series of the c.i. $X_{6,9,10} \subset \mathbb{P}(1,2,2,3,3,5,11)$. As with the examples on p. 229, this candidate is a mirage for two reasons: the equations cannot involve the variable of degree 11, and there is no variable of degree 8 to act as orbinate at the singularity (this kind of thing seems to happens fairly often with candidate models). Adding a generator of degree 8 to the ring gives a codimension 4 model $X \subset \mathbb{P}(1,2,2,3,3,5,8,11)$. We expect that this model works: we can eliminate the variable of degree 11 by a Type I projection $X \dashrightarrow X'$ corresponding to the $(2,3,8)$ blowup, as described in 3.2.4. This weighted blowup subtracts

$$\frac{t^{11}}{(1-t^2)(1-t^3)(1-t^8)(1-t^{11})}$$

from $P(T)$, and a little calculation

$$P_X(t) - \frac{t^{11}}{(1-t^2)(1-t^3)(1-t^8)(1-t^{11})} = \frac{1-t^6-t^8-t^9-t^{10}+t^{12}+t^{13}+t^{14}+t^{16}-t^{22}}{(1-t)(1-t^2)^2(1-t^3)^2(1-t^5)(1-t^8)}$$

gives the model for the projected variety $X'$ as the Pfaffian with weights

$$\begin{pmatrix} 1 & 2 & 3 & 5 \\ & 3 & 4 & 6 \\ & & 5 & 7 \\ & & & 8 \end{pmatrix} \quad \text{in} \quad \mathbb{P}(1,2,2,3,3,5,8).$$

Here $X'$ is supposed to contain $\Pi = \mathbb{P}(2,3,8) : (x = y_1 = z_1 = t = 0)$. The two ways of achieving this are: take

$$\left.\begin{array}{rl} \text{Tom:} & \text{the first } 4 \times 4 \text{ block} \\ \text{or Jerry:} & \text{the first 2 rows} \end{array}\right\} \quad \text{in the ideal} \quad I_\Pi = (x, y_1, z_1, t),$$

that is, something like

$$\begin{pmatrix} x & y_1 & z_1 & a_5 \\ & z_1 & y_1^2 & b_6 \\ & & t & c_7 \\ & & & d_8 \end{pmatrix} \quad \text{or} \quad \begin{pmatrix} x & y_1 & z_1 & x^5 \\ & z_1 & y_1^2 & y_1^3 + z_1^2 \\ & & b_6' & c_7' \\ & & & d_8' \end{pmatrix},$$

so that $X$ can be constructed either as a Tom or a Jerry unprojection (see [PR], [P1]–[P2]). As in 3.2.4, the projected variety has a line of $A_1$ singularities along the $y_2, z_2$ axis.

# References

[ABR] S. Altınok, G. Brown and M. Reid, Fano 3-folds, K3 surfaces and graded rings, in Singapore International Symposium in Topology and Geometry (NUS, 2001), ed. A. J. Berrick, M. C. Leung and X. W. Xu, to appear Contemp. Math. AMS, 2002, math.AG/0202092, 29 pp.

[GRD] Gavin Brown, Graded ring database, see www.maths.warwick.ac.uk/~gavinb/grdb.html

[Co] A. Corti, Factoring birational maps of threefolds after Sarkisov, J. Algebraic Geom. **4** (1995) 223–254

[CR] A. Corti and M. Reid, Weighted Grassmannians, in Algebraic Geometry (Genova, Sep 2001), In memory of Paolo Francia, M. Beltrametti and F. Catanese Eds., de Gruyter 2002, 141–163

[CPR] A. Corti, A. Pukhlikov and M. Reid, Birationally rigid Fano hypersurfaces, in Explicit birational geometry of 3-folds, A. Corti and M. Reid (eds.), CUP 2000, 175–258

[Ma] Magma (John Cannon's computer algebra system): W. Bosma, J. Cannon and C. Playoust, The Magma algebra system I: The user language, J. Symb. Comp. **24** (1997) 235–265. See also www.maths.usyd.edu.au:8000/u/magma

[P1] Stavros Papadakis, Gorenstein rings and Kustin–Miller unprojection, Univ. of Warwick PhD thesis, Aug 2001, pp. vi + 72, available from my website + Papadakis

[P2] Stavros Papadakis, Kustin-Miller unprojection *with* complexes, J. algebraic geometry (to appear), arXiv preprint math.AG/0111195, 23 pp.

[PR] Stavros Papadakis and Miles Reid, Kustin–Miller unprojection without complexes, J. algebraic geometry (to appear), arXiv preprint math.AG/0011094, 15 pp.

[YPG] Miles Reid, Young person's guide to canonical singularities, in Algebraic Geometry (Bowdoin 1985), ed. S. Bloch, Proc. of Symposia in Pure Math. **46**, A.M.S. (1987), vol. 1, 345–414

[Ki] Miles Reid, Graded rings and birational geometry, in Proc. of algebraic geometry symposium (Kinosaki, Oct 2000), K. Ohno (Ed.), 1–72

[qG] Miles Reid, Quasi-Gorenstein unprojection, work in progress, currently 17 pp.

[Su] Kaori Suzuki, On $\mathbb{Q}$-Fano 3-folds with Fano index $\geq 9$, math.AG/ 0210309, 7 pp.

[Su1] Kaori Suzuki, On $\mathbb{Q}$-Fano 3-folds with Fano index $\geq 2$, Univ. of Tokyo Ph.D. thesis, 69 pp. + v, Mar 2003

[T] TAKAGI Hiromichi, On the classification of $\mathbb{Q}$-Fano 3-folds of Gorenstein index 2. I, II, RIMS preprint 1305, Nov 2000, 66 pp.

Miles Reid,
Math Inst., Univ. of Warwick,
Coventry CV4 7AL, England
e-mail: miles@maths.warwick.ac.uk
web: www.maths.warwick.ac.uk/~miles

SUZUKI Kaori,
Graduate School of Mathematical Sciences,
University of Tokyo
3-8-1 Komaba, Meguro, Tokyo 153-8914, Japan
e-mail: suzuki@ms.u-tokyo.ac.jp

# On obstructions to the Hasse principle

Per Salberger

*to Sir Peter Swinnerton-Dyer*

## Introduction

A basic problem in arithmetic geometry is to decide if a variety defined over a number field $k$ has a $k$-rational point. This is only possible if there is a $k_v$-point on the variety for each completion $k_v$ of $k$. It remains to decide if there is a $k$-point on a variety with a $k_v$-point at each place $v$ of $k$. The first positive results were obtained by Hasse for quadrics and varieties defined by means of certain norm forms. A class of varieties, therefore, is said to satisfy the Hasse principle if each variety in the class has a $k$-point as soon as it has $k_v$-points for all places $v$. The corresponding property for the smooth locus is called the smooth Hasse principle. It is also natural to ask if weak approximation holds. This means that the set of $k$-points is dense in the topological space of adelic points on the smooth locus.

There are counterexamples to the Hasse principle and weak approximation already for smooth cubic curves and cubic surfaces. These counterexamples can be explained by means of a general obstruction to the Hasse principle due to Manin based on the Brauer group of the variety and the reciprocity law in class field theory. Most but not all of the known counterexamples can be explained by this obstruction (Skorobogatov [Sk]). It is likely that Manin's obstruction is the only obstruction to the (smooth) Hasse principle for rational varieties. But it has only been proved in very special cases.

It is more reasonable to study the Hasse principle for 0-cycles of degree one. For curves it is possible to relate the uniqueness of Manin's obstruction to the finiteness of the Tate–Shafarevich group of the jacobian, which has been proved for some elliptic curves by Kolyvagin and Rubin. Another fairly general result is due to the author [Sa] and concerns conic bundle surfaces over the projective line. There we proved a difficult conjecture of Colliot-Thélène and Sansuc (Conjecture B on p. 443 in [CT/S1]). It says that a new kind of Shafarevich group $Ш^1(k, M)$ defined by means of K theory vanishes for rational surfaces. This result has several consequences. One corollary concerns the size of the Chow group of degree zero cycles (cf. [CT/S1] and [Sa]). Another corollary obtained in 1987 and announced in [Sa] is the following

**0.1 Theorem** *Let $k$ be a number field and $X$ a conic bundle surface over $\mathbb{P}^1_k$. Then Manin's obstruction is the only obstruction to the Hasse principle for 0-cycles of degree one.*

The author included in [Sa] a proof when the Brauer group $H^2_{\text{ét}}(X, \mathbb{G}_m)$ of $X$ contains no other elements than those coming from the Brauer group of $k$. Then the Manin obstruction vanishes so that one obtains the simpler statement that the Hasse principle holds for 0-cycles of degree one. One of the motivations for the present paper is to present a proof of Theorem 0.1, by deducing it from our result on $Ш^1(k, M)$. This is an improved version of the proof found in 1987.

It is based on a generalization of the descent theory of Colliot-Thélène and Sansuc [CT/S2] for rational points to 0-cycles of degree one. The rest of the proof is to show that certain diagrams commute. This is done using techniques similar to those in Bloch [Bl] and [CT/S1].

The descent theory developed by Colliot-Thélène and Sansuc is an analog of the classical descent theory for elliptic curves developed by Fermat, Euler, Mordell and Weil. If $p_\alpha \colon \mathcal{T}_\alpha \to X$ is a class of such descent varieties and $K$ is an overfield of $k$, then the sets $p_\alpha(\mathcal{T}_\alpha(K))$ form a partition of $X(K)$. The descent varieties we consider are torsors over $X$ under commutative algebraic groups.

For varieties with finitely generated torsion-free Picard groups, Colliot-Thélène and Sansuc [CT/S2] introduced a special kind of descent varieties called universal torsors. These are torsors under the Néron–Severi torus of the variety having a certain universal property among other torsors. One of the most important results in their paper is the following

**0.2 Theorem** *Let $X$ be a smooth proper rational variety with a $k_v$-point $P_v$ in each completion of $k$. Suppose that the set of these $k_v$-points satisfies Manin's Brauer group condition. Then there exists a universal $X$-torsor $p \colon \mathcal{T} \to X$ under the Néron–Severi torus $T$ of $X$ (see (1.2)) such that the $k_v$-torsors under $T \times_k k_v$ at $P_v$ obtained by base extension are trivial for each place $v$ of $k_v$.*

This means that there are $k_v$-points $Q_v$ on $\mathcal{T}$ such that $p(Q_v) = P_v$ for each place $v$ of $k$. Therefore, if the universal torsors over $X$ satisfy the Hasse principle, then Manin's obstruction is the only obstruction to the Hasse principle for $X$. There are many applications of this result. For some classes of rational varieties $X$ it is possible to establish the Hasse principle for the universal torsors either directly or by means of some intermediate torsors.

The proof of Theorem 0.2 in [CT/S2] uses explicit computations of cocycles. The aim of Section 1 is to offer a proof based on simple functoriality properties of étale cohomology. It is not necessary to assume that $X$ is rational. It suffices to assume (just as in the proof in *op. cit.*) that the Picard group of $X \times_k \overline{k}$ is finitely generated and torsion-free for an algebraic closure $\overline{k}$ of $k$. Only Brauer classes in the "algebraic part" $\widetilde{H}^2_{\text{ét}}(X, \mathbb{G}_m)$ of the Brauer group of $X$ occur. This is the kernel of the functorial map from $H^2_{\text{ét}}(X, \mathbb{G}_m)$ to $H^2_{\text{ét}}(X \times_k \overline{k}, \mathbb{G}_m)$. If $X$ is smooth and rational, then $\widetilde{H}^2_{\text{ét}}(X, \mathbb{G}_m)$ is the full Brauer group of $X$.

The basic idea of the proof is to "kill" the nonconstant algebraic part of the Brauer group of $X$ by considering a fibre product $\Pi$ of a finite number of Severi–Brauer schemes over $X$ which are trivial at the specializations at the given $k_v$-points. The vanishing of Manin's obstruction for the algebraic part of the Brauer group implies that $\widetilde{H}^2_{\text{ét}}(\Pi, \mathbb{G}_m)$ contains no other elements than those coming from the Brauer group of $k$. The given $k_v$-points can be lifted to $k_v$-points on $\Pi$. It is now easy to show that there exists a universal $\Pi$-torsor which is trivial at these $k_v$-points on $\Pi$ and from this, construct the desired universal $X$-torsor. (Use (1.4) and its functoriality under $\Pi \to X$.) This gives a natural proof of Theorem 0.2.

There is no direct extension of this proof to 0-cycles of degree one since such cycles cannot be lifted to the Severi–Brauer schemes over $X$. We therefore replace the Severi–Brauer $X$-schemes by $X$-torsors under tori. This makes the proof less transparent. But the rôle of the auxiliary torsors is the same. They are used to simplify the cohomological obstructions. The $X$-torsors denoted by $\mathcal{S}$ are in fact chosen in such a way that they give rise to universal torsors over $\Pi$ after pull-back of their base with respect to the morphism $\Pi \to X$.

The advantage of this approach is that we can generalize Theorem 0.2 to a statement where the $k_v$-points $P_v$ are replaced by 0-cycles of degree one (see Theorem 1.27). Any 0-cycle $z_v$ on $X \times_k k_v$ defines a natural specialization map $\rho(z_v)$ from $H^1_{\text{ét}}(X, T)$ to $H^1_{\text{ét}}(k_v, T_v)$. Our generalization of Theorem 0.2 says that there exists a universal $X$-torsor $p\colon \mathcal{T} \to X$ such that the class $[\mathcal{T}] \in H^1_{\text{ét}}(X, T)$ of $\mathcal{T}$ belongs to the kernel of $\rho(z_v)$ for each place $v$ of $k$. This generalization is more difficult to prove and apply than Theorem 0.2, since the triviality of $\rho(z_v)([\mathcal{T}])$ in $H^1_{\text{ét}}(k_v, T_v)$ does not guarantee that $z_v$ can be lifted to a 0-cycle of degree one on $\mathcal{T}$ as in the case of $k_v$-points.

The results in Section 1 are the following. We first give precise criteria for when there exists a universal torsor for a large class of varieties over a number field $k$. One necessary condition is that there are universal torsors over the $k_v$-varieties that are obtained by base extension from $k$ to $k_v$. A second necessary condition is given by considering the elements in the Brauer group of the variety that become constant after all the base extensions to local fields. We first formulate one criterion (Proposition 1.12) without assuming that there are 0-cycles of degree one over the local fields $k_v$ and then, as an application, a second criterion (Proposition 1.26) under the assumptions that such 0-cycles exist over each completion $k_v$. Such criteria were first established in [CT/S2] in the case when the 0-cycles are $k_v$-points on $X$.

In Theorem 1.27 we then prove our generalization of Theorem 0.2 discussed above. It is worth noting that the result also applies to varieties with $H^1(X, \mathcal{O}_X) = 0$ and torsion-free Néron–Severi group, such as K3 surfaces. But the rationality assumption in [CT/S2, Section 3] remains essential for the conjecture that the universal torsors satisfy the Hasse principle. The converse (ii) $\Rightarrow$ (i) of Theorem 1.27 tells us that the universal torsors contain all the information about the obstruction coming from the algebraic part of the Brauer group.

To prove Theorem 0.1 we need a strange corollary of Theorem 1.27 (Corollary 1.45) for torsors defined over an open subset of $X$. To prove this result, we use arguments related to the "description locale des torseurs" in [CT/S2]. This corollary plays an important rôle in the proof of Theorem 0.1 in Section 2.

In Section 2 we first recall the $K$-theoretic construction of Bloch [Bl] for rational surfaces as well as some refinements in [CT/S1] and [Sa]. A fundamental tool in [Bl] is a characteristic homomorphism $\phi'$ for rational surfaces from the group $Z_0(X)^0$ of 0-cycles of degree zero to $H^1_{\text{ét}}(k, T)$ where $T$ is the Néron–Severi torus of $X$. In order to prove Theorem 0.1 we need that this map behaves well under specializations. This is not immediate for Bloch's map, but easy to show for another map $\phi$ of Colliot-Thélène and Sansuc defined by means of universal torsors. We shall therefore make use of the fact that $\phi = \phi'$ for rational surfaces. We then prove that the vanishing of $\text{Ш}^1(k, M)$ implies that the Manin obstruction is the only obstruction to the Hasse principle for 0-cycles of degree one. This is proved for rational surfaces and, more generally, for the class of varieties satisfying certain axioms (2.3) and (2.4). In particular, we deduce Theorem 0.1 from the deep arithmetical result on $\text{Ш}^1(k, M)$ for rational conic bundle surfaces in [Sa].

This paper is a slightly revised version of a manuscript from 1993 in which I prove Theorem 0.1 for a more general class of rational varieties with a pencil of Severi–Brauer varieties. There is also a proof of this more general result in the paper of Colliot-Thélène and Swinnerton-Dyer [CT/SwD]. Their approach is different and not based on descent theory.

I would like to express my gratitude to the referee for his careful reading of the paper.

# 1 Universal torsors, Brauer groups and obstructions to the Hasse principle

Let $k$ be a field, $\overline{k}$ a separable closure of $k$ and $\mathcal{G} := \mathrm{Gal}(\overline{k}/k)$ the absolute Galois group of $k$. There is a contravariant equivalence (cf. Borel [Bo]) between the categories of $k$-tori and the category of finitely generated torsion-free discrete $\mathcal{G}$-modules. If $S$ is a $k$-torus, then there is a natural $\mathcal{G}$-action on the character group $\widehat{S} := \mathrm{Hom}(\overline{S}, \mathbb{G}_{m,\overline{k}})$ of the $\overline{k}$-torus $\overline{S} = \overline{k} \times_k S$ such that $\widehat{S}$ becomes a finitely generated torsion-free discrete $\mathcal{G}$-module. Conversely, if $M$ is a finitely generated torsion-free discrete $\mathcal{G}$-module, then $D(M) := \mathrm{Hom}_{\mathbb{Z}}(M, \overline{k}^*)$ is a $\overline{k}$-torus with a natural $k$-structure induced by the $\mathcal{G}$-action on $M$, thereby defining a $k$-torus. In the sequel we identify $M$ with its bidual $\widehat{D(M)}$ and write $\mathrm{id}\colon \widehat{D(M)} \xrightarrow{\cong} M$ for the canonical $\mathcal{G}$-isomorphism.

We recall some basic notions and results from the descent theory of Colliot-Thélène and Sansuc [CT/S2]. We will consider $k$-varieties over a perfect field $k$ satisfying the following assumptions.

$$X \text{ is a smooth proper } k\text{-variety such that } \overline{X} := \overline{k} \times X \text{ is connected and } \mathrm{Pic}\,\overline{X} := H^1_{\text{ét}}(\overline{X}, \mathbb{G}_m) \text{ is finitely generated and torsion-free.} \tag{1.1}$$

Let $\pi\colon \mathcal{S} \to X$ be a $k$-morphism from a $k$-variety $\mathcal{S}$ which is faithfully flat and locally of finite type over $X$. Let $S$ be a $k$-torus. Then $\pi\colon \mathcal{S} \to X$ is said to be a (left) $X$-torsor under $S$ if there is a (left) action $\sigma\colon S \times \mathcal{S} \to \mathcal{S}$ such that the $k$-morphism

$$(\sigma, \mathrm{pr}_2)\colon S \times_X \mathcal{S} \longrightarrow \mathcal{S} \times_X \mathcal{S}$$

induced by $\sigma$ and the second projection $\mathrm{pr}_2\colon S \times_X \mathcal{S} \to \mathcal{S}$ is an isomorphism. We usually write $\mathcal{S}$ rather than $\pi\colon \mathcal{S} \to X$ for the $X$-torsor. An $X$-torsor under a $k$-torus is locally trivial in the étale topology by a theorem of Grothendieck. The isomorphism classes of $X$-torsors under $S$ correspond to elements of $H^1_{\text{ét}}(X, S)$.

Now let $\chi\colon H^1_{\text{ét}}(X, S) \to \mathrm{Hom}_{\mathcal{G}}(\widehat{S}, \mathrm{Pic}\,\overline{X})$ be the homomorphism induced by the additive pairing $H^1_{\text{ét}}(\overline{X}, \overline{S}) \times \mathrm{Hom}(\overline{S}, \mathbb{G}_{m,\overline{k}}) \to H^1_{\text{ét}}(\overline{X}, \mathbb{G}_{m,\overline{k}})$.

### 1.2 Definition

(a) Let $\mathcal{S}$ be an $X$-torsor under $S$ and $[\mathcal{S}]$ its class in $H^1_{\text{ét}}(X, S)$. Then $\chi([\mathcal{S}]) \in \mathrm{Hom}_{\mathcal{G}}(\widehat{S}, \mathrm{Pic}\,\overline{X})$ is called the *type* of $\mathcal{S}$.

(b) The *Néron–Severi torus* $T$ of $X$ is the $k$-torus $D(\mathrm{Pic}\,\overline{X})$ associated to the discrete $\mathcal{G}$-module $\mathrm{Pic}\,\overline{X}$.

(c) A *universal torsor* over $X$ is an $X$-torsor under the Néron–Severi torus $T$ whose type is $\mathrm{id}\colon \widehat{T} \to \operatorname{Pic} \overline{X}$.

By considering the spectral sequence

$$\operatorname{Ext}^p_{k_{\text{ét}}}(\widehat{S}, R^q p_* \mathbb{G}_{m,X}) \Rightarrow \operatorname{Ext}^{p+q}_{X_{\text{ét}}}(p^* \widehat{S}, \mathbb{G}_{m,X}) \tag{1.3}$$

for a $k$-torus $S$ and the structure morphism $p\colon X \to \operatorname{Spec} k$ (see [CT/S2, 1.5.1]), Colliot-Thélène and Sansuc obtained the exact sequence:

$$0 \to H^1_{\text{ét}}(k,S) \to H^1_{\text{ét}}(X,S) \xrightarrow{\chi} \operatorname{Hom}_{\mathcal{G}}(\widehat{S}, \operatorname{Pic}\overline{X}) \\ \xrightarrow{\delta} H^2_{\text{ét}}(k,S) \to H^2_{\text{ét}}(X,S). \tag{1.4}$$

The homomorphisms $H^i_{\text{ét}}(k,S) \to H^i_{\text{ét}}(X,S)$ are the functorial contravariant maps in étale cohomology. We shall not give any explicit description of $\delta$. All we need in the proofs is that the sequence (1.4) is functorial under field extensions of $k$ and homomorphisms of $k$-tori.

Let $\widetilde{H}^2_{\text{ét}}(X,S) := \operatorname{Ker}\bigl(H^2_{\text{ét}}(X,S) \to H^2_{\text{ét}}(\overline{X},\overline{S})\bigr)$. By analysing (1.3) further, one extends the end of (1.4) to an exact sequence:

$$\operatorname{Hom}_{\mathcal{G}}(\widehat{S}, \operatorname{Pic}\overline{X}) \\ \xrightarrow{\delta} H^2_{\text{ét}}(k,S) \to \widetilde{H}^2_{\text{ét}}(X,S) \to \operatorname{Ext}_{\mathcal{G}}(\widehat{S}, \operatorname{Pic}\overline{X}) \to H^3_{\text{ét}}(k,S). \tag{1.5}$$

In particular for $S = \mathbb{G}_{m,k}$, one obtains the well-known sequence:

$$H^2_{\text{ét}}(k,\mathbb{G}_{m,k}) \to \widetilde{H}^2_{\text{ét}}(X,\mathbb{G}_{m,k}) \to \operatorname{Ext}_{\mathcal{G}}(\mathbb{Z}, \operatorname{Pic}\overline{X}) \to H^3_{\text{ét}}(k,\mathbb{G}_{m,k}). \tag{1.6}$$

The next result is also in [CT/S2]. We include a proof, since *op. cit.* does not prove the implication (iii) $\Rightarrow$ (ii) *directly.*

**1.7 Proposition** *Let $k$, $X$ be as in (1.1) and let $T$ be the Néron–Severi torus of $X$. Then the following conditions are equivalent.*

*(i) $H^2_{\text{ét}}(k,T) \to H^2_{\text{ét}}(X,T)$ is injective for the Néron–Severi torus $T$.*

*(ii) $H^2_{\text{ét}}(k,S) \to H^2_{\text{ét}}(X,S)$ is injective for any $k$-torus $S$.*

*(iii) There exists a universal torsor over $X$.*

**Proof** (ii) $\Rightarrow$ (i) is trivial and (i) $\Rightarrow$ (iii) is immediate from (1.4). To prove (iii) $\Rightarrow$ (ii), let $S$ be a $k$-torus and $\chi \in \mathrm{Hom}_{\mathcal{G}}(\widehat{S}, \mathrm{Pic}\,\overline{X})$. Then there is a dual homomorphism $D(\chi)$ of $k$-tori $T \to S$ inducing a commutative diagram

$$\begin{array}{ccccccc} H^1_{\text{ét}}(X,T) & \to & \mathrm{Hom}_{\mathcal{G}}(\widehat{T}, \mathrm{Pic}\,\overline{X}) & \to & H^2_{\text{ét}}(k,T) & \to & H^2_{\text{ét}}(X,T) \\ \downarrow & & \downarrow & & \downarrow & & \downarrow \\ H^1_{\text{ét}}(X,S) & \to & \mathrm{Hom}_{\mathcal{G}}(\widehat{S}, \mathrm{Pic}\,\overline{X}) & \to & H^2_{\text{ét}}(k,S) & \to & H^2_{\text{ét}}(X,S) \end{array}$$

such that $\mathrm{id} \in \mathrm{Hom}_{\mathcal{G}}(\widehat{T}, \mathrm{Pic}\,\overline{X})$ goes to $\chi$ in $\mathrm{Hom}_{\mathcal{G}}(\widehat{S}, \mathrm{Pic}\,\overline{X})$. Hence $\delta(\chi) = 0$, thereby proving (iii) $\Rightarrow$ (ii). □

Recall that a 0-*cycle* on $X$ is a finite formal sum $z = \sum n_i P_i$ where the $P_i$ are closed points on $X$ and the $n_i$ integers. The integer $n := \sum n_i [k(P_i) : k]$ is called the *degree* of $z$. Denote by $Z_0(X)$ the free abelian group of 0-cycles on $X$. For each $k$-torus $S$ and each positive integer $i$, there is a natural additive pairing

$$\rho\colon Z_0(X) \times H^i_{\text{ét}}(X,S) \to H^i_{\text{ét}}(k,S) \tag{1.8}$$

sending a pair consisting of a closed point $P \in Z_0(X)$ and an element $\varepsilon \in H^i_{\text{ét}}(X,S)$ to the corestriction in $H^i_{\text{ét}}(k,S)$ of the pullback $\varepsilon(P)$ of $\varepsilon$ in $H^i_{\text{ét}}(k(P),S)$. It can be proved that this pairing factorizes through rational equivalence, but we do not need this.

If $z = \sum n_i P_i$ is a 0-cycle, write $\rho(z)\colon H^i_{\text{ét}}(X,T) \to H^i_{\text{ét}}(k,T)$ for the homomorphism sending $\varepsilon \in H^i_{\text{ét}}(X,T)$ to $\rho(z,\varepsilon) \in H^i_{\text{ét}}(k,T)$. This gives a retraction of the functorial map from $H^i_{\text{ét}}(k,T)$ to $H^i_{\text{ét}}(X,T)$ when $z$ is of degree one. Then by Proposition 1.7, there exists a universal torsor over $X$.

Let $T$ be the Néron–Severi torus of $X$ and $\mathcal{T}$ a universal $X$-torsor. Let $\phi_{\mathcal{T}}\colon Z_0(X) \to H^1_{\text{ét}}(k,T)$ be the homomorphism which sends $z \in Z_0(X)$ to $\rho(z, [\mathcal{T}])$ (see (1.8)), and $Z_0(X)^0$ the subgroup of $Z_0(X)$ consisting of 0-cycles of degree zero.

**1.9 Proposition** *The restriction of $\phi_{\mathcal{T}}$ to $Z_0(X)^0$ is independent of the choice of universal torsor $\mathcal{T}$.*

**Proof** Use (1.4) and the fact that

$$Z_0(X)^0 \times \mathrm{Im}(H^1_{\text{ét}}(k,T) \to H^1_{\text{ét}}(X,T)) \subseteq \mathrm{Ker}(\rho). \quad \square$$

We therefore drop the index and write $\phi$ for this map $Z_0(X)^0 \to H^1_{\text{ét}}(k,T)$. For other constructions of $\phi$ that do not depend on the assumption that a universal torsor exists, see [CT/S1, Section 1] and the next section.

The following almost trivial lemma from homological algebra will be useful.

**1.10 Lemma** *Let $L$ be a finitely generated torsion-free discrete $\mathcal{G}$-module. Then*

(a) $H^1(\mathcal{G}, \mathbb{Z}) = 0$.

(b) $H^1(\mathcal{G}, L) = \mathrm{Ext}_{\mathcal{G}}(\mathbb{Z}, L)$ *is finite.*

(c) *Let* $\varepsilon_1, \varepsilon_2, \dots, \varepsilon_r \in \mathrm{Ext}_{\mathcal{G}}(\mathbb{Z}^{(r)}, L)$ *and* $\varepsilon \in \mathrm{Ext}_{\mathcal{G}}(\mathbb{Z}^{(r)}, L)$ *correspond to* $\{\varepsilon_j\}_{j=1}^r \in \bigoplus \mathrm{Ext}_{\mathcal{G}}(\mathbb{Z}, L)$.

*Then there is an extension of discrete $\mathcal{G}$-modules*

$$0 \to L \to M \to \mathbb{Z}^{(r)} \to 0 \qquad (*)$$

*such that*

(i) $\mathrm{Ker}(\mathrm{Ext}_{\mathcal{G}}(\mathbb{Z}, L) \to \mathrm{Ext}_{\mathcal{G}}(\mathbb{Z}, M))$ *is the subgroup generated by* $\varepsilon_1, \varepsilon_2, \dots, \varepsilon_r$,

(ii) *the connecting homomorphism* $\mathrm{Hom}_{\mathcal{G}}(L, L) \to \mathrm{Ext}_{\mathcal{G}}(\mathbb{Z}^{(r)}, L)$ *induced by* $(*)$ *sends* $\mathrm{id} \in \mathrm{Hom}_{\mathcal{G}}(L, L)$ *to* $\varepsilon$.

Now let $k$ be a number field. Denote by $\Omega_k$ the set of places of $k$, and by $k_v$ the $v$-adic completion of $k$ for a place $v$. Choose an algebraic closure $\overline{k}_v$ of $k_v$ and an embedding $\overline{k} \subset \overline{k}_v$ for each $v \in \Omega_k$. We may then regard the Galois group $\mathcal{G}_v := \mathrm{Gal}(\overline{k}_v/k_v)$ as a subgroup of $\mathcal{G} = \mathrm{Gal}(\overline{k}/k)$ for each $v \in \Omega_k$.

If $M$ is a discrete $\mathcal{G}$-module and $i$ a positive integer, write

$$Ш^i(k, M) := \mathrm{Ker}\Big(H^i(\mathcal{G}, M) \to \prod_{\text{all } v} H^i(\mathcal{G}_v, M)\Big).$$

In particular, if $M$ is the group $S(\overline{k})$ of $\overline{k}$-points on a $k$-torus $S$, we write $Ш^i(k, S) := Ш^i(k, S(\overline{k}))$. Finally, set

$$Ч^1(k, S) := \mathrm{Coker}\Big(H^1(\mathcal{G}, S(\overline{k})) \to \bigoplus_{\text{all } v} H^1(\mathcal{G}_v, S(\overline{k}_v))\Big).$$

The following result from class field theory is due to Nakayama and Tate [Ta1]. It plays an important rôle in [CT/S2].

**1.11 Theorem** *Let $k$ be a number field and $S$ a $k$-torus. Then there is a perfect pairing*

$$Ш^2(k, S) \times Ш^1(k, \widehat{S}) \longrightarrow \mathbb{Q}/\mathbb{Z}$$

*which is functorial under homomorphisms of k-tori. The kernel of the induced epimorphism from* $\mathrm{Hom}(H^1(\mathcal{G}, \widehat{S}), \mathbb{Q}/\mathbb{Z})$ *to* $Ш^2(k, S)$ *is isomorphic to* $Ч^1(k, S)$. *Moreover,* $H^3_{\text{ét}}(k, S) = 0$ *for any split k-torus S.*

The next proposition generalizes a result in Section 3.3 in [CT/S2]. If we use the word "locally" for a property which holds for $X_v := k_v \times X$ for each place $v \in \Omega_k$, then we can express Proposition 1.12 in the following way. There exists a universal torsor over $X$ if and only if there exists one locally, and moreover every locally constant Azumaya algebra over $X$ is Brauer equivalent to a product of a locally trivial Azumaya algebra and a constant Azumaya algebra. We can replace $\widetilde{H}^2_{\text{ét}}$ by $H^2_{\text{ét}}$ in (i), since any "locally" constant Brauer class belongs to $\widetilde{H}^2_{\text{ét}}(X, \mathbb{G}_m)$. We prefer the formulation here since the universal torsors are related to $\widetilde{H}^2_{\text{ét}}(X, \mathbb{G}_m)$ rather than $H^2_{\text{ét}}(X, \mathbb{G}_m)$.

**1.12 Proposition** *Let k be a number field and X a smooth proper geometrically connected variety over k for which* $\operatorname{Pic} \overline{X}$ *is finitely generated and torsion-free. Then the following statements are equivalent.*

*(i) The map from* $\operatorname{Ker}(\widetilde{H}^2_{\text{ét}}(X, \mathbb{G}_m) \to \prod_{\text{all } v} \widetilde{H}^2_{\text{ét}}(X_v, \mathbb{G}_m))$ *to*

$$\operatorname{Ker}\Big(\widetilde{H}^2_{\text{ét}}(X, \mathbb{G}_m)/\operatorname{Im} H^2_{\text{ét}}(k, \mathbb{G}_m) \to \prod_{\text{all } v} \widetilde{H}^2_{\text{ét}}(X_v, \mathbb{G}_m)/\operatorname{Im} H^2_{\text{ét}}(k_v, \mathbb{G}_m)\Big)$$

*is surjective, and for each place* $v \in \Omega_k$ *there exists a universal torsor over* $X_v$.

*(ii) There exists a universal torsor over X.*

**Proof** We apply Lemma 1.10 for the $\mathcal{G}$-module $L = \operatorname{Pic} \overline{X}$ and choose a set of generators $\varepsilon_1, \varepsilon_2, \dots, \varepsilon_r$ of $\operatorname{Ker}(\operatorname{Ext}_{\mathcal{G}}(\mathbb{Z}, \operatorname{Pic} \overline{X}) \to \prod_{\text{all } v} \operatorname{Ext}_{\mathcal{G}_v}(\mathbb{Z}, \operatorname{Pic} \overline{X}))$. Let $\varepsilon \in \operatorname{Ext}_{\mathcal{G}}(\mathbb{Z}^{(r)}, L)$ correspond to $\bigoplus_{j=1}^r \varepsilon_j \in \operatorname{Ext}_{\mathcal{G}}(\mathbb{Z}, L)$. We then obtain an exact sequence of discrete $\mathcal{G}$-modules

$$0 \to \operatorname{Pic} \overline{X} \to M \to \mathbb{Z}^{(r)} \to 0 \tag{1.13}$$

such that

$$Ш^1(k, \operatorname{Pic} \overline{X}) = \operatorname{Ker}(H^1(\mathcal{G}, \operatorname{Pic} \overline{X}) \to H^1(\mathcal{G}, M)); \quad \text{and} \tag{1.14}$$

id $\in \operatorname{Hom}_{\mathcal{G}}(\operatorname{Pic} \overline{X}, \operatorname{Pic} \overline{X})$ maps to $\varepsilon$ under the connecting homomorphism $\operatorname{Hom}_{\mathcal{G}}(\operatorname{Pic} \overline{X}, \operatorname{Pic} \overline{X}) \to \operatorname{Ext}_{\mathcal{G}}(\mathbb{Z}^{(r)}, \operatorname{Pic} \overline{X})$ induced by (1.13). (1.15)

the extension (1.13) is split as a sequence of $\mathcal{G}_v$-modules for each $v \in \Omega_k$. (1.16)

Now apply $D(\dots)$ to (1.13) and consider the dual sequence of $k$-tori:

$$1 \to R \to S \to T \to 1, \tag{1.17}$$

where $T$ is the Néron–Severi torus of $X$ and $R = \prod_{j=1}^{r} \mathbb{G}_{m,k}$. From (1.14) and the arithmetical duality result in Theorem 1.11 we obtain that:

$$Ш^2(k,S) \subseteq \operatorname{Ker}(H^2_{\text{ét}}(k,S) \to H^2_{\text{ét}}(k,T)) \tag{1.18}$$

and from (1.16) that the sequences of $k_v$-tori

$$1 \to R_v \to S_v \to T_v \to 1 \tag{1.19}$$

induced from (1.17) split for all places $v$ of $k$.

**Proof of (i) ⇒ (ii)** Consider the following commutative diagram with exact rows and columns

$$\begin{array}{ccccc}
H^2_{\text{ét}}(k,R) & \to & \widetilde{H}^2_{\text{ét}}(X,R) & \to & \operatorname{Ext}_{\mathcal{G}}(\widehat{R}, \operatorname{Pic}\overline{X}) \\
\downarrow & & \downarrow & & \downarrow \\
H^2_{\text{ét}}(k,S) & \to & \widetilde{H}^2_{\text{ét}}(X,S) & \to & \operatorname{Ext}_{\mathcal{G}}(\widehat{S}, \operatorname{Pic}\overline{X}) \\
\downarrow & & \downarrow & & \downarrow \\
H^2_{\text{ét}}(k,T) & \to & \widetilde{H}^2_{\text{ét}}(X,T) & \to & \operatorname{Ext}_{\mathcal{G}}(\widehat{T}, \operatorname{Pic}\overline{X}) \\
\downarrow & & & & \\
0 & & & &
\end{array} \tag{1.20}$$

where the horizontal sequences are those in (1.5) and the vertical sequences are induced by (1.17). The complex in the second column is exact since (1.17) splits over $\overline{k}$. The map $H^2_{\text{ét}}(k,S) \to H^2_{\text{ét}}(k,T)$ is surjective since $H^3_{\text{ét}}(k,R) = 0$ for a number field $k$ (see (1.8)).

In order to prove that there is a universal torsor, it suffices by Proposition 1.7 to show that $H^2_{\text{ét}}(k,T) \to \widetilde{H}^2_{\text{ét}}(X,T)$ is injective. So let $\kappa \in \operatorname{Ker}(H^2_{\text{ét}}(k,T) \to \widetilde{H}^2_{\text{ét}}(X,T))$ and lift $\kappa$ to an element $\beta \in H^2_{\text{ét}}(k,S)$. Then, by exactness of (1.20), there exists $\gamma$ in $\operatorname{Ker}(\widetilde{H}^2_{\text{ét}}(X,R) \to \operatorname{Ext}_{\mathcal{G}}(\widehat{S}, \operatorname{Pic}\overline{X}))$ with the same image as $\beta$ in $\widetilde{H}^2_{\text{ét}}(X,S)$. Let $\gamma_v$ be the image of $\gamma$ in $\widetilde{H}^2_{\text{ét}}(X_v,R)$ and consider the following commutative diagram with exact rows and columns.

$$\begin{array}{ccccccc}
 & & 0 & & 0 & & 0 \\
 & & \downarrow & & \downarrow & & \downarrow \\
0 & \to & H^2_{\text{ét}}(k_v,R_v) & \to & \widetilde{H}^2_{\text{ét}}(X_v,R_v) & \to & \operatorname{Ext}_{\mathcal{G}_v}(\widehat{R_v}, \operatorname{Pic}\overline{X}_v) \\
 & & \downarrow & & \downarrow & & \downarrow \\
0 & \to & H^2_{\text{ét}}(k_v,S_v) & \to & \widetilde{H}^2_{\text{ét}}(X_v,S_v) & \to & \operatorname{Ext}_{\mathcal{G}_v}(\widehat{S_v}, \operatorname{Pic}\overline{X}_v)
\end{array} \tag{1.21}$$

The zeros in the columns come from the splitting property in (1.16) and (1.19), and the zeros in the rows from the existence of universal torsors over $X_v$ (see Proposition 1.7). Since $\gamma$ goes to zero in $\mathrm{Ext}_{\mathcal{G}}(\widehat{S}, \mathrm{Pic}\,\overline{X})$, we conclude from (1.21) that $\gamma_v \in \mathrm{Im}(H^2_{\text{ét}}(k_v, R_v) \to \widetilde{H}^2_{\text{ét}}(X_v, R_v))$ for each $v \in \Omega_k$. On considering the images of $\gamma$ in $\widetilde{H}^2_{\text{ét}}(X, \mathbb{G}_m)$ under the maps from $\widetilde{H}^2_{\text{ét}}(X, R)$ induced by the $r$ projections from $R = \prod_{j=1}^r \mathbb{G}_{m,k}$ to $\mathbb{G}_m$, we deduce from the first assumption in (i) that there exists $\alpha \in H^2_{\text{ét}}(k, R)$ that maps to $\gamma_v$ in $H^2_{\text{ét}}(X_v, R_v)$ for each $v \in \Omega_k$. Let $\tilde{\alpha}$ be the image of $\alpha$ in $H^2_{\text{ét}}(k, S)$. By the choice of $\gamma$, we conclude that $\beta - \tilde{\alpha}$ goes to 0 in $\prod_{\text{all } v} \widetilde{H}^2_{\text{ét}}(X_v, S_v)$ and by the injectivity of the functorial maps $H^2_{\text{ét}}(k_v, S_v) \to \widetilde{H}^2_{\text{ét}}(X_v, S_v)$ that $\beta - \tilde{\alpha} \in Ш^2(k, S)$. But then the image $\kappa$ of $\beta - \tilde{\alpha}$ in $H^2_{\text{ét}}(k, T)$ is equal to zero (see (1.18)). This completes the proof of (i) $\Rightarrow$ (ii).

**Proof of (ii) $\Rightarrow$ (i)** Let $\mathcal{T}$ be a universal torsor over $X$. Then $\mathcal{T}_v := k_v \times \mathcal{T}$ is a universal torsor over $X_v$ for each $v \in \Omega_k$. To prove the first part of (i), consider the following commutative diagram with exact rows

$$\begin{array}{ccccccc} H^1_{\text{ét}}(k, T) & \to & H^1_{\text{ét}}(X, T) & \to & \mathrm{Hom}_{\mathcal{G}}(\widehat{T}, \mathrm{Pic}\,\overline{X}) & \to & 0 \\ \downarrow & & \downarrow & & \downarrow & & \\ H^2_{\text{ét}}(k, R) & \to & \widetilde{H}^2_{\text{ét}}(X, R) & \to & \mathrm{Ext}_{\mathcal{G}}(\widehat{R}, \mathrm{Pic}\,\overline{X}) & \to & 0 \end{array} \tag{1.22}$$

Let $\gamma \in \widetilde{H}^2_{\text{ét}}(X, R)$ be the image of $[\mathcal{T}] \in H^1_{\text{ét}}(X, T)$ and $\gamma_1, \gamma_2, \ldots, \gamma_r$ the images of $\gamma$ in $\widetilde{H}^2_{\text{ét}}(X, \mathbb{G}_m)$ under the maps from $\widetilde{H}^2_{\text{ét}}(X, R)$ induced by the $r$ projections from $R = \prod_{j=1}^r \mathbb{G}_{m,k}$ to $\mathbb{G}_m$. Then $\gamma_1, \gamma_2, \ldots, \gamma_r$ have images $\varepsilon_1, \varepsilon_2, \ldots, \varepsilon_r$ in $\mathrm{Ext}_{\mathcal{G}}(\widehat{R}, \mathrm{Pic}\,\overline{X})$. Thus by the choice of $\varepsilon_j$ (see (1.6)) we get that the kernel of the map

$$\widetilde{H}^2_{\text{ét}}(X, \mathbb{G}_m)/\mathrm{Im}\, H^2_{\text{ét}}(k, \mathbb{G}_m) \to \prod_{\text{all } v} \widetilde{H}^2_{\text{ét}}(X_v, \mathbb{G}_m)/\mathrm{Im}\, H^2_{\text{ét}}(k_v, \mathbb{G}_m)$$

is generated by the images of $\gamma_1, \gamma_2, \ldots, \gamma_r$ in $\widetilde{H}^2_{\text{ét}}(X, \mathbb{G}_m)/\mathrm{Im}\, H^2_{\text{ét}}(k, \mathbb{G}_m)$. To verify the first condition in (i), it thus suffices to show that the elements $\gamma_1, \gamma_2, \ldots, \gamma_r$ belong to $\mathrm{Ker}(\widetilde{H}^2_{\text{ét}}(X, \mathbb{G}_m) \to \prod_{\text{all } v} \widetilde{H}^2_{\text{ét}}(X_v, \mathbb{G}_m))$. That is, we must prove that $[\mathcal{T}]$ belongs to the kernel of the composite map:

$$H^1_{\text{ét}}(X, T) \longrightarrow \widetilde{H}^2_{\text{ét}}(X, R) \longrightarrow \prod_{\text{all } v} \widetilde{H}^2_{\text{ét}}(X_v, R_v).$$

But $[\mathcal{T}_v] \in H^1_{\text{ét}}(X_v, T_v)$ maps to zero in $\widetilde{H}^2_{\text{ét}}(X_v, R_v)$ since the sequence $1 \to R_v \to S_v \to T_v \to 1$ splits. This completes the proof of Proposition 1.12. $\square$

Now suppose that we are given a 0-cycle $z_v$ on $X_v$ for each place $v \in \Omega_k$. If $S$ is a $k$-torus, let $S_v$ be the $k_v$-torus obtained by base extension and let

$$\rho_v \colon Z_0(X_v) \times H^i_{\text{ét}}(X_v, S_v) \longrightarrow H^i_{\text{ét}}(k_v, S_v) \tag{1.23}$$

be the pairing described in (1.8). We denote this map by $\rho_v$ for all $k$-tori $S$ and all positive integers $i$. Let $\rho_v(z_v) \colon H^i_{\text{ét}}(X_v, S_v) \to H^i_{\text{ét}}(k_v, S_v)$ be the homomorphism sending $\varepsilon_v \in H^i_{\text{ét}}(X_v, S_v)$ to $\rho_v(z_v, \varepsilon_v) \in H^i_{\text{ét}}(k_v, S_v)$.

Now recall the fundamental exact sequence of Hasse (see, for example, Tate [Ta2])

$$0 \to H^2_{\text{ét}}(k, \mathbb{G}_m) \to \bigoplus_{\text{all } v} H^2_{\text{ét}}(k_v, \mathbb{G}_m) \to \mathbb{Q}/\mathbb{Z} \to 0. \tag{1.24}$$

The map from $H^2_{\text{ét}}(k, \mathbb{G}_m)$ is the direct sum over $v \in \Omega_k$ of the functorial maps $H^2_{\text{ét}}(k, \mathbb{G}_m) \to H^2_{\text{ét}}(k_v, \mathbb{G}_m)$. The map to $\mathbb{Q}/\mathbb{Z}$ is the direct sum of the local maps $\text{inv}_v \colon H^2_{\text{ét}}(k_v, \mathbb{G}_m) \to \mathbb{Q}/\mathbb{Z}$ which are isomorphisms for non-archimedean places. The fact that the sum of all local invariants is 0 for an element of the Brauer group $H^2_{\text{ét}}(k, \mathbb{G}_m)$ of $k$ is called the reciprocity law.

Manin [Ma] noticed that the reciprocity law gives rise to the following necessary condition for the existence of a 0-cycle of degree $r$ on $X$.

$$\begin{array}{l}\text{There exists a set of 0-cycles } z_v \text{ of degree } r \text{ on } X_v \text{ indexed by} \\ v \in \Omega_k \text{ s.t. } \sum_{\text{all } v} \text{inv}_v(\rho_v(z_v))(\mathcal{A}_v) = 0 \text{ for all } \mathcal{A} \in H^2_{\text{ét}}(X, \mathbb{G}_m).\end{array} \tag{1.25}$$

We now relate the Brauer group obstruction to the Hasse principle for 0-cycles of degree one to another obstruction based on universal torsors. The following result is an immediate corollary of Proposition 1.12.

**1.26 Proposition** *Let $k$ be a number field and $X$ a smooth proper geometrically connected $k$-variety for which* $\operatorname{Pic} \overline{X}$ *is finitely generated and torsion-free. Suppose given a 0-cycle of degree one $z_v$ on $X_v$ for each place $v \in \Omega_k$. Then the following statements are equivalent.*

- *(i) Manin's reciprocity condition $\sum_{\text{all } v} \text{inv}_v(\rho_v(z_v))(\mathcal{A}_v) = 0$ holds for all $\mathcal{A} \in \operatorname{Ker}\big(\widetilde{H}^2_{\text{ét}}(X, \mathbb{G}_m) \to \prod_{\text{all } v} \widetilde{H}^2_{\text{ét}}(X_v, \mathbb{G}_m)/\operatorname{Im} H^2_{\text{ét}}(k_v, \mathbb{G}_m)\big)$.*
- *(ii) There exists a universal torsor over $X$.*

**Proof** Given 0-cycles $z_v$ of degree one on $X_v$ for each place $v \in \Omega_k$, we have to prove that the conditions (1.12i) and (1.26i) are equivalent. It was already noticed after (1.8) that the existence of a 0-cycle of degree one on $X_v$ implies the existence of a universal torsor over $X_v$. It thus suffices to show that the subgroup of

$$\operatorname{Ker}\Big(\widetilde{H}^2_{\text{ét}}(X,\mathbb{G}_m) \to \prod_{\text{all } v} \widetilde{H}^2_{\text{ét}}(X_v,\mathbb{G}_m)/\operatorname{Im} H^2_{\text{ét}}(k_v,\mathbb{G}_m)\Big)$$

generated by $\operatorname{Ker}\big(\widetilde{H}^2_{\text{ét}}(X,\mathbb{G}_m) \to \prod_{\text{all } v} \widetilde{H}^2_{\text{ét}}(X_v,\mathbb{G}_m)\big)$ and $\operatorname{Im}(H^2_{\text{ét}}(k,\mathbb{G}_m))$ equals the subgroup of elements $\mathcal{A}$ satisfying $\sum_{\text{all } v} \operatorname{inv}_v(\rho_v(z_v))(\mathcal{A}_v) = 0$. This is a formal consequence of the Hasse exact sequence of Brauer groups (1.24) and the fact that for all places $v$ of $k$, the map $\rho_v(z_v)$ defines a retraction of $H^2_{\text{ét}}(k_v,\mathbb{G}_m) \to \widetilde{H}^2_{\text{ét}}(X_v,\mathbb{G}_m)$. □

We now consider Manin's obstruction to the Hasse principle for 0-cycles of degree one given by arbitrary elements in $\widetilde{H}^2_{\text{ét}}(X,\mathbb{G}_m)$ and relate it to the existence of universal torsors with certain properties. The following result was proved in [CT/S2, 3.5.1] in the case of rational points.

**1.27 Theorem** *Let $k$ be a number field and $X$ a smooth proper geometrically connected $k$-variety for which* $\operatorname{Pic}\overline{X}$ *is finitely generated and torsion-free. Suppose given a 0-cycle $z_v$ of degree one on $X_v$ for each place $v \in \Omega_k$. Then the following statements are equivalent.*

*(i) Manin's reciprocity condition* $\sum_{\text{all } v} \operatorname{inv}_v(\rho_v(z_v))(\mathcal{A}_v) = 0$ *holds for all* $\mathcal{A} \in \widetilde{H}^2_{\text{ét}}(X,\mathbb{G}_m)$.

*(ii) There exists a universal torsor $\mathcal{T}$ over $X$ such that* $\rho_v(z_v)([\mathcal{T}_v]) = 0$ *in* $H^1_{\text{ét}}(k_v,T)$ *for each* $v \in \Omega_k$.

**Proof** We again apply Lemma 1.10 for the $\mathcal{G}$-module $L = \operatorname{Pic}\overline{X}$. Let $\varepsilon_1,\varepsilon_2,\dots,\varepsilon_r$ be generators of $\operatorname{Ext}_{\mathcal{G}}(\mathbb{Z},\operatorname{Pic}\overline{X})$, and let $\varepsilon \in \operatorname{Ext}_{\mathcal{G}}(\mathbb{Z}^{(r)},L)$ correspond to $\{\varepsilon_j\}_{j=1}^r \in \operatorname{Ext}_{\mathcal{G}}(\mathbb{Z},\operatorname{Pic}\overline{X})$. We then obtain an exact sequence of discrete $\mathcal{G}$-modules:

$$0 \to \operatorname{Pic}\overline{X} \to M \to \mathbb{Z}^{(r)} \to 0 \tag{1.28}$$

such that

$$H^1(\mathcal{G},M) = 0; \quad \text{and} \tag{1.29}$$

id $\in \operatorname{Hom}_{\mathcal{G}}(\operatorname{Pic}\overline{X},\operatorname{Pic}\overline{X})$ maps to $\varepsilon \in \operatorname{Ext}_{\mathcal{G}}(\mathbb{Z}^{(r)},\operatorname{Pic}\overline{X})$ under the connecting homomorphism induced by (1.28). (1.30)

Now apply $D(\dots)$ to (1.28) and consider the dual sequence of $k$-tori.

$$1 \to R \to S \to T \to 1, \tag{1.31}$$

where $T$ is the Néron–Severi torus of $X$ and $R = \prod_{j=1}^{r} \mathbb{G}_{m,k}$. From (1.29) and the arithmetical duality result in Theorem 1.11 we obtain

$$Ч^1(k,S) = 0 \quad \text{and} \quad Ш^2(k,S) = 0. \tag{1.32}$$

**Proof of (i) ⇒ (ii)** Consider the following commutative diagram with exact rows and columns:

$$\begin{array}{ccccc} H^1_{\text{ét}}(k,S) & \to & H^1_{\text{ét}}(X,S) & \to & \text{Hom}_{\mathcal{G}}(\widehat{S}, \text{Pic}\,\overline{X}) \\ \downarrow & & \downarrow & & \downarrow \\ H^1_{\text{ét}}(k,T) & \to & H^1_{\text{ét}}(X,T) & \to & \text{Hom}_{\mathcal{G}}(\widehat{T}, \text{Pic}\,\overline{X}) \\ \downarrow & & \downarrow & & \downarrow \\ H^2_{\text{ét}}(k,R) & \to & \widetilde{H}^2_{\text{ét}}(X,R) & \to & \text{Ext}_{\mathcal{G}}(\widehat{R}, \text{Pic}\,\overline{X}) \\ \downarrow & & \downarrow & & \downarrow \\ H^2_{\text{ét}}(k,S) & \to & \widetilde{H}^2_{\text{ét}}(X,S) & \to & \text{Ext}_{\mathcal{G}}(\widehat{S}, \text{Pic}\,\overline{X}) \end{array} \tag{1.33}$$

deduced from (1.31) and the spectral sequence in (1.3). We know from Proposition 1.26 that there exists a universal torsor over $X$. Let $[\mathcal{T}] \in H^1_{\text{ét}}(X,T)$ be the class of one such torsor $\mathcal{T}$ and consider the images $\gamma$ in $\widetilde{H}^2_{\text{ét}}(X,R)$ and $\gamma_v \in \widetilde{H}^2_{\text{ét}}(X_v,R_v)$, $v \in \Omega_k$, of $[\mathcal{T}]$. Then, since $R = \prod_{j=1}^{r} \mathbb{G}_{m,k}$, we deduce from Manin's reciprocity condition (i) and the Hasse exact sequence (1.24) that there exists $\beta \in H^2_{\text{ét}}(k,R)$ that maps to $\rho_v(z_v)(\gamma_v)$ in $H^2_{\text{ét}}(k_v,R_v)$ for each $v \in \Omega_k$. But $\rho_v(z_v)(\gamma_v) \in \text{Ker}(H^2_{\text{ét}}(k_v,R_v) \to H^2_{\text{ét}}(k_v,S_v))$ since it is the image of $\rho_v(z_v)([\mathcal{T}_v]) \in H^1_{\text{ét}}(k_v,T_v)$ in $H^2_{\text{ét}}(k_v,R_v)$. Therefore, $\beta \in \text{Ker}(H^2_{\text{ét}}(k,R) \to H^2_{\text{ét}}(k,S))$ since $Ш^2(k,S) = 0$ (cf. (1.32)). Let $\alpha \in H^1_{\text{ét}}(k,T)$ be a lifting of $\beta$ and $\alpha_v$ the image of $\alpha$ in $H^1_{\text{ét}}(k_v,T_v)$. Then $\rho_v(z_v)([\mathcal{T}_v]) - \alpha_v$ vanishes for all but finitely many $v \in \Omega_k$ and maps to 0 in $H^2_{\text{ét}}(k_v,R_v)$ for all $v \in \Omega_k$. This combined with the fact that $Ч^1(k,S) = 0$ implies that there exists $\sigma \in H^1_{\text{ét}}(k,S)$ whose image $\sigma$ in $H^1_{\text{ét}}(k_v,T_v)$ is $\rho_v(z_v)([\mathcal{T}_v]) - \alpha_v$ for each $v \in \Omega_k$. Let $\tilde{\sigma}$ be the image of $\sigma$ in $H^1_{\text{ét}}(X,T)$ and $\tilde{\alpha}$ the image of $\alpha$ in $H^1_{\text{ét}}(X,T)$. Then, since $\tilde{\alpha} + \tilde{\sigma}$ belongs to the image of $H^1_{\text{ét}}(k,T) \to H^1_{\text{ét}}(X,T)$ it follows that $[\widetilde{\mathcal{T}}] := [\mathcal{T}] + \tilde{\alpha} + \tilde{\sigma}$ is the class of a torsor $\widetilde{\mathcal{T}}$ of the same type as $\mathcal{T}$. Further, $\rho_v(z_v)([\widetilde{\mathcal{T}}_v]) = 0$ for all $v \in \Omega_k$. This completes the proof of (i) ⇒ (ii).

**Proof of (ii) ⇒ (i)** Let $\mathcal{T}$ be a universal torsor over $X$ with the property that $\rho_v(z_v)([\mathcal{T}_v]) = 0$ in $H^1_{\text{ét}}(k_v,T)$ for all $v \in \Omega_k$. We now proceed as in the proof of Proposition 1.12, (ii) ⇒ (i) and consider the image $\gamma$ of $[\mathcal{T}] \in H^1_{\text{ét}}(X,T)$ in $\widetilde{H}^2_{\text{ét}}(X,R)$ under the vertical map in (1.33), and the images $\gamma_1, \gamma_2, \ldots, \gamma_r$ of $\gamma$ in $\widetilde{H}^2_{\text{ét}}(X,\mathbb{G}_m)$ under the maps from $\widetilde{H}^2_{\text{ét}}(X,R)$ induced by the $r$ projections from $R = \prod_{j=1}^{r} \mathbb{G}_{m,k}$ to $\mathbb{G}_m$. Then $\rho_v(z_v)(\gamma_j) = 0$ for all

$j = 1, \dots, r$ and all places $v$ of $k$. This together with the reciprocity law (1.24) implies that $(z_v)_{v \in \Omega_k}$ satisfies Manin's condition for any $\mathcal{A} \in \widetilde{H}^2_{\text{ét}}(X, \mathbb{G}_m)$ in the subgroup $\Gamma$ generated by $\gamma_1, \gamma_2, \dots, \gamma_r$ and the image of $H^2_{\text{ét}}(k, \mathbb{G}_m)$. But the images $\varepsilon_1, \varepsilon_2, \dots, \varepsilon_r$ in $\text{Ext}_{\mathcal{G}}(\mathbb{Z}, \text{Pic}\,\overline{X})$ of $\gamma_1, \gamma_2, \dots, \gamma_r$ were chosen to generate $\text{Ext}_{\mathcal{G}}(\mathbb{Z}, \text{Pic}\,\overline{X})$. Thus, $\Gamma = \widetilde{H}^2_{\text{ét}}(X, \mathbb{G}_m)$, as was to be proved.

**1.34 Corollary** *Let $k$ be a number field and $X$ a smooth proper geometrically connected $k$-variety for which $\text{Pic}\,\overline{X}$ is finitely generated and torsion-free. Suppose given a 0-cycle $z_v$ of degree one on $X_v$ for each place $v$ such that Manin's reciprocity condition $\sum_{\text{all } v} \text{inv}_v(\rho_v(z_v))(\mathcal{A}_v) = 0$ holds for all $\mathcal{A} \in \widetilde{H}^2_{\text{ét}}(X, \mathbb{G}_m)$. Then for each $k$-torus $S$ and each element $\tau$ in $\text{Hom}_{\mathcal{G}}(\widehat{S}, \text{Pic}\,\overline{X})$ there exists an $X$-torsor $\mathcal{S}$ under $S$ of type $\tau$ such that $\rho_v(z_v)([\mathcal{S}_v]) = 0$ in $H^1_{\text{ét}}(k_v, S)$ for all $v \in \Omega_k$.*

**Proof** We know from (1.24) that there exists a universal torsor $\mathcal{T}$ over $X$ such that $\rho_v(z_v)([\mathcal{T}_v]) = 0$ in $H^1_{\text{ét}}(k_v, T_v)$ for each $v \in \Omega_k$. Let $\mathcal{S} := \mathcal{T} \times^T S$ be the torsor under $S$ induced from $\mathcal{T}$ by the $k$-homomorpism $D(\tau)\colon T \to S$ dual to $\tau$. Then $\mathcal{S}$ satisfies the above conditions.

**1.35 Theorem** *Let $k$ be a number field and $X$ a smooth proper geometrically connected $k$-variety for which $\text{Pic}\,\overline{X}$ is finitely generated and torsion-free. Let $T$ be the Néron–Severi torus of $X$ and $r$ an integer. Let $z_v$ be a 0-cycle of degree $r$ on $X_v$ for each place $v \in \Omega_k$ such that Manin's reciprocity condition $\sum_{\text{all } v} \text{inv}_v(\rho_v(z_v))(\mathcal{A}_v) = 0$ holds for all $\mathcal{A} \in \widetilde{H}^2_{\text{ét}}(X, \mathbb{G}_m)$. Then for each $X$-torsor under $T$ there exists another $X$-torsor $\mathcal{T}$ of the same type such that $\rho_v(z_v)([\mathcal{T}_v]) = 0$ for all $v \in \Omega_k$.*

**Proof** An examination of the proof of (i) $\Rightarrow$ (ii) in (1.24) reveals that we only used the hypothesis that $r = 1$ to prove that there exists a universal torsor $\mathcal{T}$. The rest of the arguments is valid for any $r$ and any $T$-torsor $\mathcal{T}$. □

We now make use of the ideas of [CT/S2, 2.3]. Let $k$ be a perfect field and let $X$ be as in (1.1). Let $U$ be an open $k$-subvariety of $X$ with $\text{Pic}\,\overline{U} = 0$. If $\mathcal{S}$ is an $X$-torsor, let $\mathcal{S}_U$ be the $U$-torsor obtained by restriction.

Consider the exact sequence of $\mathcal{G}$-modules for the absolute Galois group $\mathcal{G} := \text{Gal}(\overline{k}/k)$.

$$0 \to \overline{k}[U]^*/\overline{k}^* \to \text{Div}_{\overline{Z}}\overline{X} \to \text{Pic}\,\overline{X} \to 0, \tag{1.36}$$

where $Z$ is the complement of $U$ in $X$, and $\text{Div}_{\overline{Z}}\overline{X}$ the group of Weil divisors on $\overline{X}$ with support in $\overline{Z}$. On applying $D(\dots)$ we obtain a dual exact sequence of $k$-tori

$$1 \to T \to N \to V \to 1. \tag{1.37}$$

The spectral sequence (1.3) and the exact sequence (1.37) give rise to the commutative diagram

$$\begin{array}{ccc} \mathrm{Hom}_{\mathcal{G}}(\widehat{V}, \overline{k}[U]^*) & \xrightarrow{\delta} & \mathrm{Ext}_{\mathcal{G}}(\widehat{T}, \overline{k}[U]^*) \\ \downarrow\simeq & & \downarrow\simeq \\ H^0_{\text{ét}}(U, V) & \xrightarrow{\delta} & H^1_{\text{ét}}(U, T) \end{array} \tag{1.38}$$

The second vertical map is onto since $\mathrm{Pic}\,\overline{U} = 0$ (see [CT/S2, 1.5.1]).

**1.39 Proposition** *Let $\varepsilon \in H^1_{\text{ét}}(U, T)$. Then the following two conditions are equivalent.*

*(i) There exists a universal $X$-torsor $\mathcal{T}$ such that $[\mathcal{T}_U] = \varepsilon$.*

*(ii) There is a section $\sigma \in \mathrm{Hom}_{\mathcal{G}}(\widehat{V}, \overline{k}[U]^*)$ of the obvious map $\psi\colon \overline{k}[U]^* \to \overline{k}[U]^*/\overline{k}^*$ that maps to $\varepsilon$ in $H^1_{\text{ét}}(U, T)$.*

**Proof** See the "description locale des torseurs" in Section 2.3 of [CT/S2].

Now assume that (1.39ii) holds. Then for any $k$-torus $S$ there is a commutative diagram

$$\begin{array}{ccccc} \mathrm{Ext}_{\mathcal{G}}(\widehat{S}, \overline{k}^*) & \to & \mathrm{Ext}_{\mathcal{G}}(\widehat{S}, \overline{k}[U]^*) & \to & \mathrm{Ext}_{\mathcal{G}}(\widehat{S}, \overline{k}^*) \\ \downarrow\simeq & & \downarrow\simeq & & \downarrow\simeq \\ H^1_{\text{ét}}(k, S) & \to & H^1_{\text{ét}}(U, S) & \to & H^1_{\text{ét}}(k, S) \end{array} \tag{1.40}$$

defined in the following way. The vertical isomorphisms come from the spectral sequence (1.3) (see [CT/S2, 1.5.1]). The horizontal maps in the first square are the functorial maps and the horizontal map in the second square is induced by the $\mathcal{G}$-retraction $\sigma\psi/\mathrm{id}\colon \overline{k}[U]^* \to \overline{k}^*$ of the inclusion $\overline{k}^* \subset \overline{k}[U]^*$.

By completing the second square we obtain a homomorphism:

$$r_U\colon H^1_{\text{ét}}(U, S) \to H^1_{\text{ét}}(k, S) \tag{1.41}$$

which is a retraction of the functorial map from $H^1_{\text{ét}}(k, S)$ to $H^1_{\text{ét}}(U, S)$.

Let $r\colon H^1_{\text{ét}}(X, S) \to H^1_{\text{ét}}(k, S)$ be the composite of the restriction map from $H^1_{\text{ét}}(X, S)$ to $H^1_{\text{ét}}(U, S)$ and $r_U$.

**1.42 Proposition** *Let $\mathcal{T}$ be a universal $X$-torsor and $\sigma \in \mathrm{Hom}_{\mathcal{G}}(\widehat{V}, \overline{k}[U]^*)$ a section of $\psi \colon \overline{k}[U]^* \to \widehat{V}$ such that $\sigma$ maps to the class $[\mathcal{T}_U]$ of $\mathcal{T}_U$ in $H^1_{\text{ét}}(U,T)$ under the map in (1.38). Then the following hold:*

*(a) $r$ is a retraction of the functorial map from $H^1_{\text{ét}}(k,S)$ to $H^1_{\text{ét}}(X,S)$;*

*(b) $r$ is functorial under homomorphisms of $k$-tori;*

*(c) $r([\mathcal{T}]) = 0$;*

*(d) $r$ depends only on $[\mathcal{T}]$ and not on the choice of $\sigma$.*

**Proof** (c) To do this, we use the following commutative diagram:

$$
\begin{array}{ccc}
\mathrm{Hom}_{\mathcal{G}}(\widehat{V}, \overline{k}[U]^*) & \stackrel{\delta}{\longrightarrow} & \mathrm{Ext}_{\mathcal{G}}(\widehat{T}, \overline{k}[U]^*) \\
\downarrow & & \downarrow \\
\mathrm{Hom}_{\mathcal{G}}(\widehat{V}, \overline{k}[U]^*/\overline{k}^*) & \stackrel{\delta}{\longrightarrow} & \mathrm{Ext}_{\mathcal{G}}(\widehat{T}, \overline{k}[U]^*/\overline{k}^*) \\
\downarrow & & \downarrow \\
\mathrm{Hom}_{\mathcal{G}}(\widehat{V}, \overline{k}[U]^*) & \stackrel{\delta}{\longrightarrow} & \mathrm{Ext}_{\mathcal{G}}(\widehat{T}, \overline{k}[U]^*)
\end{array}
\tag{1.43}
$$

where the horizontal maps are induced by (1.37) and the vertical maps by $\psi$ and $\sigma$. Then $\delta(\sigma)$ corresponds to $[\mathcal{T}_U]$ under the isomorphism between $\mathrm{Ext}_{\mathcal{G}}(\widehat{T}, \overline{k}[U]^*)$ and $H^1_{\text{ét}}(U,T)$. Therefore, $r_U([\mathcal{T}_U]) = 0$ if and only if $\delta(\sigma)$ maps to itself under the endomorphism of $\mathrm{Ext}_{\mathcal{G}}(\widehat{T}, \overline{k}[U]^*)$ induced by $\sigma\psi$. But this is clear from the commutative diagram (1.43).

(d) Let $\mathcal{S}$ be an $X$-torsor under $S$ of type $\chi([\mathcal{S}]) \in \mathrm{Hom}_{\mathcal{G}}(\widehat{S}, \mathrm{Pic}\,\overline{X})$. Then $\mathcal{S}$ is of the same type as the $X$-torsor $\mathcal{T} \times^T S$ obtained from the $k$-homomorpism $D(\tau) \colon T \to S$ dual to $\tau = \chi([\mathcal{S}])$. Therefore, $[\mathcal{S}] - [\mathcal{T} \times^T S] \in H^1_{\text{ét}}(X,S))$ is the image of a unique element $\alpha$ in $H^1_{\text{ét}}(k,S)$ by (1.4). Also, $r([\mathcal{T} \times^T S]) = 0$ by (b) and (c). Hence, $r([\mathcal{S}]) = \alpha$ by (a), thereby completing the proof. □

Now suppose there is a 0-cycle $z$ of degree one on $X$. Then (cf. (1.8)) there is a natural retraction $\rho(z) \colon H^1_{\text{ét}}(X,S) \to H^1_{\text{ét}}(k,S)$ associated to $z$ for each $k$-torus $S$ which is functorial under homomorphisms of $k$-tori.

**1.44 Proposition** *Let $k$, $X$ be as above and suppose that there exists a universal $X$-torsor $\mathcal{T}$ such that $\rho(z)([\mathcal{T}]) = 0$. Let $S$ be a $k$-torus and $r$ the retraction from $H^1_{\text{ét}}(X,S)$ to $H^1_{\text{ét}}(k,S)$ defined by $[\mathcal{T}]$ (see (1.42d)). Then the two maps $\rho(z)$ and $r$ coincide.*

**Proof** The map $\rho(z)$ satisfies the same axioms (1.42a–c) as $r$. It therefore follows from the proof of (1.42d) that the two maps coincide. □

One can give another proof of Proposition 1.44 based on the $\mathcal{G}$-retraction from $\overline{k}[U]^*$ to $\overline{k}^*$ associated to $z$.

The following result will be used in the next section in the case $S = T$.

**1.45 Corollary** *Let $k$ be a number field and $X$ a smooth proper geometrically connected $k$-variety for which* $\operatorname{Pic} \overline{X}$ *is finitely generated and torsion-free. Let $S$ be a $k$-torus. Suppose that for each $v$ we are given a 0-cycle $z_v$ of degree one on $X_v := X \times_k k_v$ and an $X_v$-torsor $\mathcal{S}_v$ under $S_v$ such that the following hold.*

*(i) Manin's reciprocity condition $\sum_{\text{all } v} \operatorname{inv}_v(\rho_v(z_v))(\mathcal{A}_v) = 0$ holds for all $\mathcal{A} \in \widetilde{H}^2_{\text{ét}}(X, \mathbb{G}_m)$.*

*(ii) There exists an element $\eta$ of $H^1_{\text{ét}}(k(X), S \times_k k(X))$ having the same image as $[\mathcal{S}_v]$ in $H^1_{\text{ét}}(k_v(X), S_v \times_{k_v} k_v(X))$ for each $v \in \Omega_k$.*

*Then there exists an element $\alpha \in H^1_{\text{ét}}(k, S)$ with image equal to $\rho_v(z_v)([\mathcal{S}_v])$ in $H^1_{\text{ét}}(k_v, S_v)$ for every $v \in \Omega_k$.*

**Proof** Let $U$ be an open nonempty subset of $X$ and $v \in \Omega_k$ any place of $k$. We first show that there exists a 0-cycle $u_v$ of degree one on $U_v := U \times_k k_v$ with $\rho_v(u_v)(\mathcal{A}_v) = \rho_v(z_v)(\mathcal{A}_v)$ for all $\mathcal{A}_v \in H^2_{\text{ét}}(X_v, \mathbb{G}_m)$ and such that $\rho_v(u_v)([\mathcal{S}_v]) = \rho_v(z_v)([\mathcal{S}_v])$. By the additivity and functoriality of $\rho_v$ under corestrictions it suffices to do this in the case where $z_v$ is a $k_v$-point $P_v$.

Let $O_v$ be an affine open neighbourhood of $P_v$. We may then represent each element in $H^2_{\text{ét}}(X_v, \mathbb{G}_m)$ by an Azumaya algebra over $O_v$ (see [Mi, p. 149]) and consider the corresponding Severi–Brauer scheme over $O_v$ (cf. *op. cit.*). We shall only consider elements in the finite kernel of the specialization map from $\widetilde{H}^2_{\text{ét}}(X_v, \mathbb{G}_m)$ to $H^2_{\text{ét}}(k_v(P_v), \mathbb{G}_m)$. Let $\Pi_v$ be the fibre product over $O_v$ of the Severi–Brauer schemes corresponding to restrictions of these elements in $\widetilde{H}^2_{\text{ét}}(X_v, \mathbb{G}_m)$. Then $\Pi_v$ is a smooth proper $O_v$-scheme and its fibre over $P_v$ is a multiprojective space over $k_v$.

Let $W_v$ be the restriction over $O_v$ of an $X_v$-torsor of the same type as $\mathcal{S}_v$ which is trivial over $P_v$. It then follows from the $v$-adic implicit function theorem applied to the fibre product of $\Pi_v$ and $W_v$ over $O_v$ that there exists a $k_v$-point on $U_v \cap O_v$ that can be lifted to $k_v$-points on $\Pi_v$ and $W_v$. This $k_v$-point has all the desired properties. We may therefore replace $z_v$ by a 0-cycle on $U_v$ for each $v$ without changing the hypothesis in Corollary 1.45.

Now choose an open subset $U$ of $X$ such that $\operatorname{Pic} \overline{U} = 0$ and such that $\eta$ is the restriction of an element $\varepsilon \in H^1_{\text{ét}}(U, S)$. Assume, as we may, that $z_v$ is a 0-cycle on $U_v$ for each $v \in \Omega_k$.

Now apply Theorem 1.27. Then there exists a universal torsor $\mathcal{T}$ over $X$ such that $\rho_v(z_v)([\mathcal{T}_v]) = 0$ in $H^1_{\text{ét}}(k_v, T)$ for all $v \in \Omega_k$. Also, let $\sigma$ be a $\mathcal{G}$-module homomorphism from $\overline{k}[U]^*/\overline{k}^*$ to $\overline{k}[U]^*$ as in Proposition 1.39. Finally, let $r_U$ be the retraction from $H^1_{\text{ét}}(U, S)$ to $H^1_{\text{ét}}(k, S)$ in (1.41) defined by means of $\sigma$.

Then $\alpha = r_U(\varepsilon) \in H^1_{\text{ét}}(k, S)$ is the desired element with image $\rho_v(z_v)([\mathcal{S}_v])$ in $H^1_{\text{ét}}(k_v, S_v)$ for all $v \in \Omega_k$. To show this, we fix one place $v$ and change the notation so that $k = k_v$. We also omit the index $v$ for all varieties, morphisms, cohomology groups defined over $k = k_v$. Thus $U$, resp. $\rho(z)([\mathcal{S}])$, will mean $U_v$, resp. $\rho_v(z_v)([\mathcal{S}_v])$, and $\varepsilon$, $\mathcal{T}$ will now mean the images after base extension to $k_v$. We shall also make use of the functoriality of $r$ and $\rho(z)$ under extensions of the base field without further comments.

Then we get an element $\varepsilon \in H^1_{\text{ét}}(U, S)$, a 0-cycle $z$ of degree one on $U$, a universal $X$-torsor $\mathcal{T}$ with $\rho(z)([\mathcal{T}]) = 0$ and an $X$-torsor $\mathcal{S}$ under $S$ satisfying the following condition:

> The image of $\varepsilon \in H^1_{\text{ét}}(U, S)$ in $H^1_{\text{ét}}(k(X), S)$ equals that of the class $[\mathcal{S}] \in H^1_{\text{ét}}(X, S)$ in $H^1_{\text{ét}}(k(X), S)$. $(*)$

But it follows from the commutative diagram (cf. (1.40))

$$\begin{array}{ccc} \operatorname{Ext}_{\mathcal{G}}(\widehat{S}, \overline{k}[U]^*) & \rightarrow & \operatorname{Ext}_{\mathcal{G}}(\widehat{S}, \overline{k}(X)^*) \\ \downarrow\simeq & & \downarrow\simeq \\ H^1_{\text{ét}}(U, S) & \rightarrow & H^1_{\text{ét}}(k(X), S) \end{array}$$

that the restriction map from $H^1_{\text{ét}}(U, S)$ to $H^1_{\text{ét}}(k(U), S)$ is injective. Therefore, $\varepsilon = [\mathcal{S}_U]$, and hence $r_U(\varepsilon) = r([\mathcal{S}])$. Moreover, $r([\mathcal{S}]) = \rho(z)([\mathcal{S}])$ by Proposition 1.44. Hence $r_U(\varepsilon) = \rho(z)([\mathcal{S}])$, as was to be proved.

In Corollary 1.45 and some other results in this section we have assumed that the functorial maps from $\operatorname{Pic} \overline{X}$ to $\operatorname{Pic}(\overline{k}_v \times X)$ are isomorphisms for all $v \in \Omega_k$. This was used to guarantee that the base extensions of universal $X$-torsors to torsors over $X_v$ remain universal. We therefore include the following result for which we could find no reference.

**1.46 Proposition** *Let $k$ be an algebraically closed field, and let $X$ be a smooth and proper $k$-variety for which* $\operatorname{Pic} X$ *is finitely generated. Then the functorial map from* $\operatorname{Pic} X$ *to* $\operatorname{Pic}(X \times E)$ *is an isomorphism for any extension field $E$ of $k$.*

**Proof** The assumption implies that $H^1(X, \mathcal{O}_X) = 0$. Thus $\operatorname{Pic}(X \times V) = \operatorname{Pic} X \times \operatorname{Pic} V$ for any (integral) $k$-variety $V$ by the exercise on p. 292 in [Ha]. (The assumption that $X$ is projective is not necessary since Grothendieck's theorem on pp. 290–291 in *op. cit.* also holds for proper morphisms.) Now make use of the fact that $E$ is the union of its finitely generated $k$-subalgebras $A$. Therefore, there are canonical isomorphisms

$$\varinjlim \operatorname{Pic}(\operatorname{Spec} A) = \operatorname{Pic}(E) = 0 \quad \text{and}$$

$$\operatorname{Pic}(X \times E) = \varinjlim \operatorname{Pic}(X \times \operatorname{Spec} A) = \operatorname{Pic} X \oplus \varinjlim \operatorname{Pic}(\operatorname{Spec} A) = \operatorname{Pic} X,$$

as was to be proved.

# 2 K theory and obstructions to the Hasse principle

Let $k$ be a perfect field, $\overline{k}$ an algebraic closure of $k$ and $\mathcal{G} := \operatorname{Gal}(\overline{k}/k)$ the absolute Galois group of $k$. Let $X$ be a smooth proper $k$-variety such that $\overline{X} := \overline{k} \times X$ is connected.

Then there is a complex of discrete $\mathcal{G}$-modules (cf. [Bl])

$$\bigoplus_{\sigma \in \overline{X}_2} K_2(\overline{k}(\sigma)) \xrightarrow{\text{tame}} \bigoplus_{\gamma \in \overline{X}_1} \overline{k}(\gamma)^* \xrightarrow{\text{div}} \bigoplus_{\overline{X}_0} \mathbb{Z}, \tag{2.1}$$

where $\overline{X}_i$ denotes the set of points of dimension $i$. The first map is given by tame symbols and the second is the usual divisor map. Let $M$ be the cokernel of the first map and $\bigoplus^0_{\overline{X}_0} \mathbb{Z}$ the image of the second. (This notation will become natural later after (2.4).) Then (2.1) induces a short exact sequence of discrete $\mathcal{G}$-modules

$$0 \to \operatorname{Ker}(\text{div})/\operatorname{Im}(\text{tame}) \to M \to \textstyle\bigoplus^0_{\overline{X}_0} \mathbb{Z} \to 0. \tag{2.2}$$

Let $Z_i(\overline{X})$ be the free abelian group of cycles of dimension $i$ on $\overline{X}$; write $R_i(\overline{X})$ for the subgroup of $i$-cycles rationally equivalent to zero and $\operatorname{Ch}_i(\overline{X}) := Z_i(\overline{X})/R_i(\overline{X})$ for the Chow group of cycles of dimension $i$ on $\overline{X}$. The degree of a 0-cycle on $\overline{X}$ depends only on its rational equivalence class since $\overline{X}$ is proper. Let $A_0(\overline{X})$ be the subgroup of $\operatorname{Ch}_0(\overline{X})$ of 0-cycles of degree 0. Finally, define the map

$$\pi\colon \operatorname{Ch}_1(\overline{X}) \otimes_{\mathbb{Z}} \overline{k}^* \longrightarrow \operatorname{Ker}(\text{div})/\operatorname{Im}(\text{tame})$$

by the inclusions:

$$Z_1(\overline{X}) \otimes_{\mathbb{Z}} \overline{k}^* = \bigoplus_{\overline{X}_1} \overline{k}^* \subset \operatorname{Ker}(\text{div}) \quad \text{and} \quad R_1(\overline{X}) \otimes_{\mathbb{Z}} \overline{k}^* \subset \operatorname{Im}(\text{tame}).$$

Now let $k$, $\overline{k}$, $\mathcal{G}$, $X$, $\overline{X}$ be as above and assume in addition that the following holds.

### 2.3 Assumptions

(i) $\mathrm{Ch}_1(\overline{X})$ and $\mathrm{Pic}(\overline{X}) = \mathrm{Ch}_{n-1}(\overline{X})$ are finitely generated and torsion-free.

(ii) The intersection pairing $\cup\colon \mathrm{Ch}_1(\overline{X}) \times \mathrm{Ch}_{n-1}(\overline{X}) \to \mathbb{Z}$ is perfect.

(iii) $\pi\colon \mathrm{Ch}_1(\overline{X}) \otimes_{\mathbb{Z}} \overline{k}^* \to \mathrm{Ker}(\mathrm{div})/g\,\mathrm{Im}(\mathrm{tame})$ is an isomorphism.

Then $\cup$ and $\pi$ define an isomorphism between the Néron–Severi torus $T = D(\mathrm{Pic}\,\overline{X})$ and $\mathrm{Ker}(\mathrm{div})/\,\mathrm{Im}(\mathrm{tame})$. Suppose further that

$$A_0(\overline{X}) = 0. \tag{2.4}$$

Then the Galois cohomology of (2.2) gives rise to an exact sequence

$$\begin{aligned} Z_0(X)^0 \to H^1(\mathcal{G}, T(\overline{k})) \to H^1(\mathcal{G}, M) \\ \to \mathbb{Z}/\deg(Z_0(X)) \to H^2(\mathcal{G}, T(\overline{k})), \end{aligned} \tag{2.5}$$

where $Z_0(X)$ is the group of 0-cycles of degree 0. Denote by $\phi'$ the map from $Z_0(X)^0$ to $H^1_{\text{ét}}(k,T)$ obtained from (2.5) by identifying $H^1(\mathcal{G}, T(\overline{k}))$ with $H^1_{\text{ét}}(k,T)$.

**2.6 Example** Let $k$ be a perfect field and $X$ a smooth proper rational geometrically connected $k$-surface. Then Bloch [Bl] showed that (2.3) and (2.4) hold and from that deduced the map $\phi'$ described above. He also noticed that the values of $\phi'$ only depend on the rational equivalence class in $Z_0(X)$.

**2.7 Proposition** *Let $k$, $\overline{k}$, $\mathcal{G}$, $X$, $\overline{X}$ be as above and assume in addition that (2.3) and (2.4) hold. Suppose that there exists a universal torsor over $X$. Then the maps $\phi$ (see Proposition 1.9) and $\phi'$ coincide.*

**Proof** This is stated and proved in [CT/S1, Section 1] for rational surfaces, but the proof uses no other properties of rational surfaces than (2.3) and (2.4).

Now consider a discrete valuation ring $A$ containing $k$; let $K$ be its field of fractions and $F$ its residue field, and suppose that these fields are perfect. For a closed point $P$ on $X_K$, write $A(P)$ for the integral closure of $A$ in $K(P)$. The valuative criterion of properness for $X_A \to \mathrm{Spec}\,A$ implies that there is a unique $A$-morphism $g\colon \mathrm{Spec}\,A(P) \to X_A$ extending $P \to X_K$. Let

$$\mathrm{sp}\colon Z_0(X_K) \to Z_0(X_F)$$

be the specialization homomorphism that sends a closed point $P$ to the cycle associated to the 0-dimensional closed subscheme $\mathrm{Spec}\, A(P) \times_{\mathrm{Spec}\, A} F$ of $X_F$. Then extend sp to arbitrary 0-cycles by additivity.

It is easy to see that sp sends 0-cycles of degree zero to 0-cycles of degree zero. Denote by $\mathrm{sp}^0$ the associated map from $Z_0(X_K)^0$ to $Z_0(X_F)^0$. Then the obvious diagram

$$\begin{array}{ccc} Z_0(X)^0 & \xrightarrow{\mathrm{id}} & Z_0(X)^0 \\ \downarrow & & \downarrow \\ Z_0(X_K)^0 & \xrightarrow{\mathrm{sp}^0} & Z_0(X_F)^0 \end{array} \tag{2.8}$$

commutes and sp and $\mathrm{sp}^0$ have the expected functoriality properties under field extensions of $k$. It can be shown that sp induces a specialization map of Chow groups of 0-cycles, but we shall not need this.

**2.9 Proposition** *Suppose that there exists a universal torsor $\mathcal{T}$ over $X$, and let $\phi_{\mathcal{T}}$ be the map described in Proposition 1.9. Then the following holds.*

*(a) The functorial map from $H^1_{\text{ét}}(\mathrm{Spec}\, A, T_A)$ to $H^1_{\text{ét}}(K, T_K)$ is injective.*

*(b) $\phi_{\mathcal{T}}(Z_0(X_K)) \subseteq \mathrm{Im}(H^1_{\text{ét}}(\mathrm{Spec}\, A, T_A) \to H^1_{\text{ét}}(K, T_K))$.*

*(c) The following diagram commutes*

$$\begin{array}{ccc} Z_0(X_K) & \xrightarrow{\mathrm{sp}} & Z_0(X_F) \\ \Big\downarrow \phi_{\mathcal{T}} & & \Big\downarrow \phi_{\mathcal{T}} \\ \mathrm{Im}(H^1_{\text{ét}}(\mathrm{Spec}\, A, T_A) \to H^1_{\text{ét}}(K, T_K)) & \xrightarrow{\Theta} & H^1_{\text{ét}}(F, T_F) \end{array}$$

*for the functorial map $\Theta$ from $H^1_{\text{ét}}(\mathrm{Spec}\, A, T_A)$ (cf. (a)).*

**Proof** (a) See [CT/S3, Section 4].

(b) The argument is well known (see, for example, [CT/S1, p. 428]). The $X_K$-torsor $\mathcal{T}_K$ extends to an $X_A$-torsor $\mathcal{T}_A$ under $T_A$, and any closed point $P$ on $X_K$ can be extended to a morphism $\mathrm{Spec}\, A(P) \to X_A$ (see the construction of sp). Combined with the existence of corestriction maps from $H^1_{\text{ét}}(\mathrm{Spec}\, A(P), T_{A(P)})$ to $H^1_{\text{ét}}(\mathrm{Spec}\, A, T_A)$, this implies that $\phi(Z_0(X_K)) \subseteq H^1_{\text{ét}}(\mathrm{Spec}\, A, T_A)$.

(c) The horizontal maps factorize over the completion of $K$. We may thus assume that $A$ is complete and hence that $A(P)$ is discrete for each closed point $P$. By using obvious functoriality properties under corestriction of the maps involved, one reduces to prove that $\Theta(\phi_{\mathcal{T}}(P)) = \phi_{\mathcal{T}}(\sigma(P))$ for a rational point $P$. To see this, note that both composites give the pullback of $\mathcal{T}_A$ at the closed point on $X_F$ determined by $\mathrm{Spec}\, A(P) \to X_A$.

**2.10 Lemma** *Let $k$ be a field of characteristic 0, and $X$, $Y$ two smooth, proper, geometrically connected $k$-varieties. Suppose that (2.3) and (2.4) hold for $\overline{X} := X \times_k \overline{k}$ for any algebraically closed field $\overline{k}$ containing $k$, and that there exists a universal torsor over $X$. Then for any 0-cycle $y$ on $Y$, the following holds:*

*(a) The map $\rho(y)\colon H^1_{\text{ét}}(Y,T) \to H^1_{\text{ét}}(k,T)$ factorizes through a map $\rho'(y)$ from $\operatorname{Im}\big(H^1_{\text{ét}}(Y,T) \to H^1_{\text{ét}}(k(Y), T \times_k k(Y))\big)$ to $H^1_{\text{ét}}(k,T)$.*

*(b) $\phi'(Z_0(X \times_k k(Y))^0) \subseteq \operatorname{Im}\big(H^1_{\text{ét}}(Y,T) \to H^1_{\text{ét}}(k(Y), T \times_k k(Y))\big)$*

*(c) $\phi'(Z_0(X \times_k k(Y))^0)$ maps to $\phi'(Z_0(X)^0)$ under $\rho'(y)$.*

**Proof** (a) See [CT/S2, 2.7.5].

(b) By [CT/S1, p. 428], it is known that

$$\operatorname{Im}\Big(H^1_{\text{ét}}(Y,T) \to H^1_{\text{ét}}(k(Y), T \times_k k(Y))\Big)$$
$$= \bigcap_Q \operatorname{Im}\Big(H^1_{\text{ét}}(\mathcal{O}_{Y,Q}, T \times_k \mathcal{O}_{Y,Q}) \to H^1_{\text{ét}}(k(Y), T \times_k k(Y))\Big),$$

where $Q$ runs over all points of codimension one on $Y$. The desired inclusion is therefore a consequence of (2.9b) and the fact that $\phi = \phi'$ (see Proposition 2.7).

(c) Let $y = \sum n_i y_i$, where the $y_i$ are closed points on $Y$. Since $\rho$ is additive with respect to $Z_0(Y)$, it suffices to prove the statement for each $\rho'(y_i)$. By factorizing $\rho(y_i)$ through $H^1_{\text{ét}}(Y \times_k k(y_i), T \times_k k(y_i))$ and using the functoriality of $\phi$ under extensions of the base field, we reduce further to the case when $y$ is a rational point. We now use induction on $\dim Y$ and note that the case $\dim Y = 0$ is trivial. If $\dim Y \geq 1$, let $f\colon \tilde{Y} \to Y$ be the blowup at the $k$-rational point $y$, $Z = f^{-1}(y)$ and $A$ the stalk of $\mathcal{O}_{\tilde{Y}}$ at the generic point of $Z$. Then $A$ is a discrete valuation ring with field of fractions $K := k(\tilde{Y}) = k(Y)$ and residue field $F := k(Z)$. Then by (2.9c) and Proposition 2.7 there is a commutative diagram

$$\begin{array}{ccc} Z_0(X_K)^0 & \xrightarrow{\text{sp}} & Z_0(X_F)^0 \\ \downarrow \phi' & & \downarrow \phi' \\ \operatorname{Im}(H^1_{\text{ét}}(\operatorname{Spec} A, T_A) \to H^1_{\text{ét}}(K, T_K)) & \xrightarrow{\Theta} & H^1_{\text{ét}}(F, T_F) \end{array} \tag{2.11}$$

Now choose a rational $K$-point $z$ on the above $Z$. Then, since $\dim Z = \dim Y - 1$, we obtain from the induction assumption that (c) holds if we consider the pair $(Z, z)$ instead of $(Y, y)$. Further, by using the commutativity of (2.11), we deduce from this that (c) also holds for the pair $(\tilde{Y}, \tilde{y})$, which in turn implies that (c) holds for $(Y, y)$ since $K(Y) = K(\tilde{Y})$ and $f(z) = y$. This finishes the proof. □

We shall in the sequel use the following functoriality properties of (2.1). Let $k \subset k_1$ be an extension of perfect fields with algebraic closures $\overline{k} \subset \overline{k}_1$. Put $\mathcal{G} = \mathrm{Gal}(\overline{k}/k)$, $\mathcal{G}_1 = \mathrm{Gal}(\overline{k}_1/k_1)$, $X_1 = X \times_k k_1$ and $\overline{X}_1 = X \times_k \overline{k}_1$. We may then consider (2.1) as a sequence of $\mathcal{G}_1$-modules through the homomorphism $\mathcal{G}_1 \to \mathcal{G}$ obtained by restricting the $\mathcal{G}_1$-action to $\overline{k}$. This sequence is the upper row in a commutative diagram of discrete $\mathcal{G}_1$-modules where the bottom row is given by (2.1) applied to $\overline{X}_1$. Now suppose that $\overline{X}$ and $\overline{X}_1$ satisfy (2.3) and (2.4). Then we obtain the following commutative diagram with exact rows from the functoriality of (2.5) under extension of the base field:

$$\begin{array}{ccccccccc} Z_0(X)^0 & \to & H^1_{\text{ét}}(k,T) & \to & H^1(\mathcal{G},M) & \to & \mathbb{Z}/\deg Z_0(X) & \to & H^2_{\text{ét}}(k,T) \\ \downarrow & & \downarrow & & \downarrow & & \downarrow & & \downarrow \\ Z_0(X_1)^0 & \to & H^1_{\text{ét}}(k_1,T_1) & \to & H^1(\mathcal{G}_1,M_1) & \to & \mathbb{Z}/\deg Z_0(X_1) & \to & H^2_{\text{ét}}(k_1,T_1) \end{array} \tag{2.12}$$

where $T_1 = T \times_k k_1$ and $M_1$ is the cokernel of the tame symbol map in (2.1) for $\overline{X}_1$. Note that $T_1$ can be identified with the Néron–Severi torus of $\overline{X}_1$ since the functorial map gives an isomorphism from $\mathrm{Pic}(\overline{X})$ to $\mathrm{Pic}(\overline{X}_1)$ by Proposition 1.46.

From now on, let $k$ be a number field and choose algebraic closures $\overline{k}_v$ of $k_v$, and embeddings $\overline{k} \subset \overline{k}_v$ for each place $v$ of $k$. Let $\mathcal{G}_v = \mathrm{Gal}(\overline{k}_v/k_v)$, $X_v = X \times_k k_v$, $\overline{X}_v = X \times_k \overline{k}_v$, and let $M_v$ be the cokernel of the tame symbol map in (2.1) for $\overline{X}_v$. Write $\text{Ш}^1(k, M)$ for the kernel of the diagonal map from $H^1(\mathcal{G}, M)$ to $\prod_{\text{all } v} H^1(\mathcal{G}_v, M_v)$.

**2.13 Theorem** *Let $k$ be a number field and $X$ a smooth proper geometrically connected $k$-variety such that (2.3) and (2.4) hold for $\overline{X} := X \times_k \overline{k}$ for any algebraically closed field $\overline{k}$ containing $k$. Suppose that for each $v \in \Omega_k$ we are given a 0-cycle $z_v$ of degree one on $X_v$ and that Manin's reciprocity condition $\sum_{\text{all } v} \mathrm{inv}_v(\rho_v(z_v))(\mathcal{A}_v) = 0$ holds for all $\mathcal{A} \in \widetilde{H}^2_{\text{ét}}(X, \mathbb{G}_m)$.*

*Then $\text{Ш}^1(k, M)$ maps onto $\mathbb{Z}/\deg(Z_0(X))$ under the map from $H^1(\mathcal{G}, M)$ in (2.5). In particular, if $\text{Ш}^1(k, M) = 0$, then there is a 0-cycle of degree one on $X$.*

**Proof** Let $\overline{k}$ be an algebraic closure of $k$, and $\overline{K}$ an algebraic closure of the function field $\overline{k}(\overline{X})$ of $\overline{X} := X \times_k \overline{k}$. Then $\overline{K}$ is also an algebraic closure of $K := k(X)$ and we have a natural homomorphism from $\mathcal{H} := \mathrm{Gal}(\overline{K}/K)$ to

$\mathcal{G} := \mathrm{Gal}(\overline{k}/k)$. Now consider (2.12) for $k_1 = K$. Then $\mathbb{Z}/\deg(Z_0(X_1)) = 0$ since the generic point of $X$ defines a $K$-rational point on $X_1 = X_K$.

By Proposition 1.26, since Manin's reciprocity condition is satisfied, there exists a universal torsor over $X$. In turn, this implies that (cf. (2.2.5) and (2.2.8) in [CT/S2]) the map from $H^2_{\text{ét}}(k,T)$ to $H^2_{\text{ét}}(k_1,T_1)$ is injective. We thus obtain the following commutative diagram with exact rows from (2.12):

$$\begin{array}{ccccccc} Z_0(X)^0 & \to & H^1_{\text{ét}}(k,T) & \to & H^1(\mathcal{G},M) & \to & \mathbb{Z}/\deg Z_0(X) & \to 0 \\ \downarrow & & \downarrow & & \downarrow & & \downarrow & \\ Z_0(X_K)^0 & \to & H^1_{\text{ét}}(K,T_K) & \to & H^1(\mathcal{H},M_K) & \to & 0 & \end{array} \tag{2.14}$$

where $M_K$ is the cokernel of the tame symbol map (see (2.1)) for $X \times_k \overline{K}$. The assertion that $Ш^1(k,M)$ maps onto $\mathbb{Z}/\deg(Z_0(X))$ therefore reduces to the assertion that $H^1(\mathcal{G},M)$ is generated by $Ш^1(k,M)$ and the image of $H^1_{\text{ét}}(k,T)$.

For each place $v$ of $k$ there is a commutative diagram with exact rows:

$$\begin{array}{ccccccc} Z_0(X)^0 & \to & H^1_{\text{ét}}(k,T) & \to & H^1(\mathcal{G},M) & \to & \mathbb{Z}/\deg Z_0(X) & \to 0 \\ \downarrow & & \downarrow & & \downarrow & & \downarrow & \\ Z_0(X_v)^0 & \to & H^1_{\text{ét}}(k_v,T_v) & \to & H^1(\mathcal{G}_v,M_v) & \to & 0 & \end{array} \tag{2.15}$$

where the zero in the second row comes from the existence of a 0-cycle of degree one on $X_v$. Let $K_v := k_v(X_v)$ be the function field of $X_v$, and $\overline{K}_v$ an algebraic closure of the function field $\overline{k}_v(\overline{X}_v)$ of $\overline{X}_v$ containing $\overline{K}$. Then $\overline{K}_v$ is also an algebraic closure of $K_v$, and there are natural homomorphisms from $\mathcal{H}_v = \mathrm{Gal}(\overline{K}_v/K_v)$ to $\mathcal{G}_v$ and $\mathcal{H}$. Let $M_{K_v}$ be the cokenel of the tame symbol map (see (2.1)) for $X \times_k \overline{K}_v$. Then there are commutative diagrams with exact rows

$$\begin{array}{ccccccc} Z_0(X_K)^0 & \to & H^1_{\text{ét}}(K,T_K) & \to & H^1(\mathcal{H},M_K) & \to & 0 \\ \downarrow & & \downarrow & & \downarrow & & \\ Z_0(X_{K_v})^0 & \to & H^1_{\text{ét}}(K_v,T_{K_v}) & \to & H^1(\mathcal{H}_v,M_{K_v}) & \to & 0 \end{array} \tag{2.16}$$

and

$$\begin{array}{ccccccc} Z_0(X_v)^0 & \to & H^1_{\text{ét}}(k_v,T_v) & \to & H^1(\mathcal{G}_v,M_v) & \to & 0 \\ \downarrow & & \downarrow & & \downarrow & & \\ Z_0(X_{K_v})^0 & \to & H^1_{\text{ét}}(K_v,T_{K_v}) & \to & H^1(\mathcal{H}_v,M_{K_v}) & \to & 0 \end{array} \tag{2.17}$$

and (2.14–17) are parts of a three-dimensional commutative diagam that also contains the commutative diagrams

$$\begin{array}{ccc} H^1_{\text{ét}}(k,T) & \to & H^1_{\text{ét}}(k_v,T_v) \\ \downarrow & & \downarrow \\ H^1_{\text{ét}}(K,T_K) & \to & H^1_{\text{ét}}(K_v,T_{K_v}) \end{array} \quad \text{and} \quad \begin{array}{ccc} H^1(\mathcal{G},M) & \to & H^1(\mathcal{G}_v,M_v) \\ \downarrow & & \downarrow \\ H^1(\mathcal{H},M_K) & \to & H^1(\mathcal{H}_v,M_{K_v}). \end{array}$$

Now let $\mu$ be an element of $H^1(\mathcal{G}, M)$, $\mu_K$ its image in $H^1(\mathcal{H}, M_K)$, $\mu_v$ its image in $H^1(\mathcal{G}_v, M_v)$ and $\mu_{K_v}$ its image in $H^1(\mathcal{H}_v, M_{K_v})$. Lift $\mu_K$ to an element $\eta$ of $H^1_{\text{ét}}(K, T_K)$ (cf. (2.14)) and $\mu_v$ to an element $\beta_v \in H^1_{\text{ét}}(k_v, T_v)$ (cf. (2.15)), and consider the images $\eta_v$ of $\eta$ and $\beta_{K_v}$ of $\beta_v$ in $H^1_{\text{ét}}(K_v, T_{K_v})$. Then $\eta_v - \beta_{K_v} \in \mathrm{Ker}\big(H^1_{\text{ét}}(K_v, T_{K_v}) \to H^1_{\text{ét}}(\mathcal{H}_v, M_{K_v})\big)$ which by exactness of the second row in (2.17) implies that $\eta_v - \beta_{K_v} \in \phi'(Z_0(X_{K_v})^0)$. Thus by (2.10b), $\eta_v - \beta_{K_v} \in \mathrm{Im}\big(H^1_{\text{ét}}(X_v, T_v) \to H^1_{\text{ét}}(K_v, T_{K_v})\big)$, and hence so does $\eta_v$. Choose for each place $v$ an $X_v$-torsor $\mathcal{T}_v$ under $T_v$ such that $[\mathcal{T}_v] \in H^1_{\text{ét}}(X_v, T_v)$ maps to $\eta_v$ in $H^1_{\text{ét}}(K_v, T_{K_v})$. Then since $\eta_v - \beta_{K_v} \in \phi'(Z_0(X_{K_v})^0)$ we conclude from (a) and (c) of Lemma 2.10 that $\rho_v(z_v)([\mathcal{T}_v] - \beta_{K_v}) \in \phi'(Z_0(X_v)^0)$. This means that $\rho_v(z_v)([\mathcal{T}_v])$ and $\rho_v(z_v)(\beta_{K_v}) = \beta_v$ have the same image $\mu_v$ in $H^1(\mathcal{G}_v, M_v)$.

From the assumption that the 0-cycles $(z_v)_{v\in\Omega_k}$ satisfy Manin's reciprocity condition for all $\mathcal{A} \in \widetilde{H}^2_{\text{ét}}(X, \mathbb{G}_m)$, we deduce from Corollary 1.45 that there exists $\alpha \in H^1_{\text{ét}}(k, T)$ with image $\rho_v(z_v)([\mathcal{T}_v])$ in $H^1_{\text{ét}}(k_v, T_v)$ for each place $v \in \Omega_k$. Therefore, $\alpha \in H^1_{\text{ét}}(k, T)$ maps to an element in $H^1(\mathcal{G}, M)$ with the same image as $\mu$ in $H^1(\mathcal{G}_v, M_v)$ for each $v \in \Omega_k$. This completes the proof.

**2.18 Theorem** *Let $k$ be a number field and $X$ a smooth proper geometrically connected $k$-surface. Suppose that there exists a rational function $t \in k(X)$ on $X$ such that $k(X)$ is the function field of a Severi–Brauer curve over $k(t)$. Then*

*(a)* $Ш^1(k, M) = 0$,

*(b) Suppose that for each $v \in \Omega_k$ we are given a 0-cycle $z_v$ of degree one on $X_v$ such that Manin's reciprocity condition $\sum_{\text{all } v} \mathrm{inv}_v(\rho_v(z_v))(\mathcal{A}_v) = 0$ holds for all $\mathcal{A} \in H^2_{\text{ét}}(X, \mathbb{G}_m)$. Then there exists a 0-cycle of degree one on $X$.*

**Proof** (a) $H^1(\mathcal{G}, M)$ and $Ш^1(k, M)$ are $k$-birational invariants [Sa]. The assumptions on $X$ implies that it is $k$-birational to a relatively minimal conic bundle surface over $\mathbb{P}^1$. It is therefore sufficient to prove that $Ш^1(k, M) = 0$ for relatively minimal conic bundle surface over $\mathbb{P}^1$. But this is the main result of [Sa].

(b) This is a consequence of (a) and the previous theorem.

# References

[Bl] S. Bloch, On the Chow groups of certain rational surfaces, *Ann. Sci. École Norm. Sup.* **14** (1981) 41–59

[Bo] A. Borel, *Linear algebraic groups*, 2nd ed. Graduate texts in Math. 126, Springer-Verlag

[CT/S1] J.-L. Colliot-Thélène and J.-J. Sansuc, On the Chow groups of certain rational surfaces: a sequel to a paper of S. Bloch, *Duke Math. J.* **48** (1981) 421–447

[CT/S2] J.-L. Colliot-Thélène et J.-J. Sansuc, La descente sur les variétés rationnelles. II, *Duke Math. J.* **54** (1987) 375–492. (cf. *C. R. Acad. Sci. Paris* **284** Série A (1977) 1215–1218)

[CT/S3] J.-L. Colliot-Thélène and J.-J. Sansuc, Principal homogeneous spaces under flasque tori: applications, *J. Algebra* **106** (1987) 145–202

[CT/SwD] J.-L. Colliot-Thélène and Sir Peter Swinnerton-Dyer, Hasse principle and weak approximation for pencils of Severi–Brauer varieties and similar varieties, *J. reine angew. Math.* **453** (1994) 49–112

[Ha] R. Hartshorne, *Algebraic Geometry*, Springer-Verlag, 1977

[Ma] Yu.I. Manin, Le groupe de Brauer–Grothendieck en géométrie diophantienne, In: *Actes Congrès Intern. Math.* (Nice 1970), Gauthiers-Villars, Paris 1971, Tome 1, 409–411

[Mi] J.S. Milne, *Étale cohomology*, Princeton University Press, Princeton 1980

[Sa] P. Salberger, Zero-cycles on rational surfaces over number fields, *Invent. Math.* **91** (1988) 505–524

[Sk] A.N. Skorobogatov, Beyond the Manin obstruction, *Invent. Math.* **113** (1999) 399–424

[Ta1] J. Tate, The cohomology groups of tori in finite Galois extensions of number fields, *Nagoya Math. J.* **27** (1966) 709–719

[Ta2] J. Tate, Global class field theory, In: *Algebraic number theory*, J.W.S. Cassels, A. Fröhlich, eds., Academic Press 1967

Per Salberger,
Department of Mathematics,
Chalmers University of Technology,
S–412 96 Göteborg, Sweden
e-mail: salberg@math.chalmers.se

# Abelian surfaces with odd bilevel structure

G.K. Sankaran

Abelian surfaces with weak bilevel structure were introduced by S. Mukai in [14]. There is a coarse moduli space, denoted $\mathcal{A}_t^{\text{bil}}$, for abelian surfaces of type $(1,t)$ with weak bilevel structure. $\mathcal{A}_t^{\text{bil}}$ is a Siegel modular threefold, and can be compactified in a standard way by Mumford's toroidal method [1]. We denote the toroidal compactification (in this situation also known as the Igusa compactification) by $\mathcal{A}_t^{\text{bil}*}$. It is a projective variety over $\mathbb{C}$, and it is shown in [14] that $\mathcal{A}_t^{\text{bil}*}$ is rational for $t \leq 5$. In this paper we examine the Kodaira dimension $\kappa(\mathcal{A}_t^{\text{bil}*})$ for larger $t$. Our main result is the following (Theorem VIII.1).

**Theorem** $\mathcal{A}_t^{\text{bil}*}$ *is of general type for t odd and* $t \geq 17$.

It follows from the theorem of L. Borisov [2] that $\mathcal{A}_t^{\text{bil}*}$ is of general type for $t$ sufficiently large. If $t = p$ is prime, then it follows from [7] and [12] that $\mathcal{A}_p^{\text{bil}*}$ is of general type for $p \geq 37$. Our result provides an effective bound in the general case and a better bound in the case $t = p$. As far as we know, all previous explicit general type results (for instance [7, 12, 15, 8, 16]) have been for the cases $t = p$ or $t = p^2$ only.

It is for brevity that we assume $t$ is odd. If $t$ is even the combinatorial details are more complicated, especially when $t \equiv 2 \bmod 4$, but the method is still applicable. In fact the method is essentially that of [12], with some modifications.

**Acknowledgement** Part of this work resulted from conversations with my research student Alfio Marini.

# I Background

If $A$ is an abelian surface with a polarisation $H$ of type $(1,t)$, $t > 1$, then a *canonical level structure*, or simply *level structure*, is a symplectic isomorphism

$$\alpha\colon \mathbb{Z}_t^2 \xrightarrow{\cong} K(H) = \left\{\mathbf{x} \in A \;\middle|\; t_{\mathbf{x}}^*\mathcal{L} \cong \mathcal{L} \text{ if } c_1(\mathcal{L}) = H\right\}.$$

The moduli space $\mathcal{A}_t^{\mathrm{lev}}$ of abelian surfaces with a canonical level structure has been studied in detail in [11], chiefly in the case $t = p$.

A *colevel structure* on $A$ is a level structure on the dual abelian surface $\hat{A}$: note that $H$ induces a polarisation $\hat{H}$ on $\hat{A}$, also of type $(1, t)$. Alternatively, a colevel structure may be thought of as a symplectic isomorphism

$$\beta\colon \mathbb{Z}_t^2 \to A[t]/K(H)$$

where $A[t]$ is the group of all $t$-torsion points of $A$. Obviously the moduli space $\mathcal{A}_t^{\mathrm{col}}$ of abelian surfaces of type $(1, t)$ with a colevel structure is isomorphic to $\mathcal{A}_t^{\mathrm{lev}}$, and each of them has a forgetful morphism $\psi^{\mathrm{lev}}$, $\psi^{\mathrm{col}}$ to the moduli space $\mathcal{A}_t$ of abelian surfaces of type $(1, t)$. We define

$$\mathcal{A}_t^{\mathrm{bil}} = \mathcal{A}_t^{\mathrm{lev}} \times_{\mathcal{A}_t} \mathcal{A}_t^{\mathrm{col}}.$$

The forgetful map $\psi^{\mathrm{lev}}\colon \mathcal{A}_t^{\mathrm{lev}} \to \mathcal{A}_t$ is the quotient map under the action of $\mathrm{SL}(2, \mathbb{Z}_t)$ given by

$$\gamma\colon\ [(A, H, \alpha)] \mapsto [(A, H, \alpha\gamma)]$$

where $\gamma \in \mathrm{SL}(2, \mathbb{Z}_t)$ is viewed as a symplectic automorphism of $\mathbb{Z}_t^2$. The action is not effective, because $(A, H, \alpha)$ is isomorphic to $(A, H, -\alpha)$ via the isomorphism $\mathbf{x} \mapsto -\mathbf{x}$; so $-\mathbf{1}_2 \in \mathrm{SL}(2, \mathbb{Z}_t)$ acts trivially. Thus $\psi^{\mathrm{lev}}$ is a Galois morphism with Galois group $\mathrm{PSL}(2, \mathbb{Z}_t) = \mathrm{SL}(2, \mathbb{Z}_t)/\pm\mathbf{1}_2$.

A point of $\mathcal{A}_t^{\mathrm{bil}}$ thus corresponds to an equivalence class $[(A, H, \alpha, \beta)]$, where $(A, H)$ is a polarised abelian surface of type $(1, t)$, $\alpha$ and $\beta$ are level and colevel structures, and $(A, H, \alpha, \beta)$ is equivalent to $(A', H', \alpha', \beta')$ if there is an isomorphism $\rho\colon A \to A'$ such that $\rho^* H' = H$, $\rho\alpha = \alpha'$ and $\hat{\rho}^{-1}\beta = \beta'$. In particular, for general $A$, we have $(A, H, \alpha, \beta) \cong (A, H, -\alpha, -\beta)$ but $(A, H, \alpha, \beta) \not\cong (A, H, -\alpha, \beta)$. Another way to express this is to say that the wreath product $\mathbb{Z}_2 \wr \mathrm{PSL}(2, \mathbb{Z}_t)$, acts on $\mathcal{A}_t^{\mathrm{bil}}$ with quotient $\mathcal{A}_t$.

**Theorem I.1** (Mukai [14]) *$\mathcal{A}_t^{\mathrm{bil}}$ is the quotient of the Siegel upper half-plane $\mathbb{H}_2$ by the group*

$$\Gamma_t^{\mathrm{bil}} = \Gamma_t^{\natural} \cup \zeta\Gamma_t^{\natural}$$

*where*

$$\Gamma_t^{\natural} = \left\{ \gamma \in \mathrm{Sp}(4, \mathbb{Z}) \;\middle|\; \gamma - \mathbf{1}_4 \in \begin{pmatrix} t\mathbb{Z} & * & t\mathbb{Z} & t\mathbb{Z} \\ t\mathbb{Z} & t\mathbb{Z} & t\mathbb{Z} & t^2\mathbb{Z} \\ t\mathbb{Z} & * & t\mathbb{Z} & t\mathbb{Z} \\ * & * & * & t\mathbb{Z} \end{pmatrix} \right\}$$

*and $\zeta = \mathrm{diag}(1, -1, 1, -1)$, acting by fractional linear transformations.*

Thus $\Gamma_t^{\mathrm{bil}}$ should be thought of as a subgroup of the paramodular group

$$\Gamma_t = \left\{ \gamma \in \mathrm{Sp}(4,\mathbb{Q}) \;\middle|\; \gamma - \mathbf{1}_4 \in \begin{pmatrix} * & * & * & t\mathbb{Z} \\ t\mathbb{Z} & * & t\mathbb{Z} & t\mathbb{Z} \\ * & * & * & t\mathbb{Z} \\ * & \frac{1}{t}\mathbb{Z} & * & * \end{pmatrix} \right\}.$$

(The paramodular group is the group denoted $\Gamma_{1,t}^{\circ}$ in [11] and [5].)

For some purposes it is more convenient to work with the conjugate $\widetilde{\Gamma}_t^{\mathrm{bil}} = R_t \Gamma_t^{\mathrm{bil}} R_{t^{-1}}$ of $\Gamma_t^{\mathrm{bil}}$ by $R_t = \mathrm{diag}(1,1,1,t)$, and with the corresponding conjugates $\widetilde{\Gamma}_t^{\natural}$, $\widetilde{\Gamma}_t^{\mathrm{lev}}$ etcetera. These groups have the advantage that they are subgroups of $\mathrm{Sp}(4,\mathbb{Z})$ rather than $\mathrm{Sp}(4,\mathbb{Q})$, and defined by congruences mod $t$, not mod $t^2$, but their action on $\mathbb{H}_2$ is not the usual one by fractional linear transformations.

If $E_i$ are elliptic curves and $(A,H) = \big(E_1 \times E_2, c_1(\mathcal{O}_{E_1}(1)\boxtimes\mathcal{O}_{E_2}(t))\big)$, we say that $(A,H)$ is a product surface. In this case $K(H) = \{0_{E_1}\} \times E_2[t]$, so a level structure on $A$ may be thought of as a full level $t$ structure on $E_2$. The automorphism $(\mathbf{x},\mathbf{y}) \mapsto (\mathbf{x},-\mathbf{y})$ of $A = E_1 \times E_2$ induces an isomorphism $(A,H,\alpha,\beta) \to (A,H,-\alpha,\beta)$ in this case, so a product surface with a weak bilevel structure still has an extra automorphism. The corresponding locus in the moduli space arises from the fixed locus of $\zeta$ in $\mathbb{H}_2$, and will be of great importance in this paper.

The geometry of $\mathcal{A}_t^{\mathrm{bil}*}$ shows many similarities with that of $\mathcal{A}_t^{\mathrm{lev}*}$, which was studied (in the case of $t$ an odd prime) in the book [11]. In many cases where the proofs of intermediate results are very similar to those of corresponding results in [11] we omit the details and simply indicate the appropriate reference.

# II Modular groups and modular forms

We first collect some facts about congruence subgroups in $\mathrm{SL}(2,\mathbb{Z})$ and some related combinatorial information. For $r \in \mathbb{N}$ we denote by $\Gamma_1(r)$ the principal congruence subgroup of $\mathrm{SL}(2,\mathbb{Z})$. We denote the modular curve $\Gamma_1(r)\backslash\mathbb{H}$ by $X^\circ(r)$, and the compactification obtained by adding the cusps by $X(r)$.

For $m, r \in \mathbb{N}$, define

$$\Phi_m(r) = \left\{ \mathbf{a} \in \mathbb{Z}_r^m \;\middle|\; \mathbf{a} \text{ is not a multiple of a zerodivisor in } \mathbb{Z}_r \right\},$$

that is, $\mathbf{a} \in \Phi_m(r)$ if and only if $\mathbf{a} = z\mathbf{a}'$ implies $z \in \mathbb{Z}_r^*$; and put $\phi_m(r) = \#\Phi_m(r)$. We also put $\overline{\Phi}_m(r) = \Phi_m(r)/\pm 1$.

**Lemma II.1** *If the primes dividing* $r$ *are* $p_1 < p_2 < \cdots < p_n$ *then*

$$\phi_m(r) = \sum_{i=0}^{n} (-1)^i \sum_{p_{j_1},\dots,p_{j_i}} \left( r \prod_{k=1}^{i} p_{j_k}^{-1} \right)^m = r^m \prod_{p|r} (1 - p^{-m}).$$

**Proof** We first prove that $\phi_m(r)$ is a multiplicative function. Suppose first that $r = pq$, with $\gcd(p, q) = 1$. It is easy to see that $\mathbf{a} \in \Phi_m(r)$ if and only if $\mathbf{a}_p \in \Phi_m(p)$ and $\mathbf{a}_q \in \Phi_m(q)$, where $\mathbf{a}_p$ denotes the reduction of $\mathbf{a}$ mod $p$.

We divide $\mathbb{Z}_r^m$ into residue classes mod $p$: that is, we write $\mathbb{Z}_r^m$ as the disjoint union of subsets $S_\mathbf{c}$ for $\mathbf{c} \in \mathbb{Z}_p^m$, where $S_\mathbf{c} = \{\mathbf{a} \mid \mathbf{a}_p = \mathbf{c}\}$. There are $\phi_m(p)$ subsets $S_\mathbf{c}$ such that $\mathbf{r} \in \Phi_m(p)$.

The reduction mod $q$ map $S_\mathbf{c} \to \mathbb{Z}_q^m$ is bijective, since it is the inverse of the injective map $\mathbf{b} \mapsto \mathbf{c} + p\mathbf{b} \in \mathbb{Z}_r^m$. Hence in each of the $\phi_m(p)$ subsets $S_\mathbf{c}$, $\mathbf{c} \in \Phi_m(p)$ there are $\phi_m(q)$ elements whose reduction mod $q$ belongs to $\Phi_m(q)$. It follows that $\phi_m(r) = \phi_m(p)\phi_m(q)$.

Finally, we check that if $r = p^k$, $p$ prime, then $\phi_m(r) = r^m(1 - p^{-m})$. If $\mathbf{a} \notin \Phi_m(r)$, then $\mathbf{a} = p\mathbf{a}'$ for a unique $\mathbf{a}' \in \mathbb{Z}_{r/p}^m$, so there are $(p^{k-1})^m$ such elements $\mathbf{a}$. □

Note that $\phi_1$ is the Euler $\phi$ function, and $\Phi_1(r)$ the set of non-zerodivisors of $\mathbb{Z}_r$.

**Corollary II.2** *The order of* $\mathrm{SL}(2, \mathbb{Z}_t)$ *is given by*

$$|\,\mathrm{SL}(2, \mathbb{Z}_t)| = t\phi_2(t) = t^3 \prod_{p|t} (1 - p^{-2}).$$

**Proof** (See also [18, §1.6].) If $A \in \mathrm{SL}(2, \mathbb{Z}_t)$, then $A_1 = (a_{11}, a_{12}) \in \Phi_2(t)$. So by Euclid's algorithm we can find $A_2' = (a_{21}', a_{22}')$ such that $\det \begin{pmatrix} A_1 \\ A_2' \end{pmatrix} = \gcd(a_{11}, a_{12}) = r$. Replacing $A_2'$ by $A_2 = r^{-1}A_2'$, we get a matrix $A$ with $\det A = 1$. Furthermore, if $B_j = \begin{pmatrix} A_1 \\ A_2 + jA_1 \end{pmatrix}$, $j = 0, \dots, t-1$, then $\det B_j = \det A = 1$, and $B_j \neq B_{j'}$ if $j \neq j'$. So $|\,\mathrm{SL}(2, \mathbb{Z}_t)| = t\phi_2(t)$. □

For $r > 2$, put $\mu(r) = [\mathrm{PSL}(2, \mathbb{Z})\colon \Gamma_1(r)]$. By Corollary II.2 we have

$$\mu(r) = r^3 \prod_{p|r} (1 - p^{-2}).$$

We need the following well-known lemma.

**Lemma II.3** *If* $r > 2$ *then* $X(r)$ *has*

$$\nu(r) = \mu(r)/r = r^2 \prod_{p|r} (1 - p^{-2})$$

*cusps and is a smooth complete curve of genus* $g = 1 + \frac{\mu(r)}{12} - \frac{\nu(r)}{2}$.

**Proof** See [18, pp. 23–24]. □

We denote $\mu(t)$ by $\mu$ and $\nu(t)$ by $\nu$. Note that $\phi_2(1) = \nu(1) = 1$ and $\phi_2(r) = 2\nu(r)$ for $r > 2$.

Now we turn to subgroups of $\mathrm{Sp}(4, \mathbb{Q})$ and modular forms. Denote by $\mathfrak{S}_n^*(\Gamma)$ the space of weight $n$ cusp forms for $\Gamma \subseteq \mathrm{Sp}(4, \mathbb{Q})$. We need the groups $\overline{\Gamma}(1) = \mathrm{PSp}(4, \mathbb{Z})$ and, for $\ell \in \mathbb{N}$,

$$\Gamma(\ell) = \left\{\gamma \in \mathrm{Sp}(4, \mathbb{Z}) \mid \overline{\gamma} = \mathbf{1}_4 \in \mathrm{Sp}(4, \mathbb{Z}_\ell)\right\}.$$

If $t^2 \mid \ell$ then $\Gamma(\ell) \lhd \Gamma_t^{\mathrm{bil}}$, because $\Gamma(\ell) \subseteq \Gamma_t^{\mathrm{bil}}$ and $\Gamma(\ell)$ is normal in $\Gamma(1) = \mathrm{Sp}(4, \mathbb{Z})$.

By a previous calculation [19] we know that

$$\dim \mathfrak{S}_n^*\big(\Gamma(\ell)\big) = \frac{n^3}{8640}\left[\overline{\Gamma}(1) : \Gamma(\ell)\right] + O(n^2)$$

(as long as $\ell > 2$ we can consider $\Gamma(\ell)$ as a subgroup of $\mathrm{PSp}(4, \mathbb{Z})$ rather than $\mathrm{Sp}(4, \mathbb{Z})$). A standard application of the Atiyah–Bott fixed-point theorem (see [9], or in this context [12]) gives

$$\dim \mathfrak{S}_n^*\big(\Gamma_t^{\mathrm{bil}}\big) = \frac{a}{\left[\Gamma_t^{\mathrm{bil}} : \Gamma(\ell)\right]} \dim \mathfrak{S}_n^*\big(\Gamma(\ell)\big) + O(n^2)$$

where $a$ is the number of elements $\gamma \in \Gamma_t^{\mathrm{bil}}$ whose fixed locus in $\mathbb{H}_2$ has dimension 3. Thus $a$ is the number of elements of $\Gamma_t^{\mathrm{bil}}$ that act trivially on $\mathbb{H}_2$. In $\mathrm{Sp}(4, \mathbb{Z})$ there are two such elements, $\pm\mathbf{1}_4$, but if $t > 2$ then $-\mathbf{1}_4 \notin \Gamma_t^{\mathrm{bil}}$. So $a = 1$, and hence

$$\begin{aligned}
\dim \mathfrak{S}_n^*\big(\Gamma_t^{\mathrm{bil}}\big) &= \frac{1}{\left[\Gamma_t^{\mathrm{bil}} : \Gamma(\ell)\right]} \dim \mathfrak{S}_n^*\big(\Gamma(\ell)\big) + O(n^2) \\
&= \frac{n^3}{8640} \frac{\left[\overline{\Gamma}(1) : \Gamma(\ell)\right]}{\left[\Gamma_t^{\mathrm{bil}} : \Gamma(\ell)\right]} + O(n^2) \\
&= \frac{n^3}{8640}\left[\overline{\Gamma}(1) : \Gamma_t^{\mathrm{bil}}\right] + O(n^2). \qquad (1)
\end{aligned}$$

The number $\left[\overline{\Gamma}(1) : \Gamma_t^{\mathrm{bil}}\right]$ is equal to the degree of the map $\mathcal{A}_t^{\mathrm{bil}} \to \mathcal{A}_1$ (actually there are two such maps of the same degree), where $\mathcal{A}_1$ is the moduli space of principally polarized abelian surfaces. Now

$$\begin{aligned}
\left[\overline{\Gamma}(1) : \Gamma_t^{\mathrm{bil}}\right] &= \frac{1}{2}\left[\overline{\Gamma}(1) : \Gamma_t^{\natural}\right] \\
&= \frac{1}{2}\left[\overline{\Gamma}(1) : \Gamma_t^{\mathrm{lev}}\right]\left[\Gamma_t^{\mathrm{lev}} : \Gamma_t^{\natural}\right].
\end{aligned}$$

We can see directly that $\Gamma_t^{\rm lev} \supset \Gamma_t^{\natural}$ since

$$\Gamma_t^{\rm lev} = \left\{ \gamma \in \mathrm{Sp}(4,\mathbb{Z}) \;\middle|\; \gamma - \mathbf{1}_4 \in \begin{pmatrix} * & * & * & t\mathbb{Z} \\ t\mathbb{Z} & t\mathbb{Z} & t\mathbb{Z} & t^2\mathbb{Z} \\ * & * & * & t\mathbb{Z} \\ * & * & * & t\mathbb{Z} \end{pmatrix} \right\}.$$

**Lemma II.4** *The map*

$$\varphi\colon \Gamma_t^{\rm lev} \to \mathrm{SL}(2,\mathbb{Z}_t) \quad \textit{given by} \quad A \mapsto \begin{pmatrix} a_{11} & a_{13} \\ a_{31} & a_{33} \end{pmatrix}$$

*is a surjective group homomorphism, and the kernel is $\Gamma_t^{\natural}$.*

**Proof** The surjectivity follows from the well-known fact that the reduction mod $t$ map $\mathrm{red}_t\colon\ \mathrm{SL}(2,\mathbb{Z}) \to \mathrm{SL}(2,\mathbb{Z}_t)$ is surjective; the rest is obvious. □

**Lemma II.5** *For $t > 2$, the index $[\overline{\Gamma}(1) : \Gamma_t^{\rm lev}]$ is equal to $t\phi_4(t)/2$.*

**Proof** The proof is almost the same as that of [13, Lemma 0.5]. In place of the chain of groups $\Gamma_{1,p} < {}_0\Gamma_{1,p} < \Gamma' = \Gamma(1)$, we use the chain $\Gamma_t^{\rm lev} < {}_0\Gamma_{1,t} < \Gamma(1)$. Furthermore, we use the set $\Phi_4(t)$ where $\mathrm{SL}(4,\mathbb{Z}_t)$ acts. Note that $\mathrm{SL}(4,\mathbb{Z})$ still acts transitively on $\Phi_4(t)$, via

$$\begin{pmatrix} b_{11} & 0 & b_{12} & 0 \\ 0 & 1 & 0 & 0 \\ b_{21} & 0 & b_{22} & 0 \\ 0 & 0 & 0 & 1 \end{pmatrix} \quad \text{and} \quad \begin{pmatrix} B & 0 \\ 0 & {}^tB^{-1} \end{pmatrix},$$

for $B \in \mathrm{SL}(2,\mathbb{Z})$.

Following the same steps as in [13], and substituting $\phi_m(t)$ for $p^m - 1 = \phi_m(p)$, we then find that $[{}_0\Gamma_{1,t} : \Gamma_t^{\rm lev}] = t\phi_1(t)$ and $[{}_0\Gamma_{1,t} : \Gamma(1)| = \phi_4(t)/\phi_1(t)$, so $[\overline{\Gamma}(1) : \Gamma_t^{\rm lev}] = t\phi_4(t)/2$. □

**Theorem II.6** *The number of cusp forms of weight $n$ for $\Gamma_t^{\rm bil}$ (for $t > 2$) is given by*

$$\begin{aligned} \dim \mathfrak{S}_n^*(\Gamma_t^{\rm bil}) &= \frac{n^3}{34560} t^2 \phi_2(t)\phi_4(t) \\ &= \frac{n^3}{34560} t^8 \prod_{p|t} (1-p^{-2})(1-p^{-4}). \end{aligned}$$

**Proof** Immediate from equation (1), Corollary II.2 and Lemma II.5. □

# III Torsion in the modular group

We know that $\Gamma_t^{\mathrm{bil}} \subset \mathrm{Sp}(4,\mathbb{Z})$, and the conjugacy classes of torsion elements in $\mathrm{Sp}(4,\mathbb{Z})$ are known ([6, 20]). See [10] for a summary of the relevant information.

If $\gamma \in \Gamma_t^{\natural}$ then the reduction mod $t$ of $\gamma$ is

$$\overline{\gamma} = \begin{pmatrix} 1 & * & 0 & 0 \\ 0 & 1 & 0 & 0 \\ 0 & * & 1 & 0 \\ * & * & * & 1 \end{pmatrix} \in \mathrm{Sp}(4,\mathbb{Z}_t),$$

so the characteristic polynomial $\chi(\overline{\gamma})$ is $(1-x)^4 \in \mathbb{Z}_t[x]$. On the other hand, if $\gamma \in \zeta\Gamma_t^{\natural}$ then

$$\overline{\gamma} = \zeta \begin{pmatrix} 1 & * & 0 & 0 \\ 0 & 1 & 0 & 0 \\ 0 & * & 1 & 0 \\ * & * & * & 1 \end{pmatrix} = \begin{pmatrix} 1 & * & 0 & 0 \\ 0 & -1 & 0 & 0 \\ 0 & * & 1 & 0 \\ * & * & * & -1 \end{pmatrix} \in \mathrm{Sp}(4,\mathbb{Z}_t),$$

so $\chi(\overline{\gamma}) = (1-x)^2(1+x)^2 \in \mathbb{Z}_t[x]$.

The only classes in the list in [20], up to conjugacy, where the characteristic polynomials have this reduction mod $t$ ($t>2$) are I(1), where $\chi(\gamma) = (1-x)^4$, II(1)a and II(1)b. Class I(1) consists of the identity; class II(1)a includes $\zeta$ so this just gives us the conjugacy class of $\zeta$. Class II(2)b is the $\mathrm{Sp}(4,\mathbb{Z})$-conjugacy class of $\xi$, where

$$\xi = \begin{pmatrix} 1 & 1 & 0 & 0 \\ 0 & -1 & 0 & 0 \\ 0 & 0 & 1 & 0 \\ 0 & 0 & 1 & -1 \end{pmatrix} \in \Gamma_t^{\mathrm{bil}}.$$

**Proposition III.1** *Every nontrivial element of finite order in $\Gamma_t^{\mathrm{bil}}$ (for $t>2$) has order 2, and is conjugate to $\zeta$ or to $\xi$ in $\Gamma_t^{\mathrm{bil}}$ if $t$ is odd.*

**Proof** It follows from the list in [20] that the only torsion for $t>2$ is 2-torsion (this is still true if $t$ is even). The 2-torsion of the group $\Gamma_t^{\mathrm{lev}}$ was studied by Brasch [3]. There are five types but only two of them occur for odd $t$. The representatives for these conjugacy classes given in [3] are (up to sign) $\zeta$ and $\xi$; so the assertion of the theorem is that the $\Gamma_t^{\mathrm{bil}}$-conjugacy classes of $\zeta$ and $\xi$ coincide with the intersections of their $\Gamma_t^{\mathrm{lev}}$-conjugacy classes with $\Gamma_t^{\mathrm{bil}}$. This is checked in [17, Proposition 3.2] for the case $t = 6$ (the relevant cases are called $\zeta_0$ and $\zeta_3$ there), but the proof works for all $t > 2$. □

We put

$$\mathcal{H}_1 = \left\{ \begin{pmatrix} \tau_1 & 0 \\ 0 & \tau_3 \end{pmatrix} \middle| \ \mathrm{Im}\,\tau_1 > 0,\ \mathrm{Im}\,\tau_3 > 0 \right\} \subset \mathbb{H}_2 \tag{2}$$

and

$$\mathcal{H}_2 = \left\{ \begin{pmatrix} \tau_1 & \tau_2 \\ \tau_2 & \tau_3 \end{pmatrix} \middle| \ 2\tau_2 + \tau_3 = 0 \right\} \subset \mathbb{H}_2. \tag{3}$$

These are the fixed loci of $\zeta$ and $\xi$ respectively. We denote by $H_1^\circ$ and $H_2^\circ$ the images of $\mathcal{H}_1$ and $\mathcal{H}_2$ in $\mathcal{A}_t^{\mathrm{bil}}$, and by $H_1$ and $H_2$ their respective closures in $\mathcal{A}_t^{\mathrm{bil}*}$.

**Lemma III.2** *$H_i^\circ$ is irreducible for $i = 1, 2$.*

**Proof** This follows at once from Proposition III.1 together with equations (2) and (3). □

The abelian surfaces corresponding to points in $H_1^\circ$ and $H_2^\circ$ are, respectively, product surfaces and bielliptic abelian surfaces, as described in [13] for the case $t$ prime.

We define the subgroup $\Gamma(2t, 2t)$ of $\Gamma(t) \times \Gamma(t)$ by

$$\Gamma(2t, 2t) = \left\{ (M, N) \in \Gamma(t) \times \Gamma(t) \ \middle| \ M \equiv {}^{\mathsf{T}}N^{-1} \mod 2 \right\}$$

**Lemma III.3** *$H_1^\circ$ is isomorphic to $X^\circ(t) \times X^\circ(t)$, and $H_2^\circ$ is isomorphic to $\Gamma(2t, 2t)\backslash\mathbb{H} \times \mathbb{H}$.*

**Proof** Identical to the proofs of the corresponding results [11, Lemma I.5.43] and [11, Lemma I.5.45]. The level $t$ structure now occurs in both factors, whereas in [11] there is level 1 structure in the first factor and level $p$ structure in the second. In [11] the level $p$ is assumed to be an odd prime but this fact is not used at that stage: $p$ odd suffices, so we may replace $p$ by $t$. Thereafter one simply replaces all the groups with their intersection with $\Gamma_t^{\mathrm{bil}}$, which imposes a level $t$ structure in the first factor and causes it to behave exactly like the second factor. □

**Lemma III.4** *$H_1^\circ$ and $H_2^\circ$ are disjoint.*

**Proof** The stabiliser of any point of $\mathbb{H}_2$ in $\Gamma_t^{\text{bil}}$ is cyclic (of order 2), since $\Gamma_t^{\natural}$ is torsion-free and therefore has no fixed points. A point of $\mathcal{H}_1 \cap \mathcal{H}_2$ would be the image of a point of $\mathbb{H}_2$ stabilised by the subgroup generated by $\zeta$ and $\xi$, which is not cyclic. □

# IV Boundary divisors

We begin by counting the boundary divisors. These correspond to $\widetilde{\Gamma}_t^{\text{bil}}$-orbits of lines in $\mathbb{Q}^4$: we identify a line by its primitive generator $\mathbf{v} = (v_1, v_2, v_3, v_4) \in \mathbb{Z}^4$ with $\operatorname{hcf}(v_1, v_2, v_3, v_4) = 1$, which is unique up to sign. We denote the reduction of $\mathbf{v}$ mod $t$ by $\overline{\mathbf{v}} = (\overline{v}_1, \overline{v}_2, \overline{v}_3, \overline{v}_4) \in \mathbb{Z}_t^4$. To fix things we shall say, arbitrarily, that $\mathbf{v}$ is positive if the first nonzero entry $\overline{v}_i$ of $\overline{\mathbf{v}}$ satisfies $\overline{v}_i \in \{1, \ldots, (t-1)/2\}$ (remember that we have assumed that $t$ is odd). Then each line has a unique positive primitive generator.

If $\mathbf{v} = (v_1, v_2, v_3, v_4) \in \mathbb{Z}^4$, we define the $t$-divisor to be $r = \operatorname{hcf}(t, v_1, v_3)$.

**Proposition IV.1** *Positive primitive vectors* $\mathbf{v}, \mathbf{w} \in \mathbb{Z}^4$ *span lines* $\mathbb{Q}\mathbf{v}$ *and* $\mathbb{Q}\mathbf{w}$ *in the same* $\widetilde{\Gamma}_t^{\text{bil}}$*-orbit if and only if* $(\overline{v}_1, \overline{v}_3) = (\overline{w}_1, \overline{w}_3)$ *(in particular* $\mathbf{v}$ *and* $\mathbf{w}$ *have the same t-divisor, r), and* $(v_2, v_4) \equiv \pm(w_2, w_4)$ *mod r.*

**Proof** Note that if $\Gamma(t)$ is the principal congruence subgroup of level $t$ in $\mathrm{Sp}(4, \mathbb{Z})$ then $\Gamma(t) \lhd \widetilde{\Gamma}_t^{\natural}$ and the quotient is

$$\widetilde{\Gamma}_t^{\natural}(t) = \left\{ \begin{pmatrix} 1 & k & 0 & k' \\ 0 & 1 & 0 & 0 \\ 0 & l & 1 & l' \\ 0 & 0 & 0 & 1 \end{pmatrix} \in \mathrm{Sp}(4, \mathbb{Z}_t) \right\} \cong \mathbb{Z}_t^4.$$

We claim that two primitive vectors $\mathbf{v}$ and $\mathbf{w}$ are equivalent modulo $\Gamma(t)$ if and only if $\overline{v} = \overline{w}$. It is obvious that $\Gamma(t)$ preserves the residue classes mod $t$. Conversely, suppose that $\overline{v} = \overline{w}$. Then we can find $\gamma \in \mathrm{Sp}(4, \mathbb{Z})$ such that $\gamma\mathbf{v} = (1, 0, 0, 0)$ (the corresponding geometric fact is that the moduli space $\mathcal{A}_2$ of principally polarised abelian surfaces has only one rank 1 cusp). Since $\Gamma(t) \lhd \mathrm{Sp}(4, \mathbb{Z})$ this means that in order to prove the claim we may assume $\mathbf{v} = (1, 0, 0, 0)$. Then we proceed exactly as in the proof of [5, Lemma 3.3], taking $p = 1$ and $q = t$ (the assumptions that $p$ and $q$ are prime are not used at that point).

The group $\widetilde{\Gamma}_t^{\natural}(t)$ acts on the set $(\mathbb{Z}_t^4)^\times$ of nonzero elements of $\mathbb{Z}_t^4$ by $\overline{v}_2 \mapsto \overline{v}_2 + k\overline{v}_1 + l\overline{v}_3$ and $\overline{v}_4 \mapsto \overline{v}_4 + k'\overline{v}_1 + l'\overline{v}_3$: so $\overline{\mathbf{v}}$ is equivalent to $\overline{\mathbf{w}}$ if and only if $(\overline{v}_1, \overline{v}_3) = (\overline{w}_1, \overline{w}_3)$, so they have the same $t$-divisor, and $\overline{v}_2 \in \overline{w}_2 + \mathbb{Z}_t r$ and $\overline{v}_4 \in \overline{w}_4 + \mathbb{Z}_t r$. These are therefore the conditions for primitive vectors $\mathbf{v}$ and

$\mathbf{w}$ to be equivalent under $\widetilde{\Gamma}_t^\natural$. For equivalence under $\widetilde{\Gamma}_t^{\mathrm{bil}}$, we get the extra element $\zeta$ which makes $(v_1, v_2, v_3, v_4)$ equivalent to $(v_1, -v_2, v_3, -v_4)$. Since we are interested in orbits of lines, not primitive generators, we may restrict ourselves to positive generators $\mathbf{v}$. □

The irreducible components of the boundary divisor of $\mathcal{A}_t^{\mathrm{bil}*}$ correspond to the $\Gamma_t^{\mathrm{bil}}$-orbits (or equivalently to $\widetilde{\Gamma}_t^{\mathrm{bil}}$-orbits) of lines in $\mathbb{Q}^4$. We denote the boundary component corresponding to $\mathbb{Q}\mathbf{v}$ by $D_{\mathbf{v}}$. We shall be chiefly interested in the cases $r = t$ and $r = 1$. We refer to these as the standard components. They are represented by vectors $(0, a, 0, b)$ and $(a, 0, b, 0)$ respectively, in either case with $\mathrm{hcf}(a, b) = 1$, $0 \le a \le (t-1)/2$ and $0 \le b < t$. Note that there are $\nu$ of each of these.

**Corollary IV.2** *If $t$ is odd then the number of irreducible boundary divisors of $\mathcal{A}_t^{\mathrm{bil}*}$ with $t$-divisor $r$ is $\#\overline{\Phi}_2(h)\#\overline{\Phi}_2(r)$, where $h = t/r$. For $r \ne 1$, $t$, this is equal to $\frac{1}{4}\phi_2(h)\phi_2(r)$.*

**Proof** See above for the standard cases. In general, the $\Gamma_t^\natural$-orbit of a primitive vector $\mathbf{v}$ is determined by the classes of $(v_1/r, v_3/r)$ in $\Phi_2(h)$ and of $(\overline{v}_2, \overline{v}_4) \in \Phi_2(r)$. The extra element $\zeta$ and the freedom to multiply $\mathbf{v}$ by $-1 \in \mathbb{Q}$ allow us to multiply either of these classes by $-1$ and the choices therefore lie in $\overline{\Phi}_2(h)$ and $\overline{\Phi}_2(r)$. □

# V Jacobi forms

This section describes the behaviour of a modular form $F \in \mathfrak{S}_{3n}^*(\Gamma_t^{\mathrm{bil}})$ near a boundary divisor $D_{\mathbf{v}}$. The standard boundary divisors are best treated separately, since it is in those cases only that the torsion plays a role: on the other hand, the standard boundary divisors occur for all $t$ and their behaviour does not depend very much on the factorisation of $t$.

We assume at first, then, that $D_{\mathbf{v}}$ is a nonstandard boundary divisor. Since all the divisors of given $t$-divisor are equivalent under the action of $\mathbb{Z}_2 \wr \mathrm{SL}(2, \mathbb{Z}_t)$ (because the $t$-divisor is the only invariant of a boundary divisor of $\mathcal{A}_t$: see [5]), it will be enough to calculate the number of conditions imposed by one divisor of each type. That is to say, we only need consider boundary components in $\mathcal{A}_t^*$.

In view of this we may take $\mathbf{v} = (0, 0, r, 1)$ for some $r \mid t$ with $1 < r < t$. We write $(0, 0, 0, 1) = \mathbf{v}_{(0,1)}$ (for consistency with [11]) and we put $h = t/r$. Since we want to work with $\Gamma_t^{\mathrm{bil}}$ rather than $\widetilde{\Gamma}_t^{\mathrm{bil}}$ (so as to use fractional linear transformations) we must consider the lines $\mathbb{Q}\mathbf{v}R_t = \mathbb{Q}\mathbf{v}'$, where $\mathbf{v}' = (0, 0, 1, h)$ and $\mathbb{Q}\mathbf{v}_{(0,1)}R_t = \mathbb{Q}\mathbf{v}_{(0,1)}$.

Note that $\mathbf{v}'Q_r = \mathbf{v}_{(0,1)}$, where

$$Q_r = \begin{pmatrix} 1 & 1 & 0 & 0 \\ h-1 & h & 0 & 0 \\ 0 & 0 & h & 1-h \\ 0 & 0 & -1 & 1 \end{pmatrix} \in \mathrm{Sp}(4,\mathbb{Z}).$$

**Proposition V.1** *If $\mathbf{v}$ has t-divisor $r \neq t$, 1 and $F \in \mathfrak{S}_k^*(\Gamma_t^{\mathrm{bil}})$ is a cusp form of weight $k$, then there are coordinates $\tau_i^{\mathbf{v}}$ such that $F$ has a Fourier expansion near $D_{\mathbf{v}}$ as*

$$F = \sum_{w \geq 0} \theta_w^{\mathbf{v}}(\tau_1^{\mathbf{v}}, \tau_2^{\mathbf{v}}) \exp \frac{2\pi i w \tau_3^{\mathbf{v}}}{rt}.$$

**Proof** As usual (cf. [11]) we write $\mathcal{P}'_{\mathbf{v}}$ for the stabiliser of $\mathbf{v}'$ in $\mathrm{Sp}(4,\mathbb{R})$, so $\mathcal{P}'_{\mathbf{v}} = Q_r^{-1}\mathcal{P}_{\mathbf{v}_{(0,1)}}Q_r$. We take $P'_{\mathbf{v}} = \mathcal{P}'_{\mathbf{v}} \cap \Gamma_t^{\mathrm{bil}}$: this group determines the structure of $\mathcal{A}_t^{\mathrm{bil}*}$ near $D_{\mathbf{v}}$. It is shown in [11, Proposition I.3.87] that $\mathcal{P}_{\mathbf{v}_{(0,1)}}$ is generated by $g_1(\gamma)$ for $\gamma \in \mathrm{SL}(2,\mathbb{R})$, $g_2 = \zeta$, $g_3(m,n)$ and $g_4(s)$ for $m$, $n$, $s \in \mathbb{R}$, where

$$g_1(\gamma) = \begin{pmatrix} a & 0 & b & 0 \\ 0 & 1 & 0 & 0 \\ c & 0 & d & 0 \\ 0 & 0 & 0 & 1 \end{pmatrix} \quad \text{for} \quad \gamma = \begin{pmatrix} a & b \\ c & d \end{pmatrix}$$

and $g_3$ and $g_4$ are given by

$$g_3(m,n) = \begin{pmatrix} 1 & 0 & 0 & n \\ m & 1 & n & 0 \\ 0 & 1 & 0 & -m \\ 0 & 0 & 0 & 1 \end{pmatrix}, \qquad g_4(s) = \begin{pmatrix} 1 & 0 & 0 & 0 \\ 0 & 1 & 0 & s \\ 0 & 0 & 1 & 0 \\ 0 & 0 & 0 & 1 \end{pmatrix}.$$

So $P'_{\mathbf{v}}$ includes the subgroup generated by all elements of the form $Q_r^{-1}g_iQ_r$ with $a,b,c,d,m,n,s \in \mathbb{Z}$ which lie in $\Gamma_t^{\mathrm{bil}}$. In particular it includes the lattice $\{Q_r^{-1}g_4(rts)Q_r \mid s \in \mathbb{Z}\}$. If we take $Z^{\mathbf{v}} = Q_r^{-1}(Z)$ for $Z = \left(\begin{smallmatrix} \tau_1 & \tau_2 \\ \tau_2 & \tau_3 \end{smallmatrix}\right)$ then we obtain

$$Z^{\mathbf{v}} = \begin{pmatrix} h^2\tau_1 - 2h\tau_2 + \tau_3 & -h(h-1)\tau_1 + (2h-1)\tau_2 - \tau_3 \\ -h(h-1)\tau_1 + (2h-1)\tau_2 - \tau_3 & (h-1)^2\tau_1 - 2(h-1)\tau_2 + \tau_3 \end{pmatrix}.$$

One easily checks that

$$Q_r^{-1}g_4(rt)Q_r \colon Z^{\mathbf{v}} = \begin{pmatrix} \tau_1^{\mathbf{v}} & \tau_2^{\mathbf{v}} \\ \tau_2^{\mathbf{v}} & \tau_3^{\mathbf{v}} \end{pmatrix} \mapsto \begin{pmatrix} \tau_1^{\mathbf{v}} & \tau_2^{\mathbf{v}} \\ \tau_2^{\mathbf{v}} & \tau_3^{\mathbf{v}} + rt \end{pmatrix}$$

and this proves the result. $\square$

We define a subgroup $\Gamma(t,r)$ of $\mathrm{SL}(2,\mathbb{Z})$ by

$$\Gamma(t,r) = \left\{ \begin{pmatrix} a & b \\ c & d \end{pmatrix} \middle| \, a \equiv d \equiv 1 \bmod t, \; b \equiv 0 \bmod t^2, \; c \equiv 0 \bmod r \right\}.$$

**Lemma V.2** *If $D_{\mathbf{v}}$ is nonstandard then $P'_{\mathbf{v}}$ is torsion-free.*

**Proof** The only torsion in $\Gamma_t^{\mathrm{bil}}$ is 2-torsion and a simple calculation shows that if $\mathbf{1}_4 \neq g \in \mathcal{P}_{\mathbf{v}(0,1)}$ and $g^2 = \mathbf{1}_4$, then $Q_r^{-1} g Q_r \notin \Gamma_t^{\mathrm{bil}}$ for $r \neq 1, t$. $\square$

**Proposition V.3** *If $D_{\mathbf{v}}$ is nonstandard and $F \in \mathfrak{S}_k^*(\Gamma_t^{\mathrm{bil}})$ then $\theta_w^{\mathbf{v}}(r\tau_1^{\mathbf{v}}, t\tau_2^{\mathbf{v}})$ is a Jacobi form of weight $k$ and index $w$ for $\Gamma(t,r)$.*

**Proof** By direct calculation we find that $Q_r^{-1} g_1(\gamma) Q_r \in \Gamma_t^{\mathrm{bil}}$ if $\gamma \in \Gamma(t,r)$ and $Q_r^{-1} g_3(rm, tn) Q_r \in \Gamma_t^{\mathrm{bil}}$ for $m, n \in \mathbb{Z}$. Using these two elements, another elementary calculation verifies that the transformation laws for Jacobi forms given in [4] are satisfied, since

$$Q_r^{-1} g_3(rm, tn) Q_r \colon Z^{\mathbf{v}} \mapsto \begin{pmatrix} \tau_1^{\mathbf{v}} & \tau_2^{\mathbf{v}} + rm\tau_1^{\mathbf{v}} + tn \\ \tau_2^{\mathbf{v}} + rm\tau_1^{\mathbf{v}} + tn & \tau_3^{\mathbf{v}} + 2rm\tau_2^{\mathbf{v}} + r^2 m^2 \tau_1^{\mathbf{v}} \end{pmatrix}$$

$$\text{and} \quad Q_r^{-1} g_1(\gamma) Q_r \colon Z^{\mathbf{v}} \mapsto \begin{pmatrix} \gamma(\tau_1^{\mathbf{v}}) & \tau_2^{\mathbf{v}}/(c\tau_1^{\mathbf{v}} + d) \\ \tau_2^{\mathbf{v}}/(c\tau_1^{\mathbf{v}} + d) & \tau_3^{\mathbf{v}} - c{\tau_2^{\mathbf{v}}}^2/(c\tau_1^{\mathbf{v}} + d) \end{pmatrix}. \quad \square$$

**Lemma V.4** *The index of $\Gamma(t,r)$ in $\Gamma(1)$ is equal to $rt\phi_2(t)$ for $r \neq 1, t$.*

**Proof** Consider the chain of groups

$$\Gamma(1) = \mathrm{SL}(2,\mathbb{Z}) > \Gamma_0(t) > \Gamma_0(t)(r) > \Gamma(t,r)$$

and the normal subgroup $\Gamma_1(t) \lhd \Gamma_0(t)$, where

$$\Gamma_0(t) = \left\{ \gamma = \begin{pmatrix} a & b \\ c & d \end{pmatrix} \in \mathrm{SL}(2,\mathbb{Z}) \middle| \begin{array}{l} a \equiv d \equiv 1 \bmod t, \\ b \equiv 0 \bmod t \end{array} \right\},$$

$$\Gamma_1(t) = \left\{ \gamma = \begin{pmatrix} a & b \\ c & d \end{pmatrix} \in \mathrm{SL}(2,\mathbb{Z}) \middle| \begin{array}{l} a \equiv d \equiv 1 \bmod t, \\ b \equiv c \equiv 0 \bmod t \end{array} \right\},$$

$$\Gamma_0(t)(h) = \left\{ \gamma = \begin{pmatrix} a & b \\ c & d \end{pmatrix} \in \mathrm{SL}(2,\mathbb{Z}) \middle| \begin{array}{l} a \equiv d \equiv 1 \bmod t, \\ b \equiv 0 \bmod t, \; c \equiv 0 \bmod h \end{array} \right\}.$$

Thus $\Gamma_0(t)(r)$ is the kernel of reduction mod $r$ in $\Gamma_0(t)$. By Corollary II.2, $[\Gamma(1):\Gamma_1(t)] = t\phi_2(t)$. By the exact sequence

$$0 \longrightarrow \Gamma_1(t) \longrightarrow \Gamma_0(t) \longrightarrow \left\{ \begin{pmatrix} 1 & 0 \\ \bar{c} & 1 \end{pmatrix} \middle| \bar{c} \in \mathbb{Z}_t \right\} \cong \mathbb{Z}_t \longrightarrow 0$$

we have $[\Gamma_0(t):\Gamma_1(t)] = t$, and similarly

$$0 \longrightarrow \Gamma_0(t)(r) \longrightarrow \Gamma_0(t) \longrightarrow \left\{ \begin{pmatrix} 1 & 0 \\ \bar{c} & 1 \end{pmatrix} \middle| \bar{c} \in \mathbb{Z}_r \right\} \cong \mathbb{Z}_r \longrightarrow 0$$

gives $[\Gamma_0(t):\Gamma_0(t)(r)] = r$.

To calculate $[\Gamma(t)(r):\Gamma(t,r)]$, we let $\Gamma_0(t)(r)$ act on $\mathbb{Z}_t \times \mathbb{Z}_{t^2}$ by right multiplication $\gamma\colon (x,y) \mapsto (ax+cy, bx+dy)$. The stabiliser of $(1,0) \in \mathbb{Z}_t \times \mathbb{Z}_{t^2}$ is then $\{\overline{\gamma} \in \Gamma_0(t)(r) \mid a \equiv 1 \mod t, b \equiv 0 \mod t^2\}$, which is $\Gamma(t,r)$. However, the orbit of $(1,0) \in \mathbb{Z}_t \times \mathbb{Z}_{t^2}$ is $\left\{(\bar{a},\bar{b}) \in \mathbb{Z}_t \times \mathbb{Z}_{t^2} \mid \left(\begin{smallmatrix} a & b \\ c & d \end{smallmatrix}\right) \in \Gamma_0(t)(r)\right\}$: that is, the set of possible first rows of a matrix in $\Gamma_0(t)(r)$ taken mod $t$ in the first column and mod $t^2$ in the second. This is evidently equal to $\{(1,tb') \mid b' \in \mathbb{Z}_t\}$, and hence of size $t$. Thus $[\Gamma(t)(r):\Gamma(t,r)] = t$, which completes the proof. □

The standard case is only slightly different, but now there is torsion.

**Proposition V.5** *If $D_{\mathrm{v}}$ is standard and $F \in \mathfrak{S}_k^*(\Gamma_t^{\mathrm{bil}})$ then $\theta_w^{\mathrm{v}}(r\tau_1^{\mathrm{v}}, t\tau_2^{\mathrm{v}})$ is a Jacobi form of weight $k$ and index $w$ for a group $\Gamma'(t,r)$, which contains $\Gamma(t,r)$ as a subgroup of index 2.*

**Proof** Although the standard boundary components are most obviously given by $(0,0,0,1)$ for $r=t$ and $(0,0,1,0)$ for $r=1$, we choose to take advantage of the calculations that we have already performed by working instead with $(0,0,t,1)$ and $(0,0,1,1)$. Lemma V.3 is still true, but we also have $Q_t^{-1}\zeta Q_t \in \Gamma_t^{\mathrm{bil}}$ and $Q_1^{-1}(-\zeta)Q_1 \in \Gamma_t^{\mathrm{bil}}$. These give rise to the stated extra invariance. □

**Lemma V.6** *The dimension of the space $J_{3k,w}(\Gamma'(t,r))$ of Jacobi forms of weight $3k$ and index $w$ for $\Gamma'(t,r)$ is given as a polynomial in $k$ and $w$ by*

$$\dim J_{3k,w}(\Gamma'(t,r)) = \delta r t \nu \left( \frac{kw}{2} + \frac{w^2}{6} \right) + \text{linear terms}$$

*where $\delta = \frac{1}{2}$ if $r=1$ or $r=t$ and $\delta = 1$ otherwise.*

**Proof** By [4, Theorem 3.4] we have

$$\dim J_{3k,w}\big(\Gamma'(t,r)\big) \leq \sum_{i=0}^{2w} \dim \mathfrak{S}_{3k+i}\big(\Gamma'(t,r)\big). \tag{4}$$

Since $\Gamma'(t,r)$ is torsion-free, the corresponding modular curve has genus $1 + \frac{\mu(t,r)}{12} - \frac{\nu(t,r)}{2}$, where $\mu(t,r)$ is the index of $\Gamma'(t,r)$ in $\mathrm{PSL}(2,\mathbb{Z})$ and $\nu(t,r)$ is the number of cusps (see [18, Proposition 1.40]). Hence by [18, Theorem 2.23] the space of modular forms satisfies

$$\begin{aligned}\dim \mathfrak{S}_k\big(\Gamma'(t,r)\big) &= k\left(\frac{\mu(t,r)}{12} - \frac{\nu(t,r)}{2}\right) + \frac{k}{2}\nu(t,r) + O(1) \\ &= \frac{k\mu(t,r)}{12} + O(1)\end{aligned} \tag{5}$$

as a polynomial in $k$. By Lemma V.4 we have $\mu(t,r) = \frac{1}{2}rt\phi_2(t) = rt\nu$ for the nonstandard cases, $\mu(t,1) = \frac{1}{2}t\nu$ and $\mu(t,t) = \frac{1}{2}t^2\nu$. Now the result follows from equations (5) and (4). □

If $F \in \mathfrak{S}^*_{3k}(\Gamma_t^{\mathrm{bil}})$ then $F \cdot (d\tau_1 \wedge d\tau_2 \wedge d\tau_3)^{\otimes k}$ extends over the component $D_{\mathbf{v}}$ if and only if $\theta_w^{\mathbf{v}} = 0$ for all $w < k$: see [1, Chapter IV, Theorem 1]. Hence the obstruction $\Omega_{\mathbf{v}}$ coming from the boundary component $D_{\mathbf{v}}$ is

$$\Omega_{\mathbf{v}} = \sum_{w=0}^{k-1} \dim J_{3k,w}\left(\Gamma'(t,r)\right) \tag{6}$$

where $\Gamma'(t,r) = \Gamma(t,r)$ if $D_{\mathbf{v}}$ is nonstandard.

By Corollary IV.2 the total obstruction from the boundary is

$$\Omega_\infty = \sum_{r|t} \#\overline{\Phi}(h)\#\overline{\Phi}(r) \sum_{w=0}^{k-1} \dim J_{3k,w}\big(\Gamma'(t,r)\big),$$

and we may assume that $k$ is even.

**Corollary V.7** *The obstruction coming from the boundary is*

$$\Omega_\infty \leq \left(\sum_{r|t} \delta rt\nu\#\overline{\Phi}(h)\#\overline{\Phi}(r)\right)\frac{11}{36}k^3 + O(k^2).$$

**Proof** Summing the expression in Lemma V.6 for $0 \leq w < k$, as required by equation (6) gives the coefficient of $\frac{11}{36}$ and the rest comes directly from Lemma V.6 and Corollary IV.2. □

# VI Intersection numbers

We need to know the degrees of the normal bundles of the curves that generate Pic $H_1$ and Pic $H_2$. For this we first need to describe the surfaces $H_1$ and $H_2$. The statements and the proofs are very similar to the corresponding results for the case of $\mathcal{A}_p^{\mathrm{lev}}$, given in [11] and [12]. Therefore we simply refer to those sources for proofs, pointing out such differences as there are.

**Proposition VI.1** *$H_1$ is isomorphic to $X(t) \times X(t)$.*

**Proof** Identical to [11, I.5.53]. □

**Proposition VI.2** *$H_2$ is the minimal resolution of a surface $\overline{H}_2$ which is given by two* $\mathrm{SL}(2, \mathbb{Z}_2)$*-covering maps*

$$X(2t) \times X(2t) \longrightarrow \overline{H}_2 \longrightarrow X(t) \times X(t).$$

*The singularities that are resolved are $\nu^2$ ordinary double points, one over each point $(\alpha, \beta) \in X(t) \times X(t)$ for which $\alpha$ and $\beta$ are cusps.*

**Proof** Similar to [11, Proposition I.5.55] and the discussion before [12, Proposition 4.21]. $X(2)$ and $X(2p)$ are both replaced by $X(2t)$ and $X(1)$ and $X(p)$ by $X(t)$. Since $t > 3$ there are no elliptic fixed points and hence no other singularities in this case. □

**Proposition VI.3** *$H_1^\circ$ and $H_2^\circ$ meet the standard boundary components $D_{\mathbf{v}}$ transversally in irreducible curves $C_{\mathbf{v}} \cong X^\circ(t)$ and $C'_{\mathbf{v}} \cong X^\circ(2t)$ respectively. $D_{\mathbf{v}}$ is isomorphic to the (open) Kummer modular surface $K^\circ(t)$, $C_{\mathbf{v}}$ is the zero section and $C'_{\mathbf{v}}$ is the 3-section given by the 2-torsion points of the universal elliptic curve over $X(t)$.*

**Proof** This is essentially the same as [11, Proposition I.5.49], slightly simpler in fact. We may work with $\mathbf{v} = (0, 0, 1, 0)$ and copy the proof for the central boundary component in $\mathcal{A}_p^{\mathrm{lev}}$, replacing $p$ by $t$ (again the fact that $p$ is prime is not used). □

We do not claim that the closure of $D_{\mathbf{v}}$ is the Kummer modular surface $K(t)$. They are, however, isomorphic near $H_1$ and $H_2$. We remark that $H_1$ and $H_2$ do not meet the nonstandard boundary divisors, because of Lemma V.2.

**Proposition VI.4** *$\mathcal{A}_t^{\mathrm{bil}*}$ is smooth near $H_1$ and $H_2$.*

**Proof** Certainly $\mathcal{A}_t^{\text{bil}}$ is smooth since the only torsion in $\Gamma_t^{\text{bil}}$ is 2-torsion fixing a divisor in $\mathbb{H}_2$. There can in principle be singularities at infinity, but such singularities must lie on corank 2 boundary components not meeting $H_1$ nor $H_2$ (this follows again from Lemma V.2). □

**Corollary VI.5** *$H_1$ does not meet $H_2$.*

**Proof** Since $\mathcal{A}_t^{\text{bil}*}$ and the divisors $H_1$ and $H_2$ are smooth at the relevant points, the intersection must either be empty or contain a curve. However, the intersection also lies in the corank 2 boundary components. These components consist entirely of rational curves, and if $t > 5$ then $H_1 \cong X(t) \times X(t)$ contains no rational curves. Hence $H_1 \cap H_2 = \emptyset$.

With a little more work one can check that this is still true for $t \leq 5$, but we are in any case not concerned with that. □

**Proposition VI.6** *The Picard group* $\operatorname{Pic} H_1$ *is generated by the classes of* $\Sigma_1 = \overline{C}_{0010}$ *and* $\Psi_1 = \overline{C}_{0001}$. *The intersection numbers are* $\Sigma_1^2 = \Psi_1^2 = 0$, $\Sigma_1 \cdot \Psi_1 = 1$ *and* $\Sigma_1 \cdot H_1 = \Psi_1 \cdot H_1 = -\mu/6$.

**Proof** As in [12, Proposition 4.18] (but one has to use the alternative indicated in the remark that follows). □

**Proposition VI.7** *The Picard group* $\operatorname{Pic} H_2$ *is generated by the classes of* $\Sigma_2$ *and* $\Psi_2$*, which are the inverse images of general fibres of the two projections in* $X(t) \times X(t)$*, and of the exceptional curves* $R_{\alpha\beta}$ *of the resolution* $H_2 \to \overline{H}_2$. *The intersection numbers in* $H_2$ *are* $\Sigma_2^2 = \Psi_2^2 = \Sigma_2 \cdot R_{\alpha\beta} = \Psi_2 \cdot R_{\alpha\beta} = 0$, $R_{\alpha\beta} \cdot R_{\alpha'\beta'} = -2\delta_{\alpha\alpha'}\delta_{\beta\beta'}$ *and* $\Sigma_2 \cdot \Psi_2 = 6$. *In* $\mathcal{A}_t^{\text{bil}*}$ *we have* $\Sigma_2 \cdot H_2 = \Psi_2 \cdot H_2 = -\mu$ *and* $R_{\alpha\beta} \cdot H_2 = -4$.

**Proof** The same as the proofs of [12, Proposition 4.21] and [12, Lemma 4.24]. The curves $R'_{(a,b)}$ from [12] arise from elliptic fixed points so they are absent here. □

Notice that $\Sigma_2$ and $\Psi_2$ are also images of the general fibres in $X(2t) \times X(2t)$ and are themselves isomorphic to $X(2t)$.

# VII Branch locus

The closure of the branch locus of the map $\mathbb{H}_2 \to \mathcal{A}_t^{\text{bil}}$ is $H_1 \cup H_2$ and modular forms of weight $3k$ (for $k$ even) give rise to $k$-fold differential forms with poles of order $k/2$ along $H_1$ and $H_2$. We have to calculate the number of conditions imposed by these poles.

**Proposition VII.1** *The obstruction from $H_1$ to extending modular forms of weight $3k$ to $k$-fold holomorphic differential forms is*

$$\Omega_1 \leq \nu^2\Big(\frac{1}{2} - \frac{7t}{24} + t^2\Big(\frac{1}{24} + \frac{1}{864}\Big)\Big)k^3 + O(k^2).$$

**Proof** If $F$ is a modular form of weight $3k$ for $k$ even, vanishing to sufficiently high order at infinity, and $\omega = d\tau_1 \wedge d\tau_2 \wedge d\tau_3$, then $F\omega^{\otimes k}$ determines a section of $kK + \frac{k}{2}H_1 + \frac{k}{2}H_2$, where $K$ denotes the canonical sheaf of $\mathcal{A}_t^{\mathrm{bil}*}$. From

$$0 \longrightarrow \mathcal{O}(-H_1) \longrightarrow \mathcal{O} \longrightarrow \mathcal{O}_{H_1} \longrightarrow 0$$

we get, for $0 \leq j < k/2$

$$\begin{aligned} 0 \longrightarrow H^0\big(kK + (\frac{k}{2} - j - 1)H_1 + \frac{k}{2}H_2\big) \longrightarrow H^0\big(kK + (\frac{k}{2} - j)H_1 + \frac{k}{2}H_2\big) \\ \longrightarrow H^0\big((kK + (\frac{k}{2} - j)H_1 + \frac{k}{2}H_2)|_{H_1}\big). \end{aligned}$$

Thus

$$\begin{aligned} h^0\big(kK + (\frac{k}{2} - j)H_1 + \frac{k}{2}H_2\big) \;&\leq\; h^0\big(kK + (\frac{k}{2} - j - 1)H_1 + \frac{k}{2}H_2\big) \\ &\quad + h^0\big((kK + (\frac{k}{2} - j)H_1 + \frac{k}{2}H_2)|_{H_1}\big). \end{aligned}$$

Note that, by Lemma VI.5, $H_2|_{H_1} = 0$. Therefore

$$h^0\big(kK + \frac{k}{2}H_2\big) \geq h^0\big(kK + \frac{k}{2}H_1 + \frac{k}{2}H_2\big) + \sum_{j=0}^{k/2-1} h^0\big((kK + (\frac{k}{2} - j)H_1)|_{H_1}\big),$$

so

$$\Omega_1 \leq \sum_{j=0}^{k/2-1} h^0\big((kK + (\frac{k}{2} - j)H_1)|_{H_1}\big) = \sum_{j=0}^{k/2-1} h^0\big(kK_{H_1} - (\frac{k}{2} + j)H_1|_{H_1}\big). \tag{7}$$

By Lemma VI.6, $K_{H_1}$ and $H_1|_{H_1}$ are both multiples of $\Sigma_1 + \Phi_1$, and any positive multiple of $\Sigma_1 + \Psi_1$ is ample. Suppose $H_1|_{H_1} = a_1(\Sigma_1 + \Psi_1)$ and $K_{H_1} = b_1(\Sigma_1 + \Psi_1)$. Then

$$-\frac{\mu}{6} = \Sigma_1 \cdot H_1 = a\Sigma_1 \cdot (\Sigma_1 + \Psi_1) = a_1$$

and

$$\frac{\mu}{6} - \nu = 2g(\Sigma_1) - 2 = (K_{H_1} + \Sigma_1) \cdot \Sigma_1 = K_{H_1} \cdot \Sigma_1 = b_1$$

Hence, using equation (7)

$$\begin{aligned}\Omega_1 &\leq \sum_{j=0}^{k/2-1} h^0\big((\frac{k\mu}{6} - k\nu + \frac{k\mu}{12} + \frac{j\mu}{6})(\Sigma_1 + \Psi_1)\big)\\ &= \sum_{j=0}^{k/2-1} h^0\big((\frac{kt\nu}{4} - k\nu + \frac{jt\nu}{6})(\Sigma_1 + \Psi_1)\big).\end{aligned}$$

Since $t \geq 7$ (we know from [14] that $\mathcal{A}_t^{\text{bil}*}$ is rational for $t \leq 5$), we have $\frac{kt\nu}{4} - k\nu + \frac{jt\nu}{6} - \frac{t\nu}{6} + \nu > 0$ for all $j$ and hence $(\frac{kt\nu}{4} - k\nu + \frac{jt\nu}{6})(\Sigma_1 + \Psi_1) - K_{H_1}$ is ample. So by vanishing we have

$$\begin{aligned}\Omega_1 &\leq \sum_{j=0}^{k/2-1} \frac{1}{2}(\frac{kt\nu}{4} - k\nu + \frac{jt\nu}{6})^2(\Sigma_1 + \Psi_1)^2 + O(k^2)\\ &= \sum_{j=0}^{k/2-1} (\frac{kt\nu}{4} - k\nu + \frac{jt\nu}{6})^2 + O(k^2)\\ &= \nu^2\Big(\frac{1}{2} - \frac{7t}{24} + t^2\Big(\frac{1}{24} + \frac{1}{864}\Big)\Big)k^3 + O(k^2). \quad \square\end{aligned}$$

Next we carry out the same calculation for $H_2$.

**Proposition VII.2** *The obstruction from $H_2$ is*

$$\Omega_2 \leq \nu^2\Big(\Big(\frac{1}{2} + \frac{1}{72}\Big)t^2 - \Big(\frac{1}{4} + \frac{1}{24}\Big)t - \frac{7}{3} + \frac{1}{24}\Big)k^3 + O(k^2).$$

**Proof** By the same argument as above (equation (7)) the obstruction is

$$\Omega_2 \leq \sum_{j=0}^{k/2-1} h^0\big(kK_{H_2} - (\frac{k}{2} + j)H_2|_{H_2}\big).$$

In this case $H_2|_{H_2} = a_2(\Sigma_2 + \Psi_2) + c_2 R$, where $R = \sum_{\alpha,\beta} R_{\alpha\beta}$ is the sum of all the exceptional curves of $H_2 \to \overline{H}_2$, and $K_{H_2} = b_2(\Sigma_2 + \Psi_2) + d_2 R$. Since $\Sigma_2 \cong X(2t)$ we have by [18, 1.6.4]

$$2g(\Sigma_2) - 2 = \frac{1}{3}(t-3)\nu(2t) = \mu - \frac{\nu}{2}.$$

Hence

$$-\mu = \Sigma_2 \cdot H_2 = a_2\Sigma_2^2 + a_2\Sigma_2 \cdot \Psi_2 + c_2\Sigma_2 \cdot R = 6a_2;$$

so $a_2 = -\mu/6$, and

$$-4\nu^2 = R \cdot H_2 = a_2\Sigma_2 \cdot R + a_2\Psi_2 \cdot R + c_2R^2 = -2\nu^2 c_2;$$

so $c_2 = 2$. Therefore

$$H_2|_{H_2} = -\frac{\mu}{6}(\Sigma_2 + \Psi_2) + 2R.$$

Similarly

$$\mu - \frac{\nu}{2} = (K_{H_2} + \Sigma_2) \cdot \Sigma_2 = 6b_2$$

so $b_2 = \mu/6 - \nu/12$, and $0 = R.K_{H_2} = d_2R^2$ so $d_2 = 0$. Hence

$$K_{H_2} = \frac{1}{6}(\mu - \frac{\nu}{2})(\Sigma_2 + \Psi_2).$$

Moreover $L_j = (k-1)K_{H_2} - (\frac{k}{2} + j)H_2|_{H_2}$ is ample, as is easily checked using the Nakai criterion and the fact that the cone of effective curves on $H_2$ is spanned by $R_{\alpha\beta}$ and by the nonexceptional components of the fibres of the two maps $H_2 \to X(t)$. These components are $\Sigma_\alpha \equiv \Sigma_2 - \sum_\beta R_{\alpha\beta}$ and $\Psi_\beta \equiv \Psi_2 - \sum_\alpha R_{\alpha\beta}$, and it is simple to check that $L_j^2$, $L_j \cdot \Sigma_\alpha = L_j \cdot \Psi_\beta$ and $L_j \cdot R_{\alpha\beta}$ are all positive for the relevant values of $j$, $k$ and $t$. Therefore

$$\begin{aligned}
\Omega_2 &\leq \sum_{j=0}^{k/2-1} \frac{1}{2}\left(kK_{H_2} - (\frac{k}{2} + j)H_2|_{H_2}\right)^2 \\
&= \sum_{j=0}^{k/2-1} \frac{1}{2}\left(\nu(\frac{kt}{4} - \frac{k}{12} + \frac{jt}{6})(\Sigma_2 + \Psi_2) + (k+2j)R\right)^2 \\
&= \nu^2k^3\left(t^2(\frac{3}{8} + \frac{1}{8} + \frac{1}{72}) - t(\frac{1}{4} + \frac{1}{24}) + \frac{1}{24} - 2 - \frac{1}{3}\right) + O(k^2),
\end{aligned}$$

since $(\Sigma_2 + \Psi_2)^2 = 12$. □

# VIII Final calculation

In this section we assemble the results of the previous sections into a proof of the main theorem.

**Theorem VIII.1** $\mathcal{A}_t^{\mathrm{bil}*}$ *is of general type for $t$ odd and $t \geq 17$.*

**Proof** We put $n = 3k$ in Theorem II.6, and use $\phi_2(t) = 2\nu$ and the fact that

$$\phi_4(t) = t^4 \prod_{p|t}(1 - p^{-4}) = t^2\phi_2(t) \prod_{p|t}(1 + p^{-2}).$$

This gives the expression

$$\dim \mathfrak{S}_n^*(\Gamma_t^{\mathrm{bil}}) = \frac{k^3\nu^2}{320} t^4 \prod_{p|t}(1 + p^{-2}) + O(k^2).$$

From Proposition VII.1 and Proposition VII.2 we have

$$\begin{aligned} \Omega_1 &= k^3\nu^2\Big(\frac{37}{864}t^2 - \frac{7}{24}t + \frac{1}{2}\Big) + O(k^2), \\ \Omega_2 &= k^3\nu^2\Big(\frac{37}{72}t^2 - \frac{7}{24}t - \frac{55}{24}\Big) + O(k^2) \end{aligned}$$

and from Corollary V.7 and Corollary IV.2

$$\Omega_\infty = k^3\nu^2 \sum_{r|t} \frac{11}{36r} t^2 \prod_{p|(r,h)} (1 - p^{-2}) + O(k^2).$$

since $\phi_2(r)\phi_2(h) = t^2 \prod_{p|(r,h)}(1 - p^{-2})$.

It follows that $\mathcal{A}_t^{\mathrm{bil}*}$ is of general type, for odd $t$, provided

$$\frac{1}{320}\prod_{p|t}(1 + p^{-2})t^4 - \frac{481}{864}t^2 + \frac{7}{12}t + \frac{43}{24} - \sum_{r|t}\frac{11}{36r}t^2 \prod_{p|(r,h)}(1 - p^{-2}) > 0. \quad (8)$$

This is simple to check: since either $r = 1$ or $r \geq 3$, and since the sum of the divisors of $t$ is less than $t/2$, the last term can be replaced by $-\frac{11}{36}t^2 - \frac{11}{108}t^3$ and the $t$ and constant terms, and the $p^{-2}t^4$ term, can be discarded as they are positive. The resulting expression is a quadratic in $t$ whose larger root is less than 40, so we need only consider odd $t \leq 39$. We deal with primes, products of two primes and prime powers separately. In the case of primes, the expression on the left-hand side of the inequality (8) becomes $\frac{1}{320}t^4 - \frac{7433}{8640}t^2 + \frac{5}{18}t + \frac{43}{24}$, which is positive for $t \geq 17$. The expression in the case of $t = pq$ is positive if $t \geq 21$. For $t = p^2$ we get an expression which is negative for $t = 9$ but positive for $t = 25$, and for $t = p^3$ the expression is positive. □

One can also say something for $t$ even, though not if $t$ is a power of 2.

**Corollary VIII.2** $\mathcal{A}_t^{\mathrm{bil}*}$ *is of general type unless* $t = 2^a b$ *with* $b$ *odd and* $b < 17$.

**Proof** $\mathcal{A}_{nt}^{\text{bil}}$ covers $\mathcal{A}_t^{\text{bil}}$ for any $n$, and therefore $\mathcal{A}_{nt}^{\text{bil}*}$ is of general type if $\mathcal{A}_t^{\text{bil}*}$ is of general type. □

# References


[1] A. Ash, D. Mumford, M. Rapoport and Y. Tai, *Smooth compactification of locally symmetric varieties*, Math. Sci. Press, Brookline 1975

[2] L.A. Borisov, *A finiteness theorem for subgroups of* Sp(4, $\mathbb{Z}$), Algebraic geometry, 9: J. Math. Sci. (New York) **94** (1999) 1073–1099

[3] H-J. Brasch, *Branch points in moduli spaces of certain abelian surfaces*, in *Abelian varieties (Egloffstein, 1993)*, 25–54, de Gruyter, Berlin, 1995

[4] M. Eichler and D. Zagier, *The theory of Jacobi forms* Progress in Mathematics **55**, Birkhäuser, Boston 1985

[5] M. Friedland and G.K. Sankaran, *Das Titsgebäude von Siegelschen Modulgruppen vom Geschlecht 2*, Abh. Math. Sem. Univ. Hamburg **71** (2001) 49–68

[6] E. Gottschling, *Über die Fixpunkte der Siegelschen Modulgruppe*, Math. Ann. **143** (1961) 111–149

[7] V. Gritsenko and K. Hulek, *Irrationality of the moduli spaces of polarized abelian surfaces* (appendix to the paper by V. Gritsenko), in *Abelian varieties (Egloffstein, 1993)*, 83–84, de Gruyter, Berlin, 1995

[8] V. Gritsenko and G.K. Sankaran, *Moduli of abelian surfaces with a* $(1,p^2)$*-polarisation*, Izv. Ross. Akad. Nauk, Ser. Mat. **60** (1996) 19–26

[9] F. Hirzebruch, *Elliptische Differentialoperatoren auf Mannigfaltigkeiten*, Gesammelte Werke Bd. **2**, 583–608, Springer, Berlin 1987

[10] K. Hulek, C. Kahn and S. Weintraub, *Singularities of the moduli spaces of certain abelian surfaces*, Compos. Math. **79** (1991) 231–253

[11] K. Hulek, C. Kahn and S. Weintraub, *Moduli spaces of abelian surfaces: Compactification, degenerations and theta functions*, de Gruyter, Berlin 1993

[12] K. Hulek and G.K. Sankaran, *The Kodaira dimension of certain moduli spaces of abelian surfaces*, Compos. Math. **90** (1994) 1–35

[13] K. Hulek and S. Weintraub, *Bielliptic abelian surfaces*, Math. Ann. **283** (1989) 411–429

[14] S. Mukai, *Moduli of abelian surfaces, and regular polyhedral groups*, in *Moduli of algebraic varieties and the monster*, (I. Nakamura, Ed.) 68–74, Sapporo 1999

[15] K. O'Grady, *On the Kodaira dimension of moduli spaces of abelian surfaces*, Compos. Math. **72** (1989) 121–163

[16] G.K. Sankaran, *Moduli of polarised abelian surfaces*, Math. Nachr. **188** (1997) 321–340

[17] G.K. Sankaran and J. Spandaw, *The moduli space of bilevel-6 abelian surfaces*, Nagoya Math. J., **168** (2002) 113–125

[18] G. Shimura, *Introduction to the arithmetic theory of automorphic functions*, Publ. Math. Soc. Japan **11**, Iwanami Shoten, Tokyo, and Princeton University Press, 1971

[19] Y.-S. Tai, *On the Kodaira dimension of the moduli space of abelian varieties*, Invent. Math. **68** (1982) 425–439

[20] K. Ueno, *On fibre spaces of normally polarized abelian varieties of dimension 2. II, Singular fibres of the first kind*, J. Fac. Sci. Univ. Tokyo Sect. IA Math. **19** (1972) 163–199

G.K. Sankaran,
Department of Mathematical Sciences,
University of Bath,
Bath BA2 7AY, England
e-mail: gks@maths.bath.ac.uk
web: www.bath.ac.uk/~masgks

Printed in the United States
By Bookmasters